Radiolabeled Cellular Blood Elements

Pathophysiology, Techniques, and Scintigraphic Applications

NATO ASI Series

Advanced Science Institutes Series

A series presenting the results of activities sponsored by the NATO Science Committee, which aims at the dissemination of advanced scientific and technological knowledge, with a view to strengthening links between scientific communities.

The series is published by an international board of publishers in conjunction with the NATO Scientific Affairs Division

A	**Life Sciences**	Plenum Publishing Corporation
B	**Physics**	New York and London
C	**Mathematical**	D. Reidel Publishing Company
	and Physical Sciences	Dordrecht, Boston, and Lancaster
D	**Behavioral and Social Sciences**	Martinus Nijhoff Publishers
E	**Engineering and**	The Hague, Boston, and Lancaster
	Materials Sciences	
F	**Computer and Systems Sciences**	Springer-Verlag
G	**Ecological Sciences**	Berlin, Heidelberg, New York, and Tokyo

Recent Volumes in this Series

Volume 82—Receptor-Mediated Targeting of Drugs
          Edited by G. Gregoriadis, G. Poste, J. Senior, and A. Trouet

Volume 83—Molecular Form and Function of the Plant Genome
          Edited by Lous van Vloten-Doting, Gert S. P. Groot, and Timothy C. Hall

Volume 84—Behavioral Epidemiology and Disease Prevention
          Edited by Robert M. Kaplan and Michael H. Criqui

Volume 85—Primary Photo-Processes in Biology and Medicine
          Edited by R. V. Bensasson, G. Jori, E. J. Land, and T. G. Truscott

Volume 86—Wheat Growth and Modelling
          Edited by W. Day and R. K. Atkin

Volume 87—Industrial Aspects of Biochemistry and Genetics
          Edited by N. Gürdal Alaeddinoğlu, Arnold L. Demain, and
          Giancarlo Lancini

Volume 88—Radiolabeled Cellular Blood Elements
          Edited by M. L. Thakur

Series A: Life Sciences

Radiolabeled Cellular Blood Elements

Pathophysiology, Techniques, and Scintigraphic Applications

Edited by

M. L. Thakur

Thomas Jefferson University Hospital
Philadelphia, Pennsylvania

Associate Editors

M. D. Ezikowitz

Yale University School of Medicine
New Haven, Connecticut

and

M. R. Hardeman

University of Amsterdam
Amsterdam, The Netherlands

Plenum Press
New York and London
Published in cooperation with NATO Scientific Affairs Division

Proceedings of a NATO Advanced Study Institute on
Radiolabeled Cellular Blood Elements,
held August 29–September 9, 1983,
in Maratea, Italy

Library of Congress Cataloging in Publication Data

NATO Advanced Study Institute on Radiolabeled Cellular Blood Elements (1983:
Maratea, Italy)
Radiolabeled cellular blood elements.

(NATO ASI series. Series A, Life sciences; v. 88)
"Proceedings of a NATO Advanced Study Institute on Radiolabeled Cellular Blood
Elements, held August 29–September 9, 1983, in Maratea, Italy"—T.p. verso.
"Published in cooperation with NATO Scientific Affairs Division."
Includes bibliographies and index.
1. Radiolabeled blood cells—Diagnostic use—Congresses. 2. Radiolabeled blood
cells—Congresses. I. Thakur, M. L. II. Ezikowitz, M. D. III. Hardeman, Max R. IV.
North Atlantic Treaty Organization. Scientific Affairs Division. V. Title. VI. Series.
[DNLM: 1. Blood Cells—radionuclide imaging—congresses. 2. Isotope Label-
ing—methods—congresses. 3. Radioisotopes—diagnostic use—congresses.
WH 140 N279r 1983]
RC78.7.R43N37   1983   616.07'575   85-3628
ISBN-13: 978-1-4684-4924-2   e-ISBN-13: 978-1-4684-4922-8
DOI: 10.1007/978-1-4684-4922-8

PREFACE

This book is based on the contribution made by the guest lecturers and some of the participants gathered at the Advanced Study
Institute (ASI) on Radiolabeled Cellular Blood Elements held in
Maratea, Southern Italy, August 29, to September 9, 1982, under the
auspice of NATO. The first such international symposium on this subject was held in New York in September 1979, the proceedings of which
were published by Triverum Publishing Company in New York in 1980.
Since that period we have witnessed an ever increasing number of publications and investigations of the subject of radiolabeled cellular
blood elements throughout the world. The time was right, therefore
for this ASI to bring together experienced investigators, present and
potential users, and promising young scientists for exchange of knowledge and experience, informal dialogues, and relaxed and fruitful
discussions.

The guest faculty included veterans who placed emphasis on basic
cell physiology in health and disease, upon which numerous
applications of radiolabeled blood cells are based. Presentations
were also made by key investigators who have contributed for many
years, outlining problems and potential solutions and giving a
critical look at a variety of techniques used and applied in vivo.
Unlike the research articles in the proceedings of the first symposium
therefore this monograph contains chapters on basic cell physiology
and critical reviews of years of data and experience generated in the
preparation and use of radiolabeled blood cells. In addition to the
main chapters by the guest lecturers, the appendix contains a few
short papers on important contributions made by a few of the
participants. Unfortunately many could not be included owing to lack
of space.

Preparing manuscripts with figures and tables for camera reproduction was a formidable task. Care has been taken for consistency
and typographic accuracy. However, I make no claim that no errors
exist in this volume. ,I hope, however, that the reader will understand and ignore any error and find this compendious volume useful for
numerous biologic studies, physiologic explorations and clinical applications of radiolabeled cellular blood elements in years to come.

M. L. Thakur
Editor, and the ASI Director

ACKNOWLEDGEMENTS

Dr. Max Hardeman of the University of Amsterdam and Dr. Michael Ezekowitz of Yale University served as the codirectors of the Advanced Study Institute (ASI). Dr. Hardeman spent countless hours and contributed to the scientific program, took care of mailings in Europe, organized transportation in Italy, and communicated with the hotel management on numerous occasions. I cannot thank him enough!

Dr. Ezekowitz collected some manuscripts and corrected a few. I am grateful to him.

I take this opportunity once again to thank all the guest faculty, who, despite their busy schedules and time constraints, accepted my invitation and made valuable contributions to the ASI. I am also grateful to all participants, who were so friendly and were primary resources for many lively discussions. They made the ASI professionally beneficial and socially enjoyable.

Dr. diLullo, Dr. Sinclair, and Mrs. Kester of the NATO Division of Scientific Affairs in Belgium were especially helpful from the beginning of the ASI. Without their advice and understanding, this ASI could not have been so successful.

The hotel manager, Mr. Guzzardi, and his staff treated us well for two weeks. I thank them all on behalf of all of us.

It is with deep appreciation that I acknowledge the supplementary financial assistance we received from Medi-Physics, California; Amersham-International, England; Mallickrodt Inc., St. Louis, Missouri; and Beckman International, Amsterdam.

I also want to express profound gratitude to Mr. John Curry and his staff particularly, Elmina Finck, Ann Marie Webster, and Amy Yee of the American College of Radiology, Philadelphia, for the help I received in administrative and financial aspects of the ASI and in the long dragging task of correcting, typing, and retyping the manuscripts for camera reproduction.

The help I received from Patricia Vann and her editorial staff at Plenum Publishing and from my secretary, Debbie Bronson, is also appreciated.

CONTENTS

PATHOPHYSIOLOGY, BIOCHEMISTRY, AND PHARMACOLOGY OF PLATELETS

J. Fraser Mustard

Department of Pathology
Health Sciences Centre 3N26H, McMaster University
1200 Main Street West
Hamilton, Ontario, L8N 3Z5

INTRODUCTION

This subject will be presented in relation to the study of platelets and vascular disease. Platelets are involved not only in thrombosis but also in atherosclerosis, which is a key factor in the development of arterial thrombosis. The study of arterial disease and thrombosis in living subjects has been difficult.

Attempts to study vascular injury, vascular disease, and thrombosis in experimental animals and humans are limited because of our inability to monitor these processes in living subjects effectively. The development of methods for imaging internal organs and for detecting the deposition of radioactively labeled materials have made it possible to monitor some of these processes in living subjects.

Detailed reviews of the points presented in this paper can be found in the reviews listed in the References(1-9).

Among the aspects of vascular disease that could be assessed by imaging techniques are the following:

1. atherosclerotic lesions,

2. endothelial injury,

3. spasm,

4. thrombosis,

5. embolism.

ATHEROSCLEROSIS

Monitoring atherosclerotic lesions will probably require techniques such as NMR, doppler ultrasound, and positron emission tomography as well as angiographic techniques to provide effective ways of studying the progression and regression of atherosclerotic lesions. Relatively little can be done to study atherosclerotic lesions using labeled platelets, although deposition of labeled white blood cells, particularly monocytes, might be an indicator of some forms of active atherosclerosis where monocytes are playing a key role in the development of the lesions.

ENDOTHELIAL INJURY

Endothelial injury can be an initiating event in the development of atherosclerosis, in arterial spasm, and in the mechanisms involved in the formation of arterial thrombi. Therefore, a way of studying endothelial injury and endothelial alteration in intact living subjects would be desirable. At present there are no established methods for studying this, although there are some possibilities to which I will refer. Principally, these focus on the detection of products formed and secreted by the endothelial cells which may reflect stimulation or injury, for example, von Willebrand factor and plasminogen activator, and the effects of products from stimulated endothelium on blood constituents such as platelets and possibly white cells, that could influence their turnover in the circulation.

SPASM

The role of spasm in inducing the clinical complications of atherosclerosis is still controversial, but during the past decade an overwhelming body of evidence has developed that indicates that coronary artery spasm does occur, particularly in relation to variant angina, and that this can be a factor causing coronary ischemia. Spasm tends to be associated with advanced atherosclerotic disease and appears to occur most frequently at sites of extensive narrowing of the vessels as a consequence of atherosclerosis. Studying this process is difficult because the only reliable method that can be used at the present time in man is coronary artery angiography. Some of the newer techniques may, of course, make it possible to study spasm without having to inject contrast media.

THROMBOSIS

Thrombosis can be studied by some of the imaging methods and more specifically by studying the accumulation of radioactive constituents of the thrombi. The accumulation of iodine-125-labeled fibrinogen and possibly other proteins such as plasminogen, chromium-51 or indium-111-labeled platelets or chromium-51 or indium-111-labeled leukocytes can be examined. This is obviously a major focus of this meeting.

EMBOLISM

We are less successful in studying the fragments of thrombi that break off and embolize into the distal circulation. However, as I shall try to point out, this is an extremely important event that needs an approach that will allow studies in living subjects.

The next section of this paper outlines the processes that are involved in the initiation and development of vascular disease and its thromboembolic complications with particular emphasis on platelets and the implications for any in vivo monitoring techniques.

ENDOTHELIAL INJURY

Perhaps the most important point to be made in relation to endothelium is that platelets do not interact with normal endothelium; scanning electron microscopy of a normal aorta shows no platelets adherent to the endothelium. In contrast, white cells do interact with both normal and altered endothelium. The nature or importance of this white cell interaction with the endothelium has not been established, but there is some evidence that the white cells, such as polymorphonuclear leukocytes and macrophages, may be able to cause endothelial cell alteration and even loss of the endothelium and that macrophage accumulation on the subendothelium can contribute to the development of atherosclerosis. Recently it has been shown that platelets can adhere to foam cells formed from macrophages in atherosclerotic lesions when the endothelium covering the macrophage is lost.

VESSEL INJURY AND THROMBOSIS

When the endothelium is lost from the surface of the vessel, platelets generally form a thin layer over most of the surface where blood flow is mainly laminar. The adherent platelets appear to be in contact with collagen, basement membrane, and the microfibrils associated with the elastin in the subendothelium. Platelets adherent to collagen discharge their granule contents and the arachidonate pathway is activated leading to TXA_2 formation. If von Willebrand factor is available, platelets adherent to the microfibrils also undergo a release reaction. In areas where blood flow is not disturbed, no thrombi accumulate and only a single platelet layer adheres to the surface of freshly damaged normal arteries. Obviously the accumulation of platelets on a freshly injured surface could be detected using isotopically labeled platelets. However, two factors can influence this: the number of platelets that accumulate and the duration of reactivity of the injury site.

Within 30 minutes of the initial formation of the platelet layer on the subendothelium the platelet-covered surface is largely nonreactive. Platelets are gradually lost from the subendothelium,

during the first few days after loss of the endothelium. The surface
that is exposed to the circulating blood is nonreactive to circulating
platelets and does not activate the coagulation process. Thus, in
studies of injury in arteries, if the radioactive platelets are given
before the injury occurs, radioactive platelets will accumulate on the
surface and may possibly be detected by gamma scanning. However, if
the radioactive platelets are given after the injury has occurred,
they will not displace the initial layer of nonradioactive platelets
that has formed on the surface and therefore it is very unlikely that
one will be able to detect the injury sites by infusing radioactive
platelets.

At present, we do not know what makes these vessels nonreactive,
but we do know that it is not PGI_2, and it is probably not related to
the adsorption of plasma proteins onto the surface of the vessel.

When a normal vessel is injured with a balloon catheter,
platelets accumulate on the subendothelium and release their granule
contents, including the platelet-derived growth factor (PDGF), which
is mitogenic for smooth muscle cells. Smooth muscle cell migration
and proliferation leads to a thickening of the vessel wall within
seven days following removal of the endothelium. The surface of this
smooth muscle cell intimal thickening exposed to blood is
nonthrombogenic.

When a diseased vessel or previously damaged vessel is injured,
the process that occurs on the surface is different from that which
occurs on the freshly exposed subendothelium of a normal blood vessel.
When the neointima is damaged, platelet-fibrin thrombi form on much of
the surface of the vessel. In some areas there are few thrombi, but
these surfaces are covered with a platelet layer such as seen on the
subendothelium. Thus, the interaction of the constituents of the
blood with a damaged neointima appears to be more complicated than
with the subendothelium. Results from a number of experiments
indicate that in these circumstances, platelet-fibrin thrombi are
probably produced by activation of the coagulation mechanism, with
resultant thrombin generation. The thrombin induces both fibrin
formation and platelet aggregation, and a complicated thrombus forms
in which the platelets are not in contact with connective tissue of
the vessel wall but are mainly in contact with fibrin which is
adherent to the vessel wall. The formation of these platelet fibrin
thrombi can be inhibited by heparin, but, in contrast, the adherence
of the platelets to the subendothelium is not influenced by the doses
of heparin that block the formation of the platelet fibrin thombi on
the injured neointima. The injured neointima behaves in a similar
manner to the exposed subendothelium. That is, the platelet-fibrin
thrombi are lost from the surface of the vessel over the next 24 to 48
hours and there is little fresh platelet or fibrin accumulation on the
surface.

Again, assessment of thrombosis at such injury sites will be dependent upon the radioactive material being given before the injury occurs. If radioactive material is given after the injury, relatively little radioactive material will accumulate at the injury site, making detection of the injury site difficult. Thus, we can conclude from studies in experimental animals that thrombus formation in arteries, regardless of whether the mechanism initiating the accumulation of thrombi is mediated by connective tissue or mediated by thrombin, is a limited process that will be difficult to study using radioactive materials unless the materials are present before the injury occurs or there is repeated or continuous injury to the vessel wall.

In contrast to vessel walls, prosthetic grafts show a different behavior toward platelets. On the basis of the published evidence it appears that many prosthetic surfaces tend to remain reactive to circulating platelets and coagulation for a long period of time after their insertion. Thus the administration of radioactive platelets to subjects or animals with prosthetic grafts will detect platelet accumulation on the surface over a long period of time. On some of these grafts there is considerable fibrin formation as well, and therefore the process can be monitored by examining the accumulation of radioactive platelets or radioactive fibrinogen on the grafts or measuring platelet survival which may be shortened.

We should keep in mind that a nonthrombogenic surface can be an endothelial surface, a smooth muscle cell surface, a subendothelium covered with the initial platelet layer after it has undergone the initial reaction, the subendothelium after the platelets have been lost from the surface, or a damaged smooth muscle cell layer after the initial reaction has occurred. Thus, it is not surprising to find ulcerated atherosclerotic lesions with little evidence of thrombus formation on them.

If one examines the carotid artery bifurcation where atherosclerosis commonly occurs and where thromboembolic events are believed to initiate in many subjects, one can find a range of morphological appearances. There may be advanced atherosclerosis with thrombosis, advanced atherosclerosis without ulceration or thrombosis, or advanced atherosclerosis with ulceration and no thrombosis. At present we do not know what makes a surface of an ulcerated lesion nonthrombogenic. The study of arterial thrombosis is enormously difficult and since at present we have no simple way of predicting when an injury will occur that will initiate thrombosis, it is difficult to monitor thrombotic processes by means of circulating radioactive materials. It is important to keep in mind that this differs to some extent from venous thrombosis. When a thrombus occurs in a vein, the thrombus tends to persist for a considerable period of time and the fibrinogen in the thrombus turns over, making it possible to give radioactive fibrinogen and demonstrate its accumulation in the thrombus.

Occlusive thrombi in arteries tend to occur in areas where there is marked narrowing of the blood vessels in association with disturbances in blood flow. As such sites, the shear stresses may not only injure the wall but obviously augment platelet interaction with the wall and with each other. If marked narrowing is produced in a normal coronary artery, platelet aggregates will form spontaneously and may occlude the lumen for short intervals at such sites. Under these circumstances, it is possible to observe repeated formation and dissolution of platelet thrombi. Obviously the formation and dissolution of these thrombi could be monitored by the use of radioactive platelets. It is not entirely clear what occurs at such sites, but two points are important. First, when the rate of blood flow is greatly increased, it creates high shear rates that can disrupt the endothelium. This tends to occur in areas where the lumen is very narrow and induces strong movement of platelets to the vessel wall, leading to the rapid formation of platelet aggregates. This is probably one of the main mechanisms that is involved in hemostasis in small blood vessels. Thus, at stenotic lesions, it is possible that the effects of flow could induce endothelial injury and the formation of platelet thrombi and platelet-fibrin thrombi. Presumably these sites will remain at risk for thrombosis as long as the flow effects cause platelet accumulation and/or endothelial injury caused by the flowing blood occurs repeatedly. Thus one could have situations in which there is frequent and repeated induction of the thrombotic process as a result of hemodynamic forces. It should be possible to detect such thrombi with isotopic techniques. Areas in the human arterial tree that are exposed to conditions that could allow for the study of recurrent thrombotic episodes are a stenotic carotid artery and areas of stenosis in the peripheral arteries of the lower limb particularly at bifurcations and branches involving the femoral and popliteal arteries where marked atherosclerotic narrowing occurs.

SPASM AND THROMBOSIS

When a vessel is injured, it goes into spasm. This is well documented in studies of hemostasis. Thus, it is not surprising that arteries with advanced atherosclerosis can go into spasm. It has been demonstrated that the smooth muscle cells that proliferate in the intima following endothelial injury appear to be more sensitive to agents such as thromboxane A_2 and serotonin that induce spasm than the smooth muscle cells in the media. The problems here are how to study spasm in relation to ischemic events, and how to assess its contribution to thrombosis.

When spasm occurs it can alter blood flow and can facilitate thrombus formation. Spasm may cause sufficient narrowing to produce some ischemia and then the vessel may relax, allowing normal blood flow to resume. However, if a thrombus is formed, it may continue to block the lumen of the blood vessel after the vessel relaxes. Alternatively, the thrombus may be dislodged when blood flow resumes

and there may be no permanent injury. The dislodged thrombi may
fragment and embolize the microcirculation. At present, the only
technique we have for studying the process of spasm is during coronary
artery catheterization and angiography. It may, however, be possible
to study the thrombus formation that occurs in association with sites
that go into spasm using radioactive constituents such as platelets or
fibrinogen.

PATHWAYS IN THROMBOSIS AND ANTITHROMBOTIC AGENTS

Since there is considerable interest in the pathways in
thrombosis and the effect of drugs on these pathways, I shall briefly
summarize what we now know about this process.

As indicated in an earlier section, when a normal subendothelium
is exposed on a normal blood vessel by removal of the endothelium with
a balloon catheter, a thin layer of platelets accumulates on the
subendothelium. In an area of disturbed flow, extensive platelet
aggregation with fibrin formation can occur. The platelets stick to
the subendothelium (particularly to the collagen, microfibrils, and
the basement membrane). The arachidonate pathway is activated,
leading to freeing of arachidonic acid which is converted by
cyclo-oxygenase and thromboxane synthetase to thromboxane A_2. ADP and
other granule constituents, including serotonin, are released from the
granules of platelets adherent to collagen and microfibrils in the
presence of von Willebrand factor. The thromboxane A_2, ADP, and
serotonin cause platelets to change shape so they can stick to each
other and to the platelets adherent to the wall. Fibrinogen is
involved in the aggregation process by linking the platelets through
receptors (glycoproteins IIb and IIIa) that are not normally available
but become available when the platelets are stimulated with agonists
such as ADP or thromboxane A_2. These platelet aggregating agents act
synergistically so that very low concentrations are effective. The
fibrinogen receptor rapidly becomes unavailable again, and the
platelets deaggregate unless other mechanisms come into play to
stabilize the aggregates. Thrombin forms around the aggregated
platelets in part as a result of changes in the platelet membrane
allowing clotting factors to interact with membrane phospholids and
accelerates the coagulation process. Thrombin can induce platelet
aggregation independently of the release reaction, it can activate the
arachidonate pathway with the formation of thromboxane A_2, and it can
cause further release of granule contents. In all these ways it
causes further platelet aggregation. Thrombin also causes fibrin
formation around the aggregated platelets.

The mechanisms involved in platelet aggregation include the
induction of the receptors for fibrinogen and platelet adherence to
these receptors. All agents that cause platelets to aggregate appear
to induce a change in glycoproteins IIb and IIIa that allow fibrinogen
to bind to them causing crosslinking of the platelets and platelet

aggregation. The receptors are not usually maintained in a state that allows fibrinogen binding, and, therefore, when the receptor configuration reverses to the nonreactive state, the fibrinogen is lost from the platelets and the platelets deaggregate. This is probably an important factor in the reversibility of platelet thrombi and the formation of platelet emboli that may show the microcirculation distal to the thrombus and cause disturbances in the microcirculation of the affected organ.

When thrombin appears to be the primary initiating pathway at a site of repeated vessel injury or a diseased vessel wall, the pathways in thrombosis have a different emphasis. In this case, damage to cells lining the vessel probably activates the extrinsic pathway of coagulation. Thrombin formed at the surface of these damaged cells causes fibrin formation and the thrombin also causes the platelets to aggregate and adhere to the polymerizing fibrin on the wall. The formation of this type of thrombus is difficult to inhibit because thrombin induces platelet aggregation and the release reaction through pathways that are independent of ADP and thromboxane A_2. Because of this, aspirin has very little inhibitory effect on this type of thrombotic process, whereas it has some effect on thrombi induced by exposed connective tissue. In contrast, inhibitors of coagulation are highly effective in blocking thrombin-mediated thrombi. As indicated earlier, it seems most likely that both types of thrombotic processes are occurring in human subjects, that is, thrombosis initiated by platelet interactions with connective tissue contituents and thrombosis initiated by thrombin. This makes the treatment of arterial thrombosis difficult. Also, it would be of some help if we could determine the frequency of the different types of thrombi in living subjects.

One approach to handling this therapeutic problem has been to use drugs that inhibit coagulation in combination with drugs that inhibit some of the reactions of platelets. Such a combination is an anticoagulant with a nonsteroidal anti-inflammatory drug such as aspirin. However, the problem with this is that hemostasis is also significantly impaired and excessive bleeding occurs. This problem has been known for years by those involved in the treatment of hemophiliacs; aspirin is a dangerous drug for hemophiliacs. In recent controlled trial at the Mayo Clinic in which aspirin and oral anticoagulants were given to patients who received valve prostheses, the high incidence of bleeding forced the discontinuation of the administration of this combination of drugs.

Some drugs inhibit platelet adhesion to the subendothelium. PGI_2 infusions and dipyridamole inhibit platelet adhesion to the subendothelium. If drugs such as dipyridamole or PGI_2 are used at a high-enough concentration to inhibit platelet adherence to the subendothelium and their administration is continued for 6 to 8 hours, platelet accumulation does not occur when the therapy is stopped.

When dipyridamole is given over an 8-hour period after injury, platelet accumulation on the subendothelium of rabbit aortas is significantly less than for controls. Few additional platelets accumulate in an 8-hour period when the dipyridamole is discontinued. These observations have some interesting applications in the management of some aspects of arterial disease and in maintaining the patency of bypass grafts.

The group at the Mayo Clinic is now giving dipyridamole before the grafts are put in place and then adding aspirin to the drug therapy after insertion of the grafts. This regimen has proved effective in reducing the incidence of graft occlusion, probably because it inhibits the interaction of platelets when the grafts are inserted so that smooth muscle cell proliferatin is not stimulated to as great an extent as it would be without dipyridamole. This theory is supported by the experimental observation that if animals are made thrombocytopenic before the endothelium is removed, smooth muscle cell proliferation does not occur; but if the animals are made thrombocytopenic 4 hours after the initial inhibited. As indicated earlier, it is the initial platelet interaction with the damaged wall which is important and subsequent to this the vessel remains relatively nonreactive unless it is activated by some injury process.

PLATELET SURVIVAL

Platelet survival is influenced by a number of factors, one of which is vessel injury. We now know that a single injury to the aorta, such as removing the endothelium, or injury to the neointima of an aorta with a ballon catheter, does not shorten platelet survival. This is certainly understandable since the wall is reactive for such a short time that it is does not produce a significant effect on the circulating platelet population. However, when experiments are carried out using techniques that cause repeated or continuous injury to the aorta, platelet survival is shortened. It seems likely that experimental studies of injury to the vessel wall associated with shortened platelet survival have been done in circumstances in which some form of repeated or continuous injury to the wall has occurred. With repeated injury to the vessel wall, the surface repeatedly becomes reactive to the circulating platelets so that platelet interaction with the surface continues during the period of repeated injuries. In these circumstances, it would be possible to monitor platelet accumulation on a vessel wall being subjected to repeated injuries. Thus, in the study of arterial disease, finding continuous platelet accumulation at a site in the vascular system may indicate that the site is undergoing some form of continuous stimulation or injury. Of course, in diseased vessels, we do not know whether passivation of the surface always occurs, and it could be that an injury site may tend to remain reactive. Some of the vascular grafts can remain reactive in this way.

At present we do not know why the platelet survival is shortened with repeated vessel injury, but a hypothesis can be proposed that the turnover of the platelets on the vessel wall is responsible. Platelets that interact with the vessel wall ultimately disappear from the vessel wall possibly because of the effect of proteolytic enzymes that cleave the proteins or glycoproteins that bind adherent platelets to injury sites. It has been shown that removal of glycopeptides from the surface of platelets by treatment with proteolytic enzymes leads to a rapid clearance of the platelets when they are returned to the circulation.

In addition, we found in studies in which repeated injury to the vessel wall of experimental animals was caused by indwelling aortic catheters that the administration of epsilon aminocaproic acid, an inhibitor of plasmin activity, lengthened the shortened platelet survival. It is is possible that the indwelling aortic catheter stimulated or altered the endothelial cells leading to increased plasminogen activator activity in the plasma; plasmin could be formed which could cause the platelets to be freed from the vessel wall. Thus platelets would have altered membrane glycoproteins so that their survival is shortened.

Drugs that inhibit platelet adherence to the vessel wall, such as dipyridamole, prolong shortened platelet survival that has resulted from repeated vessel injury. However, this does not explain all the observations, because ticlopidine also prolongs shortened platelet survival but does not appear to inhibit platelet interaction with the damaged vessel wall. Its mechanism of action has not been established.

It seems reasonable to conclude, therefore, that changes in platelet survival do reflect repeated vessel injury and that the determination of platelet survival can be used as one approach for studying repeated vessel injury.

Continuous vessel injury may be one of the reasons shortened platelet survival is associated with cigarette smoking. One of the theories is that cigarette smoking alters, stimulates, or injures the endothelium. Although there is no conclusive evidence on this point, a recent study of the effect of smoking in rats has demonstrated that smoking causes platelets to adhere to the surface of the rat aorta, particularly around vessel orifice.

Hemocysteine has also been implicated in shortening platelet survival in patients with homocysteinemia. Homocysteine is known to cause endothelial injury. Here the situation is complicated, however, by the fact that consistent findings have not been reported by all the investigators. About half the studies indicate that patients with homocysteinemia have shortened platelet survival, whereas the other half have failed to demonstrate this. There is no known explanation for this difference.

Diabetes is another condition in which shortened platelet survival is observed; in this condition, also, there is a reasonable probability of change to the vessel wall. In diabetics, the plasma contains abnormally high amounts of von Willebrand factor activity; this may be a reflection of endothelial cell alteration or stimulation since von Willebrand factor is produced by endothelial cells.

While platelet survival may be a measure of repeated injury to a vessel wall, it is not specific for this and is difficult to do routinely because of the duration of the study. Other methods are needed. One approach is to measure in the blood products formed or released by the endothelium; von Willebrand factor is such substance, and elevated plasma levels have been demonstrated in association with endothelial cell stimulation and injury.

SUMMARY

Study of vascular disease and its complication is difficult in living subjects. The key factors that need to be monitored are the growth and regression atherosclerotic lesions and endothelial cell injury. It seems likely that if endothelial injury does not occur, the extent of atherosclerosis and its complications, such as spasm or thrombosis, will be minimal.

Current isotopic techniques are of relatively little use for studying atherosclerotic lesions or endothelial injury. The reason they are limited in value in studying endothelial injury is that the time of reactivity of the blood constituents with the surface at the injury site is relatively limited, making it difficult to catch the actual point in time when the injury occurs.

The study of thrombosis using the labeled platelets and fibrinogen could be useful, however, in certain circumstances in which a repeated stimulus for thrombus formation is occurring. Such events may occur in markedly stenosed lesions in the coronary arteries, in, association with diseased vessels in the peripheral vascular system, and in the carotid arteries.

Attempts at imaging thrombosis using our present approaches with radioactive labels in the arterial system are unlikely to be successful in conditions other than those in which repeated, frequent injury occurs. I suspect that these techniques will be applicable only in a small proportion of clinical situations in comparison with the overall problem of arterial disease and complications in the peripheral vascular, coronary, and cerebral circulations.

However, it is important that we do develop effective approaches to studying vessel injury, platelets, atherosclerosis, and thrombosis in living subjects. It will be virtually impossible to develop and secure new therapies for management of arterial thrombosis unless we

can develop methods for studying arterial thromboembolism in living
subjects. Large scale clinical trials with death as the end point are
far too expensive and unjustified now for establishing the possible
benefit of all new drugs. Only drugs that have been shown to be
effective in inhibiting arterial thrombosis in small scale trials with
a specific end point should be put into large scale clinical trials in
man.

REFERENCES

1. S. M. Schwartz, C. M. Gajdusek, S. C. Selden III, Vascular wall
   growth control: The role of the endothelium, Arteriosclerosis
   1:107 (1981).
2. R. Ross, Atherosclerosis: a problem of the biology of arterial wall
   cells and their interactions with blood components,
   Arteriosclerosis 1:293 (1981).
3. R. L. Kinlough-Rathbone, M. A. Packham, J. F. Mustard, Vessel
   injury, platelet adherence and platelet survival,
   Arteriosclerosis 3:(in press) (1983).
4. B. B. Weksler, Prostacyclin, in: "Progress in Hemostasis and
   Thrombosis," vol. 6:113, T. H. Spaet, ed., Grune & Stratton, New
   York (1982).
5. A. Maseri, S. Severi, P. Marzullo, Role of coronary arterial spasm
   in sudden coronary ischemic death, Ann NY Acad Sci 382:204
   (1982).
6. J. F. Mustard, R. L. Kinlough-Rathbone, M. A. Packham, Aspirin in
   the treatment of cardiovascular disease: a review, in: "Am J
   Med Proceedings of a Symposium - New Perspectives on Aspirin
   Therapy," L. Lasagna and F. G. McMahon, eds. (1983).
7. J. F. Mustard, R. L. Kinlough-Rathbone, M. A. Packham, Vessel
   injury, thrombosis, and the progression and regression of
   atherosclerotic lesions, in: "Clinical Diagnosis and
   Atherosclerosis. Quantitative Methods of Evaluation," M. G.
   Bond, W. Insull, Jr., S. Glagov, A. B. Chandler, F. Cornhill,
   eds., Springer-Verlag, New York (1983).
8. J. F. Mustard, R. L. Kinlough-Rathbone, M. A. Packham, The vessel
   wall and thrombosis, in: "Hemostasis and Thrombosis," R. W.
   Colman, J. Hirsh, V. Marder, E. W. Salzman, eds., Lippincott,
   Philadelphia (1982).
9. J. F. Mustard, M. A. Packham, R. L. Kinlough-Rathbone, Mechanisms in
   thrombosis, in: " Haemostasis and Thrombosis," A. L. Bloom, D. P.
   Thomas, eds., Churchill Livingstone, New York (1981).

PHYSIOLOGY AND PATHOPHYSIOLOGY OF HUMAN NEUTROPHILS

Ron S. Weening and Dirk Roos*

*Central Laboratory of the Netherlands
Red Cross Blood Transfusion Service and
Laboratory for Experimental and Clinical Immunology
of the University of Amsterdam

Department of Pediatrics
Academic Medical Center
Amsterdam, The Netherlands

INTRODUCTION

Circulating blood phagocytes, including neutrophils (polymorphonuclear leukocytes [PMN]) as well as monocytes, can recognize, ingest,
kill, and, to a certain extent, digest microbes. These blood cells,
therefore, are of crucial importance in host defense against invading
microorganisms. Disorders in this line of defense are accompanied by
serious, recurrent bacterial and fungal infections. In the last
decade, a number of these often inherited disorders have been
classified and their origin recognized.

For a good understanding of PMN function, we will briefly discuss
the series of events leading to the final elimination of microorganisms by these cells. These events include granulopoiesis and
maturation, chemotaxis, opsonization, adherence, phagocytosis and
intracellular killing (Fig. 1). (For a handbook, see Klebanoff and
Clark [1].) Abnormalities in any of these events may result in an
undue susceptibility to bacterial infections.

Granulopoiesis and Maturation

Mature PMN are highly-specialized, nondividing, short-living cells
(circulation half-time, 6 to 7 hours; survival in tissues, 4 to 5
days), with a diameter of 12-15 μm, a multilobulated nucleus, and many
cytoplasmic granules. These cells are produced from a hematopoietic
stem cell in the bone marrow in approximately 15 days. The first
stage, the myeloblast, can be recognized by its large nucleus, some

nucleoli, and the absence of cytoplasmic granules. By cell division, the promyelocyte is formed. In this cell, the first type of cytoplasmic granules, the primary, or azurophil, granules are found. From the following division, the myelocyte derives. At this stage, a second type of granule is formed: the secondary, or specific, granule (2). Beyond the myelocyte stage, the cells do not divide anymore but differentiate via the metamyelocyte and band form within approximately 10 days to mature PMN. These mature PMN thus contain at least two classes of granules, each containing different proteins and enzymes (3) (Table 1). As discussed later, fusion of these granules and the subsequent liberation of their contents into the phagocytic vacuoles are important processes in the destruction of phagocytosed material.

Most likely, humoral factors are regulating the number of neutrophil precursors and the rate of neutrophil production. Among such factors are the "colony stimulating factor" (4) and factors derived

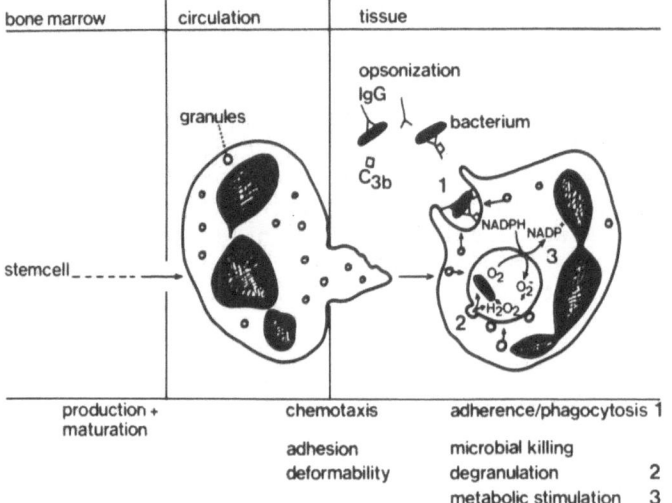

Fig. 1.   The neutrophilic response to microbial invasion. After maturation in the bone marrow, the neutrophils are released into the circulation. In response to a chemotactic stimulus, the cells migrate into the tissues. This process involves adhesion of the neutrophils to the blood vessel walls and deformation, in order to crawl between the endothelial cells and through the basal membranes. In the tissues, bacteria opsonized by IgG and/or C3b are bound, ingested, and killed. The last process involves generation of antimicrobial oxygen species and release of lysosomal enzymes into the phagosomes.

Table 1. Characteristics of Granules in Human PMN

	Azurophil	Specific
% of total granules	33	67
density (determined in 0.34 M sucrose) (g/ml)	1.23	1.19
size (μm)	0.3	0.13
electron density	dense	translucent

Enzymes

acid hydrolases (e.g., β-glucuro- nidase)	+	−
myeloperoxidase	+	−
cationic proteins	+	−
neutral proteases	+	−
lysozyme	+	+
lactoferrin	−	+
vitamin-B12-binding protein	−	+

from granulocyte breakdown. Maturation and release of neutrophils may also be regulated by humoral factors, which possibly affect cell stickiness and deformability (5) as well as the permeability of the marrow vascular space. The release of neutrophils from the bone marrow (normally 1.6×10^9 PMN/kg body weight per day) can be stimulated by endotoxin (releasing the marginating pool), etiocholanolone, and hydrocortisone. Our knowledge of the mechanisms controlling the production and the release of PMN from the bone marrow is far from complete. Sufficient numbers of circulating granulocytes are necessary for an adequate elimination of microorganisms. Several conditions are known in which the granulocyte concentration in the circulation is below 1500 per cubic mm (neutropenia).

Chemotaxis

Neutrophils can move in an amoeboid fashion. These movements are nondirectional and have been defined as random migration. The reaction by which the speed of these nondirection movements is enhanced is called chemokinesis. If a directional movement of cells towards the source of a chemical gradient is occurring, the term chemotaxis is used. By this movement, the cells will finally accumulate at the source of the chemical stimulus. The factors that induce chemotaxis of cells are defined as chemotactic factors.

For the accumulation of PMN at inflammatory sites, a complicated series of events is needed (see Fig. 1). First, before leaving the vascular compartment, the cells have to sense the stimulus and have to give direction to their movement (chemotaxis). Second, the cells have to adhere to the endothelial lining of a blood vessel. Third, for diapedesis through the intercellular pores, the cells have to deform. Furthermore, feedback mechanisms (e.g., inhibitors of chemotaxis) regulate the influx of cells at the inflammatory site.

Electron-microscopic examinations have revealed a radial area of microfilaments (a mesh work of actin bundles of 4-6 nm with associated contractile proteins) and microtubules (tubular structures mainly consisting of tubulin) in the leading part of the cell (6). Microfilaments are the contractile elements in these cells, essential for locomotion because disruption of microfilaments by cytochalasin B inhibits cell movement (7). However, the action of this drug is rather complex and nonspecific (membrane transport mechanisms, for instance, are also affected). Therefore, the most convincing evidence that microfilaments are essential for cell mobility has been provided by the observations of Boxer et al (8) in a male infant with repeated infections. The neutrophils of this patient did not migrate in a chemotactic gradient, and the ingestion of bacteria was markedly depressed. The explanation for these findings proved to be an abnormal polymerization of actin in the filaments. It has been suggested that locomotion is the result of sliding of interdigitating actin and myosin filaments. Actin-binding protein, a Ca++ ion-dependent regulator protein (gelsolin), and some other proteins (acumentin, profilin) regulate the assembly and gelation of actin, whereas myosin contracts these gels.

Because inhibitors of microtubule assembly, such as high concentrations of colchicine and vinblastine, inhibit chemotaxis, it has been postulated that microtubules are important for the direction of the movement (9). Moreover, agents that induce microtubule assembly (e.g., C5a) have been found to promote chemotaxis (10). Most likely, the chemotactic response is energy-dependent, because this process is inhibited by inhibitors of glycolysis and by a low ATP content (e.g., in hypophosphatemia).

The requirement of ions for chemotactic responsiveness has long been disputed. Exchange (influx as well as efflux) of Ca++ ions has been reported (10, 11). Probably only redistribution of intracellular calcium is necessary and may be related to the assembly of microtubules and/or microfilaments. These ionic movements, induced by chemotactic stimuli, most probably lead to a decrease in the charge of the cell surface. In fact, Gallin (12) has proposed that the decrease in the net negative surface charge initiates the activation of cell movement.

There are a number of reports that deal with the role of cyclic AMP and cyclic GMP in leukocyte mobility. Agents that increase intracellular cyclic AMP (e.g., cAMP, β-adrenergic agents, theophylline, prostaglandin El, cholera toxin, histamine) inhibit chemotaxis (13, 14). In agreement with the Yin-Yang hypothesis (15), which proposes opposite actions of the two cyclic nucleotides, agents that increase intracellular cyclic GMP (e.g., imidazole, carbamylcholine, cGMP) enhance cell mobility (16). However, a strict quantitative relationship between the change in cyclic nucleotide levels and the chemotactic response has not always been established.

Additional evidence for the involvement of these nucleotides in cell locomotion is derived from the observation of Boxer et al (16) and Weening et al (17) in patients with the Chediak-Higashi syndrome. In this rare, autosomal recessive condition, a diminished chemotactic responsiveness of the PMN toward chemoattractants has been found. It has been demonstrated that the PMN in this condition have abnormally high intracellular cyclic AMP levels. Ascorbic acid corrected the cyclic AMP levels as well as the functional defects (16, 17).

Electron microscopic studies have shown that agents that increase intracellular cyclic GMP lead to the appearance of microtubules. Therefore, cyclic nucleotides exert their effects on cell locomotion possibly by regulating microtubule assembly. The mechanism by which cyclic nucleotides cause these morphologic and functional changes is largely unknown, however. Because tubulin is the most prominent protein in microtubules, one may speculate that cyclic GMP promotes microtubule assembly by activating certain protein kinases.

In conclusion, it is assumed that chemotactic factors interact with sites on the cell surface, perhaps by interaction with serine esterases, inducing ionic movements and electropotential changes over the cell membrane. This signal is translated into cell movement by the action of microtubules and microfilaments. This energy-dependent process may be regulated by alterations in cytoplasmic calcium concentrations and cyclic nucleotide levels.

Chemoattractants

It will be clear from the foregoing that efficient generation of chemotactic factors in the host is a prerequisite for the chemotactic response of the phagocytic leukocytes. Among these factors, complement components activated by antigen-antibody complexes or bacteria (producing the chemotactic split product C5a), for example, are of major importance. Furthermore, chemotactic factors may derive from bacterial products (e.g., N-formyl-L-methionyl-L-leucyl-L-phenylalanine [FMLP]). Activation of the fibrinolytic system (e.g., plasminogen-activator, plasmin), the kinin-generating system (e.g., kallikrein), or the coagulation process (fibrino-peptides) also produces chemoattractants. The inflammatory process is limited by the release of inhibitors of chemotaxis. These mediators react either directly on the cells or by inactivation of the chemoattractants.

Adherence and Aggregation

PMN that are activated by soluble stimuli (e.g., FMLP, C5a) will become stickier in the presence of Ca++ ions than unstimulated cells. Probably a decrease in the charge of the cell surface is responsible for the increased adherence and margination. Ionic movements across the plasma membrane and/or binding of certain proteins to the plasma membrane (e.g., lactoferrin released from specific granules) may enhance this stickiness. This phenomenon will also lead to an enhanced adherence of the cells to each other (aggregation of cells), which can be measured in the aggregometer. This process is of clinical importance, because it explains why cells are sequestered in the lung after contact, for example, with a dialysis membrane.

Opsonization and Phagocytosis

One of the most intriguing and still unsolved questions in phagocyte physiology is the process of recognition, binding, and ingestion of foreign particles and old or damaged – but not undamaged – autologous cells by phagocytes. Possibly the degree of surface hydrophobicity and the overall surface charge of the particle are important for recognition and binding. Van Oss and Gillman (18) have suggested that only particles that are more hydrophobic and/or more positively charged than phagocytes will be ingested. These investigators have also provided evidence that the presence of opsonins (certain serum factors bound to the surface of particles) may promote phagocytosis by increasing the hydrophobicity of these particles (19). However, this cannot be the sole explanation, because from this point of view it is not clear why the phagocytic cells sometimes bind particles without subsequent pseudopod formation and engulfment. Obviously, this whole series of events is more complex than the physical theory suggests.

Opsonins and Opsonization

Binding of factors from fresh human serum to the surface of particles (opsonization) promotes their uptake by PMN (see Fig. 1). Serum immunoglobulins (heat-stable opsonins) and complement components (heat-labile opsonins) are involved in this process. Specific antibodies of the IgG class (especially of the IgG1 and IgG3 subclass) will bind with their F(ab) regions to the particles, whereas the Fc regions attach to the PMN surface. Such antibodies may also activate the classical complement pathway (Cl, C4, C2, and C3), and generate the opsonically active C3B fragment. Although IgG-coated micro-organisms can be phagocytosed without complement activation, C3b attachment enhances synergistically the rate of ingestion. Moreover, IgM antibodies are only opsonically active in the presence of C3b.

In the absence of specific antibodies, the complement system can be activated by bacteria via the alternative complement pathway. This pathway, which depends on the presence of Mg++ ions, properdin, Factor D, Factor B, and other proteins, also leads to C3b formation. Therefore, the presence of C3 is of major importance of the heat-labile opsonic activity. It is not clear whether C5 may also have some opsonic activity. Miller and Nillson (20) have described an opsonic defect for yeast particles in patients with Leiner's disease, due to an abnormal function of C5. This finding could not be confirmed, however, in a patient with hereditary C5 deficiency (21). Other phagocytosis-promoting serum factors are also known. Among these, a fraction of the Fc portion of a leukophilic γ-globulin (tuftsin) is of interest, because hereditary deficiency of this peptide has been described (22).

Adherence and Phagocytosis

As pointed out before, IgG and C3b promote the binding of particles to the plasma membrane of PMN (immune adherence). It has recently been established that specific receptors (Fc and C3b receptor) exist on the plasma membrane of PMN. Indeed, there are different kinetics of phagocytosis for both opsonins (23), and only few types of particle are bound (and phagocytosed) in the absence of serum (nonspecific adherence). Furthermore, monoclonal antibodies against the Fc receptor block the uptake via the Fc receptor, whereas the uptake via the C3b receptor is not affected. The opposite phenomena are found with antibodies against the C3b receptor.

The moment particles adhere to the plasma membrane of PMN, these cells are activated by an unknown mechanism: Redistribution of intra-cellular Ca++ seems to be important for transmitting the activation signal. Whether methylation of phospholipids is important is still unproven. No clear evidence exists to support a role for cyclic nucleotides. After adherence to the plasma membrane, pseudopods (parts of the plasma membrane) circumferentially surround opsonized particles by creeping from one opsonizing molecule to another (in this way it is conceivable why particles with opsonins on a part of their surface are bound but not ingested). This process has been called the "zipper" mechanism of phagocytosis (24). After ingestion, a phagocytic vacuole, the phagosome, is formed. The movement of pseudopods most likely depends on the presence of microfilaments in these extrusions, because disturbances in actin polymerization (see discussion on chemotaxis) decrease ingestion rates. Microtubules do not appear to be crucial to the phagocytic process. Engulfment is an energy-dependent process. In PMN, the energy for this process in the form of ATP is largely supplied by glycolysis (25). In hypophosphatemia (intravenous hyperalimentation), low ATP levels in PMN result in decreased phagocytosis. Restoration of ATP levels by incubation in the presence of adenosine or phosphate enhances the ingestion rate (26).

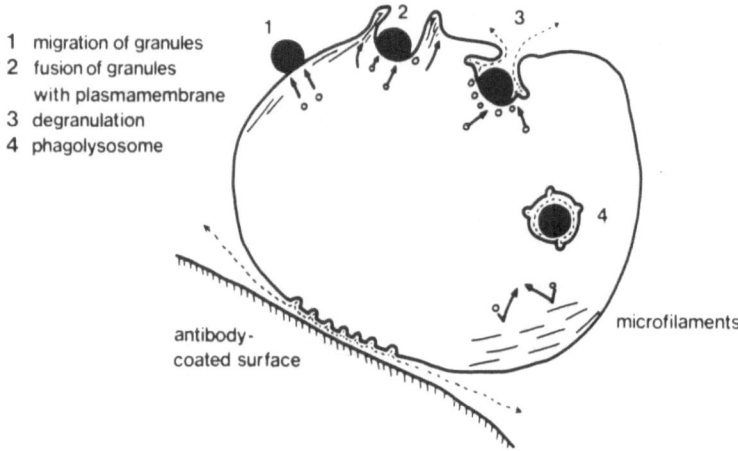

1 migration of granules
2 fusion of granules
    with plasmamembrane
3 degranulation
4 phagolysosome

antibody-
coated surface

microfilaments

Fig. 2.   Release of lysosomal enzymes. 1. Adherence of particles and
subsequent migration of granules. 2. Pseudopod formation.
3. Fusion of granules with the plasma membrane,
degranulation, and release of granular constituents into the
phagosomes. 4. Phagolysosome.

Degranulation

     As soon as the plasma membrane is perturbed by particle binding,
cytoplasmic granules move to the binding site, fuse with that part of
the plasma membrane that surrounds the phagocytic vacuole, and deliver
their enzyme content into the phagosome (degranulation). As shown in
Table 1, these granules contain a variety of digestive enzymes, which
contribute to the killing and degradation of microorganisms.

     During phagocytosis, lysosomal enzymes are also spilled extracel-
lularly, because degranulation occurs before the phagocytic vacuole is
closed (Fig. 2).

     This "regurgitation during feeding" may contribute considerably to
the tissue injury observed during inflammation. This phenomenon also
enables one to quantitatively measure this process in vitro.
Lysosomal enzyme release is not due to nonspecific cell lysis, because
cytoplasmic enzymes (e.g., lactate dehydrogenase) are not liberated
during phagocytosis. It has been shown that specific granules
degranulate earlier than azurophilic granules (27).

Most likely granule movement is initiated by membrane perturbation, because particle adherence suffices for degranulation (28).
Moreover, soluble agents (e.g., phorbol-myristate acetate [PMA] or
C5a) also induce lysosomal enzyme release (28). Although the exact
nature of the cytoplasmic granule movement is unknown, that process
may depend on microtubule assembly, because agents that disrupt
microtubules (e.g., colchicine or vinblastine) inhibit the release of
these enzymes. As described in the section on chemotaxis, assembly of
microtubules may be regulated by cyclic nucleotide levels. Indeed, it
has been shown that agents that increase the intracellular level of
cyclic GMP enhance lysosomal enzyme release, whereas cyclic AMP-
increasing drugs have an opposite effect (29). It has been proposed
that the microfilaments prevent the fusion of granules with the plasma
membrane in the resting cell (see Fig. 2). During ingestion,
microfilaments, closely linked to the plasma membrane, appear in the
pseudopods but dissapear from the original contact place.
Cytochalasin B, which supposedly disrupts microfilaments, enhances
lysosomal enzyme release (30). Moreover, in the patient with abnormal
actin polymerization (8), lysosomal enzyme release was enhanced.

The Respiratory Burst*

During phagocytosis, PMN are metabolically stimulated. Figure 3
shows a schematic representation of the reactions that take place
during ingestion.

Within 30 seconds after the initiation of phagocytosis, a sharp
increase in oxygen uptake by the cells occurs. This increase in cell
respiration (known as the "respiratory burst") is cyanide-insensitive
and thus nonmitochondrial (25). The consumed oxygen is converted
enzymatically into hydrogen peroxide (H_2O_2) and perhaps other reactive
oxygen metabolites (e.g., superoxide and hydroxyl radicals). It was
also found that glucose oxidation by the hexose monophosphate shunt
(HMP shunt) is stimulated severalfold during phagocytosis.

The reducing equivalents (NADPH + H+) for the reduction of oxygen
to hydrogen peroxide are probably derived from glucose via stimulation
of the HMP shunt (see Fig. 3). Additional stimulation of the HMP
shunt occurs by detoxification of excess H_2O_2 via the glutathione
cycle in the cytoplasm (Fig. 4).

The respiratory burst is initiated by, but not dependent on,
phagocytosis, because membrane perturbation either by particle binding
or by soluble agents (e.g., FMLP, PMA, C5a) will start these reactions
(28). Which process actually activates the oxygen-consuming enzyme(s)
is unknown. Segal et al (32) have found that the phagocytic vacuoles
of human neutrophils contain a cytochrome-b-like protein. The
characteristic spectrum of this protein is missing in the neutrophils
from most patients with chronic granulomatous disease (CGD).

---

*For a review, see Babior (31)

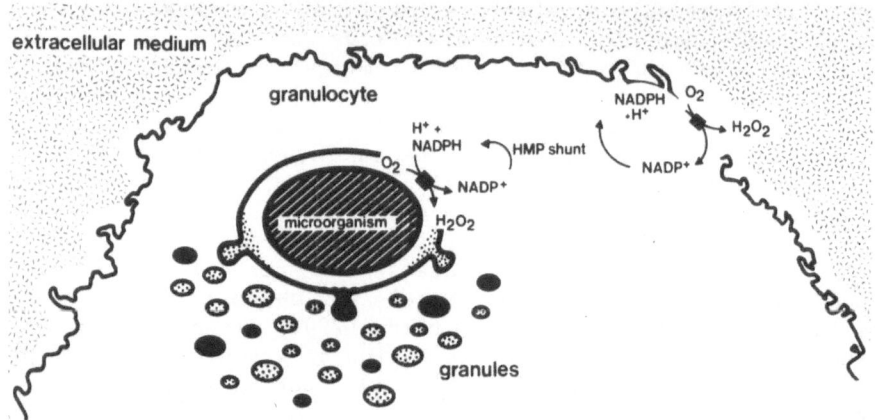

Fig. 3. Degranulation and metabolic activation. During and after
formation of the phagosomes, specific (light) and azurophil (dark)
granules fuse with the phagosomal membrane; the granular enzymes
diffuse into the phagosomal space. At the same time, NADPH oxidase
(dark area in membrane) is activated to produce H_2O_2. Reprinted with
permission of the copyright holder and authors from: D. Roos, M. N.
Hamers, R. van Zwieten and R. S. Weening, Acidification of the
phagocytic vacuole: a possible defect in chronic granulomatous
disease? <u>in</u>: "Advances in Host Defense Mechanisms, vol. 3, Chronic
Granulomatous Disease," J. I. Gallin and A. S. Fauci, eds., Raven
Press, New York (1983).

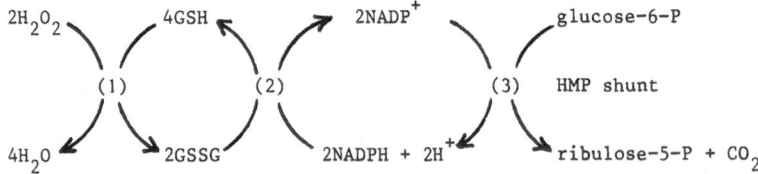

(1) Glutathione peroxidase
(2) Glutathione reductase
(3) Glucose-6-phosphate dehydrogenase + 6-phosphogluconate dehydrogenase

Fig. 4. Glutathione cycle.

Possibly, this cytochrome system is coupled to an NADPH-oxidizing flavoprotein (see Fig. 5). This figure also show a proton translocation from the cytoplasm into the extracellular medium and possibly into the phago(lyso)-somes. This movement of protons is strictly coupled to the oxidase system (33). The acidification of the phago(lyso)somes might (in part) be due to this proton translocation.

As shown in Figure 5, a model has been designed in which the oxidase system, located in the plasma membrane, is accessible for NADPH from the inside and for molecular oxygen (O_2) from the outside of the cell. This enzyme system is activated upon particle attachment and releases hydrogen peroxide and possibly other activated oxygen species at the outside of the membrane. During internalization of particles, the oxidative products will be released into the phagosome, in close contact with the ingested material. Excess hydrogen peroxide will be degraded in the cytoplasm by catalase and the glutathione system.

Oxidative Microbicidal Mechanisms

The decreased microbicidal activity of PMN to certain microorganisms under conditions of anaerobiosis and of PMN with a defect in the oxidative metabolism (e.g., CGD) has firmly established the importance of the oxygen-dependent microbicidal mechanisms. However, certain other strains of bacteria are easily killed by PMN under anaerobic conditions and by CGD cells; this indicates the simultaneous presence of oxygen-independent microbicidal mechanisms.

The oxygen-dependent antimicrobial systems can be divided into those that require myeloperoxidase (MPO) and those that do not (Table 2).

Table 2. Microbicidal Mechanisms

---

1. Oxygen-dependent, MPO-mediated
   Hypochlorous acid (HOCl)

2. Oxygen-dependent, MPO-independent
   Hydrogen peroxide ($H_2O_2^-$)
   Superoxide anion ($O2.-$)
   Hydroxyl radical (OH.)

3. Oxygen-independent
   Acid
   Lysozyme
   Lactoferrin
   Cationic proteins

Oxygen-Dependent, MPO-Mediated Killing Mechanisms

Hydrogen Peroxide, in combination with isolated MPO and a
halide (I^-, Br^-, or Cl^-), has potent microbicidal activity (34).
In addition, MPO-deficient PMN kill microorganisms less
efficiently than do normal cells (35). This illustrates the
importance of MPO in the oxidative killing reactions.

During ingestion, H_2O_2 and MPO are delivered into the
phagosome, whereas halides (especially Cl^-) are available
throughout the cell. In the MPO-mediated reaction, hypochlorous
acid (HOCl) is produced, which can perforate the bacterial wall
(this process can be measured in vitro [36]). In the presence of
chloride, decarboxylation of either free amino acids from the
coingested medium or decarboxylation of bacterial amino acids may

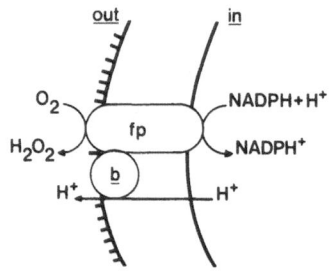

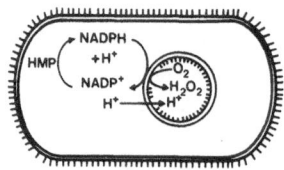

Fig. 5. Configuration of NADPH oxidase in the plasma membrane of
         phagocytes (top) and in the phagosomal membrane (bottom): fp,
         flavo-protein; b, cytochrome-b; out, extracellular medium;
         in, cytoplasm. The model shows proton translocation from the
         cytoplasm into the extracellular medium and into the
         phago(lyso)some. Reprinted with permission of the copyright
         holder and the authors from: D. Roos, M. N. Hamers, R. van
         Zwieten and R. S. Weening, Acidification of the phagocytic
         vacuole: a possible defect in chronic granulomatous disease?
         in: "Advances in Host Defense Mechanisms, vol. 3, Chronic
         Granulomatous Disease," J. I. Gallin and A. S. Fauci, eds.,
         Raven Press, New York (1983).

also produce toxic chloramines and aldehydes (37). Furthermore, hydroxyl radicals (OH.) may be produced in MPO-mediated reactions. So far, it is clear that at least high concentrations of H_2O_2 and hypochlorous acid can kill microorganisms.

Oxygen-Dependent, MPO-Independent Killing Mechanisms

Hydrogen peroxide itself has microbicidal activity, but only at relatively high concentrations. The antimicrobial properties of H_2O_2 may be potentiated by ascorbate (abundantly present in PMN) in the presence of certain metal ions (38) or lysozyme (39).

Superoxide anions (O_2^-), produced enzymatically by the one-electron reduction of oxygen, are highly reactive. They can act both as oxidants and as reductants. Although superoxide radicals would be expected to be strongly bactericidal, the presence of superoxide dismutase in virtually every aerobic organism probably explains why most of these organisms are resistant to this agent.

Possibly, hydroxyl radicals (OH.) are also formed in the respiratory burst. Recently, evidence for the production of OH. was derived from the finding that stimulated phagocytes liberate ethylene from methional (3-methylthiopropionaldehyde) and from 2-keto, 4-methylthiobutyric acid (40, 41). Absolute proof for the production of this radical is lacking by want of a specific identification system. The relative contribution of OH. radicals to bacterial killing in PMN is completely unknown.

Oxygen-Independent Microbicidal Mechanisms

Acid

From studies measuring ingestion of particles stained with indicator dyes, it has been derived that the intraphagosomal pH drops to approximately 4-6 within 3 to 15 minutes (42). The mechanisms of acidification are not known. The low pH may be microbicidal or microbiostatic to certain ingested microorganisms. Furthermore, the optimum of the MPO-mediated microbicidal reactions and of most of the lysosomal hydrolases is at acid pH.

Lysozyme

The enzyme lysozyme (present in azurophilic and specific granules) hydrolyses β-1,4-links between N-acetyl-muramic acid and 2-acetyl-2-deoxy-D-glucose residues in mucopolysaccharides or mucopeptides. Some kinds of bacterium are lysed by lysozyme. Most organisms, however, resist the action of lysozyme unless pretreated in some way (e.g., by proteases). Therefore, the function of lysozyme in vivo might be the digestion of bacteria that have been killed in other ways (43).

Lactoferrin

Iron is required as an essential nutrient for microbial growth; therefore, lactoferrin (present in specific granules) most probably has a microbiostatic effect by binding iron.

Granular Cationic Proteins

Among the cationic proteins, at least seven electrophoretically distinct proteins have been separated. These proteins were found to have antimicrobial activity against both gram-positive and gram-negative organisms (44). Little is known regarding their precise role in intact cells.

Lysosomal Hydrolases

In general, the lysosomal hydrolases have a digestive rather than a microbicidal function. Some of these enzymes (e.g., Cathepsin D) may also be toxic to certain microorganisms. Patients are known with a selective deficiency of specific granules in the PMN, with half-normal lysozyme levels, and undetectable lactoferrin activity (45). These patients have recurrent bacterial infections and a killing defect of PMN in vitro.

Clinical Relevance of Defective Microbicidal Mechanisms

There are a number of hereditary defects in microbial killing that illustrate the importance of the oxygen-dependent killing mechanisms in PMN.

Patients with chronic granulomatous disease (CGD) are very susceptible to severe infections, especially by staphylococcus aureus, enterobacteriaceae, and fungi. Predominantly, lymph nodes, skin, lung, and liver are affected (for reviews, see Klebanoff and Clark [1] and Johnston and Newman [46]).

The killing of certain microorganisms in vitro by CGD neutrophils as well as by CGD monocytes is defective. Upon incubation with bacteria, these cells fail to show the normal respiratory burst, probably because of a defect in the oxygen-consuming, hydrogen-peroxide-generating system. The causal relationship between the killing defect and the lack of H_2O_2 production is suggested by the partial restoration of the killing capacity in CGD cells during the simultaneous ingestion of an H_2O_2 generating system together with the bacteria (47).

In addition, H_2O_2 releasing bacteria (catalase-negative), such as lactobacilli, pneumococci, and certain streptoccocci, are readily killed by CGD cells (48). Presumably, the bacterial H_2O_2 is used for the destruction of these bacteria.

CGD is seen predominantly in male children. In most instances, an X-linked transmission has been demonstrated (classical form). The description of CGD in females and in siblings of opposite sex suggests either unequal inactivation of the X chromosome (Lyon hypothesis) or an autosomal transmitted variant.

REFERENCES

1. S. J. Klebanoff and R. A. Clark, "The Neutrophil: Function and Clinical Disorders," North-Holland Publishing Company, Amsterdam/New York/Oxford (1978).
2. D. F. Bainton, J. L. Ullyot, M. G. Farquhar, The development of neutrophilic polymorphonuclear leukocytes in human bone marrow. Origin and content of azurophil and specific granules, J Exp Med 134:907 (1971).
3. U. Bretz and M. Baggiolini, Biochemical and morphological characterization of azurophil and specific granules of human neutrophilic polymorphonuclear leukocytes, J Cell Biol 63:251 (1974).
4. P. A. Chervenick and D. R. Boggs, Bone marrow colonies: stimulation in vitro by supernatant from incubated human blood cells, Science 169:691 (1970).
5. M. A. Lichtman and R. I. Weed, Alteration of the cell periphery during granulocyte maturation: relationship to cell function, Blood 39:301 (1972).
6. J. I. Gallin, E. K. Gallin, H. L. Malech, E. B. Cramer, Structural and ionic events during leukocyte chemotaxis, in: "Leukocyte Chemotaxis: Methods, Physiology and Clinical Implications," J.I. Gallin and P.G. Quie, eds., Raven Press, New York (1978).
7. E. L. Becker, A. T. Davis, R. D. Estensen, P. G. Quie, Cytochalasin B. IV. Inhibition and stimulation of chemotaxis of rabbit and human polymorphonuclear leukocytes, J Immunol 108:396 (1972).
8. L. A. Boxer, E. T. Hedley-Whyte, T. P. Stossel, Neutrophil actin dysfunction and abnormal neutrophil behavior, New Engl J Med 293:1093 (1974).
9. E. L. Becker and H. J. Showell, The ability of chemotactic factors to induce lysosomal enzyme release. II. The mechanism of release, J Immunol 112:2055 (1974).
10. J. I. Gallin and A. S. Rosenthal, The regulatory role of divalent cations in human granulocyte chemotaxis: evidence for an association between calcium exchanges and microtubule assembly, J Cell Biol 62:594 (1974).
11. M. M. Boucek and R. Snyderman, Calcium influx requirement of human neutrophil chemotaxis: inhibition by lanthanum chloride, Science 194:905 (1976).
12. E. K. Gallin and J. I. Gallin, Interaction of chemotactic factors with human macrophages: induction of transmembrane potential changes, J Cell Biol 75:277 (1977).

13. I. Rivkin, J. Rosenblatt, E. L. Becker, The role of cyclic AMP in
     the chemotactic responsiveness and spontaneous motility of
     rabbit peritoneal neutrophils. The inhibition of neutrophil
     movement and the elevation of cyclic AMP levels by
     catecholamines, prostaglandins, theophylline, and cholera
     toxin, J Immunol 115:1126 (1975).
14. H. R. Hill, R. D. Estensen, P. G. Quie, N. A. Hogan, N. D.
     Goldberg, Modulation of human neutrophil chemotactic responses
     by cyclic 3',5'-guanosine monophosphate and cyclic
     3',5'-adenosine monophosphate, Metabolism 24:447 (1975).
15. N. D. Goldberg, M. K. Haddox, S. E. Nicol, D. B. Glass, C. H.
     Sanford, F. A. Kuehl, Jr., R. Estensen, Biologic regulation
     through opposing influences of cyclic GMP and cyclic AMP: the
     Yin-Yang hypothesis, in: "Advances in Cyclic Nucleotide
     Research," G. I. Drummond, P. Greengard, and G. A. Robison,
     eds., Raven Press, New York (1975).
16. L. A. Boxer, A. M. Watanabe, M. Rister, H. R. Besch, Jr., J.
     Allen, R. L. Baehner, Correction of leukocyte function in
     Chediak-Higashi syndrome by ascorbate, New Engl J Med 295:1041
     (1976).
17. R. S. Weening, E. P. Schoorel, D. Roos, M. L. J. van Schaik, A.
     A. Voetman, A. A. M. Bot, A. M. Batenburg-Plenter, Ch.
     Willems, W. P. Zeijlemaker, A. Astaldi, Effect of ascorbate on
     abnormal neutrophil, platelet, and lymphocyte function in a
     patient with the Chediak-Higashi syndrome, Blood 57:856
     (1981).
18. C. J. van Oss and C. F. Gillman, Phagocytosis as a surface
     phenomenon. I. Contact angles and phagocytosis of non-
     opsonized bacteria, J Reticuloendothel Soc 12:283 (1972).
19. C. J. van Oss and C. F. Gillman, Phagocytosis as a surface
     phenomenon. II. Contact angles and phagocytosis of
     encapsulated bacteria before and after opsonization by
     specific antiserum and complement, J Reticuloendothel Soc
     12:497 (1972).
20. M. E. Miller and U. S. Nillson, A familial deficiency of the
     phagocytosis-enhancing activity of serum related to a dysfunc-
     tion of the fifth component of complement (C5), New Engl J Med
     282:354 (1970).
21. S. I. Rosenfeld, J. Baum, R. T. Steigbigel, J. P. Leddy,
     Hereditary deficiency of the fifth component of complement in
     man. II. Biological properties of C5-deficient human serum,
     J Clin Invest 57:1635 (1976).
22. A. Constantopoulos, V. A. Najjar, J. W. Smith, Tuftsin deficien-
     cy: a new syndrome with defective phagocytosis, J Pediatr
     80:564 (1972).
23. T. P. Stossel, Phagocytosis: recognition and ingestion, Semin
     Hematol 12:83 (1975).
24. F. M. Griffin, Jr., J. A. Griffin, J. E. Leider, S. C.
     Silverstein, Studies on the mechanism of phagocytosis. I.
     Requirement for circumferential attachment of particle-bound

ligands to specific receptors on the macrophage plasma membrane, J Exp Med 142:1263 (1975).

25. A. J. Sbarra and M. L. Karnovsky, The biochemical basis of phagocytosis. I. Metabolic changes during the ingestion of particles by polymorphonuclear leukocytes, J Biol Chem 234:1355 (1959).

26. P. R. Craddock, Y. Yawata, L. van Santen, S. Gilberstadt, S. Silvis, H. S. Jacob, Acquired phagocyte dysfunction. A complication of the hypophosphatemia of parenteral hyperalimentation, New Engl J Med 290:1403 (1974).

27. D. F. Bainton, Sequential degranulation of the two types of polymorphonuclear leukocyte granules during phagocytosis of microorganisms, J Cell Biol 58:249 (1973).

28. I. M. Goldstein, D. Roos, H. B. Kaplan, G. Weissman, Complement and immunoglobulins stimulate superoxide production by human leukocytes independently of phagocytosis, J Clin Invest 56:1155 (1975).

29. G. Weissmann, I. Goldstein, S. Hoffstein, G. Chauvet, R. Robineaux, Yin-Yang modulation of lysosomal enzyme release from polymorphonuclear leukocytes by cyclic nucleotides, Ann NY Acad Sci 256:222 (1975).

30. P. Davies, A. C. Allison, R. I. Fox, M. Polyzonis, A. D. Haswell, The exocytosis of polymorphonuclear-leucocyte lysosomal enzymes induced by cytochalasin B, Biochem J 128:78 (1972).

31. B. M. Babior, Oxygen-dependent microbial killing by phagocytes, New Engl J Med 298:659 (1978).

32. A. W. Segal, D. Webster, O. T. G. Jones, A. C. Allison, Absence of a newly-described cytochrome b from neutrophils of patients with chronic granulomatous disease, Lancet ii:446 (1978).

33. R. van Zwieten, R. Wever, M. N. Hamers, R. S. Weening, D. Roos, Extracellular proton release by stimulated neutrophils, J Clin Invest 68:310 (1981).

34. S. J. Klebanoff, Iodination of bacteria: a bactericidal mechanism, J Exp Med 126:1063 (1967).

35. R. I. Lehrer and M. J. Cline, Leukocyte myeloperoxidase deficiency and disseminated candidiasis: the role of myeloperoxidase in resistance to candida infection, J Clin Invest 48:1478 (1969).

36. H. J. Sips and M. N. Hamers, Mechanism of the bactericidal action of myeloperoxidase: increase permeability of the Escherichia coli cell envelope, Infect Immun 31:11 (1981).

37. J. M. Zgliczynski, R. J. Selvaraj, B. B. Paul, T. Stelmaszynska, P. K. F. Poskitt, A. J. Sbarra, Chlorination by the myeloperoxidase-H_2O_2-Cl- antimicrobial system at acid and neutral pH, Proc Soc Exp Biol Med 154:418 (1977).

38. Y. Ericsson and H. Lundbeck, Antimicrobial effect in vitro on the ascorbic acid oxidation. I. Effect on bacteria, fungi, and viruses in pure culture, Acta Pathol Microbiol Scand 37:493 (1955).

39. T. E. Miller, Killing and lysis of gram-negative bacteria through the synergistic effect of hydrogen peroxide, ascorbic acid, and lysozyme, J Bacteriol 98:949 (1969).

40. S. J. Weiss, G. W. King, A. F. LoBuglio, Evidence for hydroxyl radical generation by human monocytes, J Clin Invest 60:370 (1977).

41. A. I. Tauber and B. M. Babior, Evidence for hydroxyl radical production by human neutrophils, J Clin Invest 60:374 (1977).

42. M. S. Jensen and D. F. Bainton, Temporal changes in pH within the phagocytic vacuole of the polymorphonuclear neutrophilic leukocyte, J Cell Biol 56:379 (1973).

43. D. M. Chipman and N. Sharon, Mechanism of lysozyme action. Lysozyme is the first enzyme for which the relation between structure and function has become clear, Science 65:454 (1969).

44. H. Odeberg and I. Olsson, Anti-bacterial activity of cationic proteins from human granulocytes, J Clin Invest 56:1118 (1975).

45. J. K. Spitznagel, M. R. Cooper, A. E. McCall, L. R. DeChatelet, I. R. H. Welsh, Selective deficiency of granules associated with lysozyme and lactoferrin in human polymorphs (PMN) with reduced microbicidal capacity, J Clin Invest 51:93a (1972).

46. R. B. Johnston, Jr. and S. L. Newman, Chronic granulomatous disease. Pediatric Clinics of North-America, vol. 24/2:365, WB Saunders, Philadelphia (1977).

47. R. B. Johnston, Jr. and R. L. Baehner, Improvement of leukocyte bactericidal activity in chronic granulomatous disease, Blood 35:350 (1970).

48. G. L. Mandell and E. W. Hook, Leukocyte bactericidal activity in chronic granulomatous disease: correlation of bacterial hydrogen peroxide production and susceptibility to intra-cellular killing, J Bacteriol 100:531 (1969).

MIGRATORY PROPERTIES AND RADIOSENSITIVITY

OF LYMPHOCYTES

Jonathan Sprent

Department of Immunology
Scripps Clinic and Research Center
1066 North Torrey Pines Road
La Jolla, CA 92037

INTRODUCTION

Radioactive isotopes have proven of enormous importance in studying nearly all aspects of lymphocyte physiology, including the origin, immune functions, migratory pathways, and life-spans of these cells. In this article I shall attempt to give a brief overview of lymphocyte physiology and function and then discuss the pros and cons of the various radioisotopes that have been used to study lymphocyte migration in vivo. Detailed reviews of lymphocyte physiology have appeared elsewhere (1-4).

A true understanding of lymphocyte physiology depends crucially on an appreciation of lymphocyte function. A short summary of the main immune functions of T and B lymphocytes and their subsets is given below.

SUBSETS OF LYMPHOCYTES AND THEIR FUNCTIONS

The vast majority of lymphocytes are small resting cells. Although these cells display striking homogeneity in terms of mor- phology, it has become apparent in the last 15 years that lymphocytes represent an extremely complex network of cells with a bewildering variety of different functions. Lymphocytes display marked heterogeneity with respect to their cell surface glycoproteins, and specific antisera and monoclonal antibodies to these surface markers have proved of crucial importance in defining the various functional subsets of lymphocytes (6, 7).

T Lymphocyte Subsets

Functionally, T cells fall into at least three broad categories: T helper (Th) cells, T killer (Tk) cells, and T suppressor (Ts) cells. Each of these three groups of cells can be subdivided into further functional subsets. For example, there appear to be at least three distinct types of Ts cells displaying different functions and surface markers (8). In man, cell-surface T4 and T8 molecules are useful markers for distinguishing between Th and Tk cells: most Th cells are T4$^+$, T8$^-$ whereas Tk cells are T4$^-$, T8$^+$.

Although T cells are exquisitely antigen specific, paradoxically these cells are unable to bind or recognize free antigen (9). In this respect they differ radically from B cells, which readily bind free antigen. How, then, do T cells display antigen-specificity without being able to recognize free antigen? The answer is that T cells "see" antigen only when it is presented in the context of certain "self" components, namely the gene products of the major histocompatibility complex (MHC). The MHC--termed the HLA complex in man--consists of multiple loci (10-12). These loci encode two types of cell surface glycoproteins, termed class I and class II molecules. Both types of molecules display extensive genetic polymorphism and are the main targets for transplantation immunity.

Class I molecules are transmembrane glycoproteins with a molecular weight of 65,000; these molecules are expressed on all cells and are noncovalently associated with β2 microglobulin. In man, class I molecules are encoded by three different loci: HLA-A, B, C. Although class I molecules probably have several biological functions, their chief immunological role appears to be to bind or associate with exogenous antigens in such a way that these antigens are accessible to Tk cells. Precisely how this association occurs is unclear. Perhaps the simplest view is that small polypeptide segments of the exogenous antigen bind to class I molecules so as to create a complex molecule expressing new antigen determinants. Tk cells ignore both the exogenous antigen and the self class I molecule per se but are able to recognize the "altered self" determinants created by the association of the two ligands.† Interestingly, Tk cells are unable to recognize the exogenous antigen in question when it is presented in the context of a class I molecule of another individual--a phenomenon termed "MHC restriction"; this implies that exogenous antigen associates with the polymorphic segments of the class I molecule. In this respect, Tk cells also discriminate between the individual's own class I molecules--another example of MHC-restriction. Most individuals are

---

† It should be emphasized that although this idea is favored by many investigators and fits most of the available facts, direct evidence of antigen-MHC association is conspicuously lacking.

heterozygous for HLA-A, B, and C and thus express 6 different
molecules, 3 paternal and 3 maternal. In theory, a particular antigen
could associate with all 6 of these molecules and generate 6 different
altered self-determinants recognized by 6 discrete subsets of Tk
cells.

Class II molecules consist of two polypetide chains with molecular
weights of 30-35,000. These chains are noncovalently associated and
are coded by pairs of closely-linked genes situated in the MHC. In
man, there are at least two sets of class II molecules, termed HLA-DR
and HLA-DC.* These are the products of two closely linked loci, each
containing a pair of genes encoding the two polypeptide chains of each
molecule. Unlike class I molecules, class II molecules are expressed
on only certain cell types, e.g., B cells, a set of macrophage-like
cells, and dendritic cells. The main function of class II molecules
appears to be to guide the function of Th cells. Thus Th cells fail
to recognize antigen unless it is presented by macrophage-like cells
expressing class II molecules. Macrophages are believed to break down
antigens in such a way as to allow small pieces of the antigen to
associate with class II molecules on the cell surface. Th cells then
recognize the association of antigen plus self class II molecules,
i.e., in much the same way as Tk cells recognize antigen plus self
class I molecules. As for Tk cells, MHC-restriction affecting Th
cells is highly specific and there is little or no cross reactivity
for antigen presented in the context of other class II molecules.

After contact with MHC-associated antigen, both Tk and Th cells
become activated, proliferate extensively and differentiate into
effector cells. Tk cells function by lysing cells expressing the
requisite altered self molecules, e.g., virus-infected cells. Th
cells have several different functions, perhaps the most important of
which is to interact with specific B cells. Although B cells are able
to bind antigen, the activation and differentiation of B cells to the
stage of antibody formation usually, though not invariably, requires T
cell help. After activation by antigen plus class II molecules on
macrophages, Th cells recognize the same association of antigen and
self class II molecules on antigen-specific B cells; as a result of
this recognition, the Th cells deliver an activation signal to the B
cells.† After activation, the B cells become susceptible to various
"growth factors" (some of which are T cell-derived); these factors
cause the B cells to proliferate extensively and then differentiate

---

* Recent evidence (12) suggests that there is a third class II
  molecule in man, HLA-SB.

† This applies to small resting B cells; other B cells do not require
  direct cell contact with Th cells.

into antibody-forming cells and plasma cells (13). Another important
function of Th cells is to interact with macrophage-like cells to
enable the latter to destroy intracellular parasites, e.g., myco-
bateria (9). Delayed-type hypersensitivity is a manifestation of this
interaction.

Ts cells comprise a complex network of cells and function by
inhibiting the normal immune response of other T cells and B cells
(8). Ts cells probably play a major role in preventing autoimmune
disease. In mice, MHC-linked molecules expressing "I-J determinants"
play an important function in guiding the interactions between the
various members of the suppressor cell network.

B Lymphocyte Subsets

Unlike T cells, B cells appear to recognize free antigen rather
than antigen seen in association with MHC molecules. In the case of
"T-dependent" antigens such as proteins or hapten-protein conjugates,
B cells bind antigen and then differentiate to antibody-forming cells
under the control of Th cells and various growth factors (see above).
The response of B cells to certain antigens, however, appears to be
relatively independent of Th cells. These "thymus-independent" (TI)
antigens fall into two categories, termed TI-1 and TI-2 antigens.
This subdivision is of particular interest because the response to
these antigens seems to involve different subsets of B cells (14).
These B cell subsets show distinct differences in terms of surface
markers, the ratio of surface IgM:IgD molecules and distribution in
the lymphoid tissues. A key, as yet unsolved, issue is whether these
B cell subsets represent distinct cell lineages or, alternatively, the
same lineage of cells seen at a different stage of differentiation and
maturation.

ONTOGENY OF LYMPHOCYTES

T cells

The vast majority of, and perhaps all, typical T cells arise in
the thymus. This organ is derived from the third and fourth
pharyngeal pouches and is initialy alymphoid. In mice, the thymus
becomes populated from the blood by large "pre-T" stem cells at about
the 10th day of gestation; this process occurs early in the first
trimester in the human. The stem cells differentiate rapidly into
small lymphocytes, and the thymus becomes compartmentalized into the
cortex and medulla. The interrelationship between the cortex and
medulla is poorly understood. The cortex--which accounts for about
90% of the thymic mass--consists of immature T cells and shows a high
rate of cell turnover; the cells in the medulla, by contrast, are more
akin to the mature T cells of the spleen and lymph nodes (LN) and show
a much slower turnover than cortical thymocytes.

The question of which cells leave the thymus to give rise to the mature T cells of the secondary lymphoid organs is still controversial, although the most recent studies suggest that the cells exit from the thymus with a mature phenotype (16). There is also debate as to the number of cells leaving the thymus. Estimates based on the number of labeled cells accumulating in the spleen and LN after injection of radioactive isotopes or fluorescent dyes into the thymus suggest that cell migration from the thymus is comparatively limited (16). If so, one is left with the paradox that the vast majority of cells generated in the thymus die in situ. Although this might seem curiously wasteful, a considerable amount of evidence suggests that extensive proliferation in the thymus is a manifestation of T cells "learning" to recognize their own MHC molecules as self (9). In this respect thymus-grafting studies have demonstrated that the capacity of peripheral T cells to recognize antigen in association with self MHC molecules depends on prior confrontation with these molecules in the thymus, presumably on thymic epithelial cells.

It is well-known that congenitally athymic "nude" mice and patients with di George's syndrome in man show a profound deficiency of T cells. A similar deficiency occurs in neonatally thymectomized mice or adult thymectomized mice subjected to whole body irradiation and reconstitution with bone marrow cells. In recent years a lot of attention has been directed to the fact that small numbers of T cells, e.g., 1% of normal, persist in these situations. Many workers interpret such findings to indicate that some T cells can differentiate in the absence of the thymus. An oppposing viewpoint is that the T cells found in such situations are in fact thymus-processed and reflect T cell differentiation in an ectopic thymus, inadequate T cell depletion of marrow cells, incomplete destruction of T cells by irradiation, etc.

B Cells

In the fetus, B cell differentiation takes place predominantly in the liver. At the time of birth in rodents and during the second trimester in man, B cell differentiation shifts to the marrow and to a lesser extent to the spleen. In certain species such as the sheep, rabbit, and, possibly, man, some B cell differentiation occurs in Peyer's patches and the appendix; B cells do not differentiate in these organs in rodents.

The immediate precursors of B cells appear to be large, cycling cells expressing intracytoplasmic immunoglobulin (Ig). These cells give rise to noncycling small lymphocytes which rapidly express surface Ig, Fc receptors, class II MHC molecules, and various other surface markers. These immature B cells exit from the marrow, probably in large numbers, and migrate to the secondary lymphoid tissues, i.e., the spleen, LN, and Peyer's patches. After further differentiation small numbers of the young B cells join the

recirculating lymphocyte pool as long-lived recirculating cells (see
below). It seems likely, though as yet unproven, that many B cells
exiting from the marrow or other centers of B cell production are
doomed to die within a few days. Precise estimates of the rate of B
cell production and destruction, however, are not available.

LYMPHOCYTE MIGRATION

    The bulk of lymphocytes in the secondary lymphoid organs reside in
the recirculating lymphocyte pool (RLP). Recirculating lymphocytes
migrate continuously from the blood through the lymphoid tissues into
the thoracic duct lymph and then back into the blood.

Spleen

    All blood-borne cells enter the spleen via the small tributaries
of the splenic artery. With the exception of lymphocytes, blood cells
empty into the sinusoids of the red pulp and then eventually leave the
spleen via the splenic vein. Lymphocytes enter the spleen at the
junction of the white and red pulps; T and B cells then migrate into
the white pulp and surround the central arterioles which traverse the
white pulp. The areas immediately surrounding the central arterioles
are termed periarteriolar lymphocyte sheaths (PALS) and are the main
conduits for T cell traffic. B cells move rapidly from the PALS and
make their way to the peripheral regions of the white pulp. These
B-dependent regions consist of both primary follicles (accumulations
of small B lymphocytes) and secondary follicles (primary follicles
containing germinal centers--the hallmark of an ongoing humoral
response). After varying periods, T and B cells migrate from the
white pulp by ill-defined pathways into the red pulp and then leave
the spleen via the splenic vein. Studies on rat spleens perfused in
vitro have suggested that T cells take about 5 hours to migrate
through the spleen (1). B cell migration through the spleen is much
slower and probably takes in excess of 24 hours.

Lymph Nodes

    As in the white pulp of the spleen, T and B lymphocytes percolate
through different regions of LN. T cells are found predominantly in
the lower areas of the cortex, termed the deep cortex or paracortex.
The B-dependent areas comprise two regions: 1) primary and secondary
follicles lying just beneath the subcapsular sinus, and 2) the
medulla. The medulla consists of cords of plasma-cell-enriched
lymphoid tissue which interdigitate into the medullary sinus; the
latter drains into the efferent lymphatic vessel.

    In contrast to the spleen, entry of cells into LN is highly
specific. A few lymphocytes and macrophages reach LN via the afferent
lymphatic vessels which empty into the subcapsular sinus. However,
the vast majority of lymphocytes enter LN by passing through the walls

of post-capillary venules (PCV), specialized small vessels situated in
the paracortex. Post-capillary venules are characterized by a lining
of high endothelial cells overlying a basement membrane. Unlike other
blood-borne cells, lymphocytes have receptors for the inner surface of
PCV (19). After transitory binding, T and B cells penetrate PCV by
passing between the endothelial cells. The cells then cross the
basement membrane and gain entry to the paracortex; T cells remain in
situ while B lymphocytes subsequently migrate to the B-dependent
areas, e.g., the primary follicles. T and B cells leave LN by
migrating via poorly defined routes to the medullary sinus and enter
the efferent lymphatic vessels. The latter either empty directly into
the thoracic duct or form afferent lymphatic vessels for other LN.
The contents of the thoracic duct pass into the venous system via the
left subclavian vein in the neck.

Tempo of Lymphocyte Migration

Thoracic duct lymph contains an almost pure population of
lymphocytes, of which about 90% are typical small lymphocytes. The
remainder are a mixture of large lymphocytes, including plasmablasts;
macrophages are very rare in normal thoracic duct lymph. Chronic
thoracic duct cannulation is a convenient method for obtainng detailed
information about lymphocyte migration. This procedure has been
performed successfully in many different species, including man.
Continuous collection of thoracic duct lymph from rats and mice is
possible for periods of several weeks, and much of the information on
the relative rates of T versus B cell recirculation has been obtained
from studies on these species.

The number of lymphocytes entering thoracic duct lymph is enormous
and an average mouse will yield greater than 10^8 cells over a period
of 24 hours; this represents about one third of the total number of
lymphocytes in the secondary lymphoid organs. During the first day of
collection, T cells predominate over B cells by a ratio of 2-4:1.
Thereafter, T and B cells are present in approximately equal
proportions. Numbers of thoracic duct lymphocytes (TDL) decline
rapidly after the first day and reach a low plateau of less than 10%
of normal levels after 7-10 days. In terms of absolute numbers,
outputs of small T lymphocytes fall precipitately within the first few
days of cannulation. Outputs of small B lymphocytes decline much more
slowly, and even after drainage for 1 week B cell outputs are reduced
by only 2-3 fold. These data imply that, whereas T cells can be
mobilized from the lymphoid tissues into the central lymph very
rapidly, B cells are mobilized only quite slowly. In support of this
notion, cellular depletion of the lymphoid tissue of mice subjected to
chronic thoracic duct drainage is initially more prominent in the
T-dependent areas than in the B-dependent areas. As will be discussed
later, most T and B cells entering the lymph during thoracic duct
drainage show a slow turnover, even during late stages of drainage.

In keeping with the rapid disappearance of T cells from the RLP during chronic thoracic duct drainage, studies with purified populations of T cells labeled with appropriate radioisotopes have shown that T cells recirculate rapidly into the lymph after intravenous injection with a mean transit time of 12-18 hours. B cells, by contrast, take many days to enter the lymph, particularly in mice.

Perhaps the simplest approach for demonstrating quantitative and qualitative differences in T and B cell migration is to inject purified populations of radiolabeled T and B cells intravenously and then examine the distribution of the labeled cells in the lymphoid tissues at various times post-injection. With the aid of autoradiography one can trace the migratory paths of T and B cells within the lymphoid tissues, demonstrating that these cells follow their respective T- and B-dependent routes. To determine the proportional localization of cells in one organ compared with another, however, the most quantitative approach is to remove the organs concerned, count the amount of radioactivity in the organs and express these values as a percentage of the injected counts. With this approach one can show that both T and B cells localize predominantly in the spleen within 30 minutes of injection; in rodents up to 40% of the injected counts may reach the spleen. Counts in the liver are low (5-10%) when fully-viable cells are injected; higher localization in the liver reflects the presence of damaged cells. Localization of cells in the lungs is usually low unless the cells are clumped.

Localization of cells in the spleen is usually maximal at 1-4 hours post-injection; thereafter the counts begin to fall. In the case of T cells, counts in the spleen decline by a factor of 2-3 fold between 1-4 hours and 24 hours. This decline is associated with a gradual increase in counts in the LN, and by 24 hours counts in LN exceed those in spleen. This rapid movement of T cells from the spleen to the LN presumably reflects the fact that T cells take only a short time (5 hours) to pass through the spleen. In this respect the decline in counts in the spleen between 1-4 hours and 24 hours is much less marked with injection of B cells. Likewise, the gradual accumulation of counts in LN after B cell injection is much slower, and even by 24 hours or later LN counts are always much less than counts in the spleen.

LYMPHOCYTE TURNOVER

Lymphocytes in the primary lymphoid organs--the thymus and marrow--show a very rapid turnover, and the vast majority of cells in these organs incorporate tritiated thymidine (H-3-TdR) into their nuclei during a 5-day course of injections. The turnover of lymphocytes in the secondary lymphoid tissues occurs much more slowly. In the case of LN and TDL, nearly all of the small lymphocytes incorporate H-3-TdR at slow linear rate, implying that the cells in these tissues are homogeneous with respect to lifespan. However, the use of surface markers has shown that, at least in TDL, B cells turnover slightly more rapidly than T cells. In mice the half-lives of recirculating lymphocytes entering thoracic duct lymph are in the order of 3 months for T cells and 3 weeks for B cells.

Although small cells with a rapid turnover are very rare in LN and TDL, these cells may comprise 10-30% of the lymphocytes in blood and the spleen. Some of the recently divided cells in these organs probably represent young cells emerging from the primary lymphoid organs; others are presumably the progeny of cells engaged in immune responses to various environmental antigens.

ACTIVATED LYMPHOCYTES

When lymphocytes encounter antigen in vivo, specifically reactive T and B cells leave the recirculating pool and become transiently sequestered in the lymphoid tissues at sites of antigen concentration (1, 2). Here, the cells undergo activation and proliferate extensively. After 2-3 days, the progeny of the stimulated lymphocytes--T blasts and B blasts (plasmablasts)--enter the circulation in large numbers. What happens to these cells?

In the case of T blasts, the cells migrate in large numbers to the spleen and the gut, particularly to Peyer's patches. Some of the blasts migrate from the T-dependent areas of Peyer's patches (the interfollicular areas) and penetrate the epithelial lining of the gut; here the cells die or enter the gut lumen. Other T blasts are carried throughout the body in the bloodstream and percolate into areas where they re-encounter the stimulating antigen. Most T blasts appear to die within a few days, although a few can differentiate to long-lived recirculating memory cells.

Most B blasts rapidly differentiate into plasma cells, release antibody for several days, and then die; a small percentage of plasma cells, however, appear to survive for much longer periods, i.e., several months in mice. In the case of IgA-secreting plasmablasts, most of the cells home to mucosal surfaces, e.g., to the lungs and the lamina propria of the gut. Cells secreting immunoglobulins of other isotypes, e.g., IgM and IgG, do not migrate to mucosal surfaces and lodge largely in the spleen and LN.

RADIOSENSITIVITY OF LYMPHOCYTES

Before considering the problems associated with labeling lymphocytes with radioisotopes, it is important to mention that these cells are extremely radiosensitive. Most cell types are comparatively radioresistant and die only during cell division, i.e., as a result of chromosome damage. Lymphocytes, however, are unique in that they are susceptible to death in interphase after irradiation. Thus, despite the fact that the vast majority of lymphocytes are long-lived cells,

exposure of mice to comparatively low doses of irradiation, e.g., 600
rads, induces gross depletion of the lymphoid tissues within 24 hours.
Likewise, when lymphocytes are exposed to irradiation in vitro, the
cells die within a few hours. In terms of both cell survival and
immune function, D37 value for T cells may be less than 100 rads
for T cells and even lower for B cells.

Precisely why lymphocytes die in interphase after irradiation is
unclear, but the prevailing view is that, because of their inactive
metabolic state, small lymphocytes are unable to repair irradiation-
induced membrane damage. In this respect, activated lymphocytes are
far more resistant to irradiation.

LYMPHOCYTE LABELING WITH RADIOISOTOPES

Ideally, radioisotopes used for studying cell migration should be
nontoxic, easily measured, remain bound to cells for long periods, and
not be reutilized upon cell death; in the case of studies on large
animals and man, there is the additional requirement that the isotope
be suitable for external imaging. In practice, no radioisotope meets
all of the above specifications. Nevertheless, many of the existing
radioisotopes have provided much valuable information on lymphocyte
migration and turnover. The advantages and disadvantages of the most
commonly used isotopes are considered below. For a useful review on
this subject, see Rannie and Donald (22).

BETA-EMITTING RADIONUCLIDES

In rodents and other small animals used for experimental purposes,
the short path length of β-emitting isotopes makes these labels
particularly suitable for autoradiography, both on cell suspensions
and tissue sections.

H-3- OR C-14-LABELED AMINO ACIDS

For cytoplasmic labeling C-14- or H-3-uridine and C-14- or
H-3-leucine are currently the most popular β-emitting labels. These
isotopes have proved of great importance in tracing the migratory
paths of T and B cells within the lymphoid tissues and in defining the
relative rates at which T and B cells recirculate from blood to lymph
or pass through the isolated spleen perfused in vitro (1, 2). With
the aid of liquid scintillation counting, one can also use β-emitting
radioisotopes to estimate the proportion of cells localizing in one
tissue compared with another. Gamma-emitting radioisotopes are
preferable for this purpose, however, because liquid scintillation
counting is time-consuming and usually requires correction for
quenching induced by hemoglobulin and other pigments. The chief
disadvantage of β-emitting cytoplasmic labels such as C-14 or
H-3-uridine is that these labels, particularly H-3, tend to elute from
cells quite rapidly and become incorporated by other cells. This

becomes a major consideration when one attempts to trace the fate of cells for periods longer than 48 hours. Within this time period, however, elution and reutilization of label is usually not a problem, particularly if autoradiography rather than scintillation counting is used.

It may be mentioned that the different energies of the emission from C-14 and H-3 means that radioactivity from these two lables can be counted separately. Ford et al (24) made use of this fact to design an elegant experiment to estimate what proportion of rat lymphocytes are reactive to MHC alloantigens of a rat of another strain. The approach was to label TDL of rat strain a with H-3 and TDL of rat strain b with C-14. The two populations of cells were mixed together and transferred intravenously into an irradiated strain b rat. By measuring the ratio of H-3/C-14 in the mixed population before injection and then comparing this value to the ratio obtained when the cells entered the central lymph of the recipient, the authors showed that, relative to the strain b cells, about 10% of the strain a lymphocytes failed to recirculate into the lymph of the strain b recipient. Since contact with antigen induces alloreactive lymphocytes to become selectively sequestered in the spleen for several days, the value of 10% provided an accurate indication of the proportion of strain a cells that responded to the strain b alloantigens.

H-3- OR C-14-THYMIDINE

In the case of nuclear labels, β-emitters such as C-14- or H-3-thymidine are the labels of choice for studying lymphocyte turnover. Indeed, virtually all of the quantitative information on T and B cell turnover has come from the use of these labels, particularly H-3-thymidine. This isotope is relatively nontoxic, and when taken up by lymphocytes which then revert to a resting state, the label is retained for periods of weeks or months. H-3-thymidine is rapidly released from dead cells, however, and then incorporated by other cells. This problem of H-3-thymidine reutilizaton can be largely overcome by administering large quantities of cold thymidine, e.g., in studies on the rate at which labeled cells disappear in vivo after cessation of a prolonged course of H-3-thymidine injection.

GAMMA-EMITTING RADIONUCLIDES

The chief advantage of γ-emitting isotopes is their long path length. This property means that counting total levels of radioactivity in cells labeled with these isotopes, or tissues of recipients of labeled cells, is relatively simple. The reciprocal disadvantage is that γ-emitting isotopes generally do not give good quality autoradiographs because the silver grains extend far beyond the boundaries of the cell.

Chromium-51

In rodents and other experimental animals, Cr-51 (Na$_2$CrO$_4$) is
perhaps the best all-purpose isotope for studying the interorgan
distribution of lymphocytes, i.e., the relative localization of
lymphocytes in one organ compared with another. The isotope is also
much used for measuring CML in vitro. Chromium-51 has a moderately
long half-life (27d), is comparatively nontoxic, is not taken up by
dead cells, and once incorporated, the label is retained for prolonged
periods. One of the great advantages of Cr-51 is that although dead
cells release Cr-51 rapidly, the label is not reincorporated by other
cells (but see below).

In the light of the above properties, it might be thought that
Cr-51 is the "ideal" isotope for studying lymphocyte migration. This
is nearly, but not quite, true. As a label for tracing human
lymphocytes, Cr-51 has two disadvantages. Firstly, the long half-life
of the isotope means that patients injected with Cr-51-labeled cells
are exposed to irradiation for prolonged periods. Secondly, the
emission from Cr-51 is not sufficiently strong to allow effective
external imaging, the only practical approach for tracing human
lymphocytes. These problems do not arise when Cr-51 is used in
rodents, i.e., where organs can be removed from recipients of labeled
cells.

Studies in mice have shown one additional problem when
Cr-51-labeled cells are transferred in vivo. It was mentioned above
that when Cr-51-labeled cells die in vitro, the label is released
rapidly and not reincorporated. A different situation arises in vivo.
Here, phagocytes, such as macrophages, engulf dead and dying
Cr-51-labeled lymphocytes and retain the label for prolonged periods.
This can lead to confusing results. As an example, one can take the
case of Cr-51-labeled lymphocytes exposed to a high dose of radiation
(1,000 rads) before intravenous injection. If one places irradiated
lymphocytes in vitro, the cells appear normal for a few hours but then
begin to disintegrate (21). What happens to the cells after injection
in vivo? If Cr-51-labeled cells are injected intravenously soon after
irradiation, the distribution of radioactive counts at 4 hours in the
spleen, LN, liver, etc., is indistinguishable from normal (Table 1,
compare groups A and D) (25). Thereafter the counts in these organs
decline very slowly, and at 10 days post-transfer the total counts in
the lymphoid organs are no lower than in recipients of nonirradiated
labeled cells. At face value such data might be taken to imply that
irradiation has no effect on lymphocyte migration and, in marked contrast to culture in vitro, does not kill lymphocytes when the cells
are transferred in vivo. Careful scrutiny of the data in Table 1, however, shows that irradiation totally abolishes the normal movement of
cells from the spleen to the LN between 4 hours and 24 hours post-
injection. Furthermore, if the labeled cells are removed from the spleen
at 24 hours post-transfer and then injected into other hosts, the

Table 1. Distribution of Radioactivity in Recipients of Spleen Cells taken from Mice injected 1 to 10 days previously with Cr-51-labeled Syngenic Normal TDL or Irradiated TDL

Group	^{51}Cr-labeled Cells Transferred[a]	Time of Sacrifice (hr)	Liver	Spleen	Mesenteric Lymph Node	Small Intestine	Large Intestine	Lung	Spleen cpm / Liver cpm (at 4 hr)
				Percent of Injected Radioactivity in:					
A	TDL	4	9.7[b]	34.0	7.6	5.7	1.8	7.7	3.51
		24	10.4	19.7 (62%)[c]	18.1	7.9	2.7	1.6	
		240	9.9	7.0 (52%)	5.3	1.8	0.8	0.5	
B	Spleen from mice given labeled TDL 24 hr before	4	13.1	30.2	10.8	4.6	1.6	2.2	2.31
		24	11.6	15.2	19.5	5.8	3.5	3.5	
C	Spleen from mice given labeled TDL 10 days before	4	24.0	30.4	4.0	2.4	1.2	4.9	1.27
		24	24.7	13.6	11.7	4.6	1.5	1.4	
D	Irradiated TDL	4	10.2	35.2	7.1	4.2	1.4	7.5	3.45
		24	11.6	30.9 (25%)	6.8	3.6	1.0	0.5	
		240	8.7	13.9 (36%)	2.1	1.1	0.5	0.2	
E	Spleen from mice given labeled irradiated TDL 24 hr before	4	56.3	1.8	0.1	0.0	0.0	4.2	0.02
		24	51.7	1.8	0.1	0.0	0.0	1.8	

a   TDL were labeled with Cr-51 at 20 μCi/ml. Irradiated TDL received 1,000 rads in vitro just before transfer. Repassaged labeled cells in spleen were washed once before secondary transfer.

b   Arithmetic mean, three mice per group.

c   Proportion of radioactivity in the intact organ recovered after preparing single cell suspension. Adapted

Table 2. Distribution of Radioactivity in Liver, Spleen and Gastrointestinal Tract of Mice injected with Activated T cells (T.TDL)
Incubated in Vitro with various concentrations of I-125 UdR

Labeled cells injected[a]	Concentration of I-125-UdR for labeling (µCi/ml)	Time of Sacrifice (hr)	Percent of injected radioactivity in:				Percent of radioactivity recovered (including lung, LN and stomach)
			Liver	Spleen	Small intestine	Large intestine	
Viable T.TDL	10.0	4	23.4[c]	14.2	17.4	2.1	62.0
		24	0.7	0.8	7.0	3.1	13.0
		48	0.2	0.2	4.9	1.3	7.6
		120	0.1	0.1	0.5	0.2	1.0
Viable T.TDL	1.0	1	19.7	10.5	18.0	2.9	54.9
		4	18.5	14.9	19.6	3.1	61.4
		24	2.2	4.0	21.5	7.1	36.6
		48	0.5	1.1	8.2	3.3	13.5
		120	0.0	0.2	2.0	0.4	2.8
Viable T.TDL	0.1	4	21.7	14.9	18.6	3.2	61.5
		24	5.3	11.1	22.3	6.9	46.8
		48	3.2	3.6	19.0	6.8	33.0
		120	0.3	0.7	4.6	1.2	6.8
Heat-killed T.TDL	1.0	1	15.6	0.7	3.0	1.3	34.1
		4	1.7	0.5	3.0	1.4	22.0
		24	0.3	0.3	1.5	0.6	3.9
Viable T.TDL incubated with anti-H2 alloantiserum[b]	1.0	1	28.8	1.8	2.7	1.2	44.1
		4	3.3	1.0	2.9	1.1	18.1

a   To prepare activated T cells, CBA thymus cells were transferred I.V. into irradiated (CBA xC57BL)F$_1$ mice. The donor CBA cells collected in thoracic duct lymph of the recipients 4 days later consisted almost entirely of T blast cells with reactivity to the host H-2 alloantigens. These cells (T.TDL) were incubated with I-125-UdR in vitro at 5 x 10^7 cells/ml and then washed thoroughly before transfer to normal syngeneic (CBA) mice.

b   Cells incubated (after labeling) with C57BL anti-CBA alloantiserum (1:4 dilution) at 10^7 cells/ml for 20 min at 4°C in the absence of complement. The coating of antibody on the cells causes the cells to be opsonized by the liver on transfer.

c   Arithmetic mean of data; three mice per group. Adapted from ref. 25.

radioactive counts now lodge largely in the liver (compare groups B and E); counts in the lymphoid organs are negligible. A similar distribution occurs when Cr-51-labeled phagocytic cells are injected intravenously. These data thus suggest that irradiation takes several hours to affect lymphocyte migration. After the initial localization in the spleen and LN, the irradiated cells die rapidly in situ, and by 24 hours most of the label has been incorporated by phagocytes. Upon further transfer the phagocytes lodge in the liver.

Iodine-125 UdR

Like thymidine, I-125-5-iodo-2'-deoxyuridine (I-125 UdR) is taken up by the DNA of dividing cells. Unlike thymidine, however, I-125 UdR is not reutilized and is rapidly excreted from the body. A key advantage of I-125 UdR is that, in contrast to Cr-51, phagocytes do not retain label when they ingest I-125 UdR-labeled dead cells.

Being a γ-emitter, I-125 UdR is a useful label for tracing the fate of lymphocytes in S phase (25, 26). The main drawback of I-125 UdR as a label is its toxicity. This is illustrated in Table 2 where it is shown that activated T cells (see Table 2 footnote) labeled with I-125 UdR home predominantly to the small intestine, spleen and liver by 4 hours after intravenous injection, irrespective of whether the cells are labeled at a concentration of 10, 1 or 0.1 μCi/ml/50 million cells. Thereafter, the counts in these organs decrease rapidly, the rate of decay being proportional to the dose of isotope used. High doses of isotope, e.g., 10μCi/ml are clearly toxic. Although the lowest does of 0.1 μCi/ml provided the best results in the experiment illustrated, it should be pointed out that the levels of radioactivity incorporated by cells treated at this dose were extremely low, indeed so low that the organs of the cell recipients had to be counted for 10 minutes to obtain reliable data. A dose of 1 μCi/ml is quite convenient for tracing cells for 24 hours or less, but is obviously toxic if cells are followed for longer periods (Table 2).

It can be seen from Table 2 that even doses of I-125 UdR as low as 0.1 μCi/ml led to less than 10% recovery of the injected counts by day 5 post-transfer (Table 2). Does this mean that the isotope is still toxic at this dose, or do the data signify that activated T cells are short-lived cells? Although the former possibility is difficult to exclude, studies involving double adoptive transfer of activated T cells labeled with Cr-51 (at non-toxic doses) have indicated that most activated T cells do die within a few days of transfer (25).

Se-75-L-selenomethionine

This cytoplasmic label is a useful alternative to Cr-51. It has a long half-life (121 days) and, like Cr-51, is relatively non-toxic. The fact that the γ-emission spectra of Cr-51 and Se-75-seleno-methionine (Se-75) can be distinguished means that the two labels can

be used in tandem. Disadvantages of Se-75 are its rapid rate of
elution from lymphocytes and reutilization by other cells.
Reutilization tends to be uneven and can be quite high in certain
tissues, particularly the intestines (22, 27, 28).

LABELS FOR HUMAN LYMPHOCYTES

In addition to being nontoxic and retained by cells for prolonged
periods, ideal labels for human lymphocytes should have a short
half-life and provide good external imaging. Technetium-99m-sodium
pertechnetate has the latter two properties but seems to be too toxic
for lymphocytes to be used clinically (22, 27). At the present time,
In-111 appears to have the most promise as a label for human
lymphocytes (27, 29-35). Indeed this label is starting to give
invaluable information on lymphocyte migration, both in normal
subjects and in patients with various disease states.

Indium-111-oxine

Indium-111 (In-111) chelated with oxine has a convenient, and
clinically-acceptable, half-life (2.8 days) and gives excellent
external images. When used at certain defined concentrations (see
below), In-111 elutes only quite slowly from labeled cells. By these
three parameters, In-111 might seem to be the ideal isotope for
clinical studies. Sparshot et al, (34) working with rat lymphocytes,
caution, however, that several problems can arise with In-111,
including chemical toxicity by minor contaminants in the preparation,
radiation damage, and a paradoxical rapid loss of In-111 in vivo when
cells are treated with low doses of the isotope (1 μCi/100 million
cells). These workers obtained the best results when lymphocytes
(TDL) were treated with In-111 at about 10 μCi/ml/100 million cells.
With this dose the lymphocytes survived well for at least 48 hours, as
manifested by effective blood to lymphocyte recirculation. At higher
doses, the lymphocytes sustained appreciable radiation damage. Thus,
at a dose of 40 μCi/ml, the lymphocytes recirculated poorly within the
first 48 hours and by 7 days most of the labeled cells had died and
been ingested by phagocytes. (As with Cr-51, phagocytes retain label
from In-111-labeled dead cells for prolonged periods.) Sparshot et al
calculate that lymphocytes incubated with In-111 at 20 μCi/ml are
subjected to a total dose of irradiation (Auger electrons) of about
160 rads/cell over the first 24 hours and 280 rads over 48 hours. As
considered earlier, such doses of irradiation, at least when given as
a single exposure, can destroy a large proportion of resting T cells
(>50%) and the vast majority of B cells. Even at the "best" doses of
In-111, i.e., about 10 μCi/ml, radiation damage to lymphocytes will
ultimately be considerable. It is worth pointing out, however, that
lymphocytes tend to equilibrate in the lymphoid tissues within 24
hours of injection. Hence it would seem pointless to follow
lymphocytes in patients for periods longer than 24 hours. Within this
time period it seems unlikely that treating lymphocytes, particularly

T cells, with In-111 at a dose of 10 μCi/ml would cause more than minimal radiation damage.

Bearing in mind that good imaging of LN requires injecting in the order of 100 μCi of In-111, treating human lymphocytes with In-111 at a dose of 10 μCi/ml necessitates labeling quite large numbers of cells, i.e., about 10^9 PBL (34). Because of the problems of obtaining autologous lymphocytes in such numbers, some workers have elected to treat human lymphocytes with very high doses of In-111, e.g., 250 μCi/ml. Such doses would undoubtedly destroy lymphocytes within a few hours of injection and thereby prevent secondary migration of cells from the spleen to the LN. However, such a dose would probably not interfere with the initial distribution of the cells, i.e., the high localization in the spleen seen within 1-2 hours of injection. Indeed, Goodwin (30) has used this approach successfully to demonstrate rapid migration of In-111-labeled lymphocytes to the joints of patients with rheumatoid arthritis.

A rather worrying finding with In-111-labeled lymphocytes is that the cells rapidly manifest severe chromosomal abnormalities when cultured in vitro (35). (Whether In-111 is more dangerous than other radioisotopes in this regard is not clear.) Bearing in mind the evidence that pathological changes in chromosomes, particularly translocations, might lead to the activation of "oncogenes" in lymphocytes (36), the above effects of In-111 must be viewed with some concern. One resolution to this problem might be to label cells with In-111 at a very high dose, i.e., sufficient to kill virtually all of the labeled cells within a few days. The problem with this approach is that after injection there might be significant radiation damage to adjacent unlabeled lymphocytes. An alternative approach would be to label lymphocytes with In-111 at a low dose and then expose the cells to heavy ionizing irradiation (1,000 rads) immediately before injection. This treatment would ultimately kill the cells but would not interfere with the initial homing of the cells to the lymphoid organs. (The relationship of the use of In-111-lymphocytes in humans and the potential lymphoid risk is further discussed in the chapter by Dr. John Wagstaff.)

ACKNOWLEDGMENT

This study was supported by USPHS grants AI-15393, CA-15822, AI-1096, and CA-33958.

REFERENCES

1. W. L. Ford, Lymphocyte migration and immune response, Prog Allergy 19:1 (1975).

2. J. Sprent, Recirculating lymphocytes, in: "The Lymphocyte: Structure and Function", (J. J. Marchalonis, ed., Marcel Dekker, New York, (1977).

3. M. De Sousa, Lymphocyte Circulation: Experimental and Clinical Aspects, John Wiley & Sons, New York (1981).

4. D. M. V. Parrott and D. M. V. Wilkinson, Lymphocyte locomotion and migration, Prog Allergy 28:193 (1981).

5. L. E. Hood, I. L. Weissman, W. B. Wood, "Immunology," second ed., Benjamin/Cummings, CA 1984.

6. I. F. C. McKenzie and T. Potter, Murine lymphocyte surface antigens. Adv Immunol 27:181 (1979).

7. E. Rheinherz and S. Schlossman, The differentiation and function of human lymphoctyes. Cell 19:821 (1980).

8. B. Pernis and H. J. Vogel, eds., "Regulatory T Lymphocytes", Academic Press, New York (1980).

9. R. N. Zinkernagel and P. C. Doherty, MHC-restricted T cells: studies on the biological role of polymorphic major transplantation antigens determining T-cell restriction--specificity, function, and responsiveness, Adv Immunol 27:51 (1979).

10. D. C. Shreffler and C. S. David, The H-2 major histocompatibility complex and the I immune response region: genetic variation, function, and organization, Adv Immunol 20:125 (1975).

11. J. Klein and Z. A. Klein, MHC restriction and Ir genes, Adv Cancer Res 37:233 (1982).

12. J. Trowsdale, J. Lee, A. McMichael, HLA-DR bouillabaisse, Immunol Today 4:31 (1983).

13. N. Klinman, D. Mosier, I. Scher, E. Vitetta, eds., "B Lymphocytes in the Immune Response," Elsevier (1981).

14. I. Sher, The CBA/N mouse strain: an experimental model illustrating the influence of the X-chromosome on immunity, Adv Immunol 33:1 (1982).

15. H. Cantor and I. L. Weissman, Development and function of thymocytes and T lymphocytes, Prog Allergy 20:1 (1976).

16. R. Scollary, Thymus cell migration: cells migrating from the thymus to peripheral lymphoid organs have a "mature" phenotype, J Immunol 128:1566 (1982).

17. D. G. Osmond, The contribution of the bone marrow to the economy of the lymphoid system, Monogr Allergy 16:57 (1980).

18. M. E. Smith and W. L. Ford, The recirculating lymphocyte pool of the rat: a systematic description of the migratory behaviour of recirculating lymphocytes, Immunology 49:83 (1983).

19. W. M. Gallatin, I. L. Weissman, E. C. Butcher, A cell-surface molecule involved in organ-specific homing of lymphocytes, Nature 304:30 (1983).

20. D. Guy-Grand, C. Griscelli, P. Vassalli, The mouse gut T lymphocyte, a novel type of T cell. Nature, origin and

traffic in mice in normal and graft-versus-host conditions, J Exp Med 148:1661 (1978).

21. R. E. Anderson and N. L. Warner, Ionizing irradiation and the immune response, Adv Immunol 24:215 (1976).

22. G. H. Rannie and K. J. Donald, Estimation of the migration of thoracic duct lymphocytes to nonlymphoid tissues: a comparison of the distribution of radioactivity at intervals following I.V. transfusion of cells labelled with H-3, Se-75, Tc-99m, I-125 and Cr-51 in the rat, Cell Tissue Kinet 10:523 (1977).

23. W. L. Ford and M. E. Smith, in: "In Vivo Immunology," A. A. van den Broek and M. G. Hanna, eds., Adv Exp Med Biol 149:139 (1982).

24. W. L. Ford, S. J. Simmonds, R. C. Atkins, Early events in systemic graft-versus-host reaction. II. Autoradiographic estimates of the frequency of donor lymphoctyes which respond to each Ag-B-determined antigenic complex, J Exp Med 141:681 (1975).

25. J. Sprent, Fate of H-2-activated T lymphocytes in syngeneic hosts. 1. Fate in lymphoid tissues and intestines traced with H-3-thymidine, I-125-deoxyuridine and Chromium-51, Cell Immunol 21:278 (1976).

26. J. G. Hall, An essay on lymphocyte circulation and the gut, Monogr Allergy 16:100 (1980).

27. G. H. Rannie, M. L. Thakur, W. L. Ford, An experimental comparison of radioactive labels with potential application to lymphocyte migration studies in patients, Clin Exp Immunol 29:509 (1977).

28. M. L. Rose and H. S. Micklem, SeL-75-selenomethionine: a new isotopic marker for lymphocyte localization studies, J Immunol Methods 9:281 (1976).

29. J. Wagstaff, C. Gibson, N. Thatcher, W. L. Ford, H. Sharma, D. Crowther, Human lymphocyte traffic assessed by Indium-111-oxine labeling: clinical observations, Clin exp Immunol 43:443 (1981).

30. D. A. Goodwin, Cell labeling with oxine chelates of radioactive metal ions: techniques and clinical implications, J Nucl Med 19:557 (1978).

31. T. Issekutz, W. Chin, J. B. Hay, Measurement of lymphocyte traffic with Indium-111, Clin exp Immunol 39:215 (1980).

32. H. Frost, P. Frost, C. Wilcox, Lymph node scanning in sheep with Indium-111-labeled lymphocytes, Int J Nucl Med Biol 6:60 (1979).

33. J. P. Lavender, J. M. Goldman, R. B. Arnot, M. L. Thakur, Kinetics of Indium-111-labeled lymphocytes in normal subjects and patients with Hodgkin's disease, Br Med J ii:797 (1977).

34. S. M. Sparshot, H. Sharma, J. D. Kelley, W. L. Ford, Factors influencing the fate of Indium-111-labeled lymphocytes after transfer to syngeneic rats, J Immunol Methods 41:303 (1981).

35. R. J. M. ten Berge, A. T. Natarajan, M. R. Hardeman, E. A.
     van Royen, P. T. A. Schellekens, Labeling with Indium-111
     has detrimental effect on human lymphocytes: concise
     communication, <u>J Nucl Med</u> 24:615 (1983).
36. D. Forman and J. Rowly, Chromosomes and cancer, <u>Nature</u>
     300:403 (1982).

CELL ISOLATION TECHNIQUES: A CRITICAL REVIEW

M. R. Hardeman

Department of Internal Medicine
Academic Medical Center
University of Amsterdam
The Netherlands

INTRODUCTION

Labeling blood cells with a radioactive tracer is not a new technique. In 1940 Hahn and Hevesy measured the blood volume of a rabbit by labeling the animal's erythrocytes with radioactive phosphorus. Since then many experimental and clinical investigations using radioactive labeled cellular blood elements have been described in the literature. Their application has become increasingly popular during the past few years, due to the development of very efficient labeling techniques and the availability of suitable radionuclides. The following criteria need to be fulfilled by an ideal blood cell labeling method.

a. high labeling efficiency

b. no elution of the label, either in vitro or in vivo

c. no re-utilization of the label after cell destruction

d. no exchange of the label with unlabeled compounds in the body stores

e. availability of a suitable radionuclide

f. maintenance of cell viability, i.e., the ability to stay in the circulation with a normal survival

g. preservation of cell function(s) thought to be relevant for the type of study

h. short, technically simple procedure

i. no toxic substances

This list can be used as a reference for the comparision of various labeling techniques.

It seems logical to investigate cell-specific labeling methods using one or more function(s) of the cell that are unique. Some examples are:

-Phagocytosis of Tc-99m-Sn-colloid, labels preferentially monocytes

-Binding of P-32-Di-Isopropylfluorophosphate (DFP) to esterases, predominantly for granulocytes

-Active uptake of C-14-serotonin for platelets

-Binding of C-11-0 for erythrocytes

An advantage of such techniques is undoubtedly that the labeling procedure can be performed in whole blood and eventually also in vivo. Up until now, however, none of these methods has satisfactorily fulfilled the criteria mentioned earlier. Thus, the techniques of blood cell labeling used most frequently at present lack specificity. It is clear, therefore, that the numerically abundant erythrocytes in whole blood would bind the major part of the added radioactivity. This gives an unacceptable high background and diminishes the sensitivity of the method when labeled leukocytes or thrombocytes are under study. Thus, unless we want to label red cells, the bulk of these cells has to be removed. This introduces technical problems for cell isolation and an additional risk of cell damage in performing such a procedure.

QUALIFICATIONS FOR CELL PREPARATIONS

In general, cell preparation techniques should follow the guidelines listed below.

1.   Sterility - mainly of importance for the testing of procedures.

2.   Functionality - using in vitro test(s) relevant to the type of study.

3.   Viability - i.e., the capacity to circulate with normal disappearance kinetics.

4.   Purity - mainly for kinetic studies.

It is a matter of philosophy whether and to what extent one should try to purify the cell suspension for given scintigraphic studies. It is clear that subsequent pelleting and resuspension and/or so-called selective hypotonic lysis procedures include a high risk of cell damage. In my opinion one should prefer the maintenance of viability over the purity of the cell type and therefore avoid as much as possible any mechanical or osmotic trauma.

Since we are extremely short of meaningful in vitro tests reflecting the in vivo behavior of the cells, for the time being, one should avoid handling the cells as much one can!

As stated before, for most of the studies that we are interested in, the bulk of the erythrocytes should be removed. This can easily be performed by slow centrifugation, yielding a supernatant plasma, rich and pure in platelets. Unfortunately, about half the original platelets are lost in the red cell fraction. Simple sedimentation of erythrocytes leads to a mixed cell supernatant, containing platelets, leukocytes, and light erythrocytes. Although there are rather large differences in the life-spans of erythrocytes (120 days), platelets (8-10 days) and granulocytes (6-8 hours), further purification of the granulocytes is needed for cell kinetic studies. With respect to the use of radioactive labeled granulocytes for scintigraphic studies it is still debatable whether a "crude" cell suspension would be sufficient or whether further purification is needed.

Although a discussion on this subject is beyond the scope of this presentation, I should like to refer to the study of Hawker and Hall (1) who demonstrated that platelets are actively involved in abscesses. In this respect it might be that the platelets, metabolizing arachidonic acid via the lipoxygenase pathway, yield products which have chemotactic activity (2).

Extending the discussion about highly pure cell suspensions to the detection of thrombotic processes with radioactive labeled platelets, a similar study of Charkes can be recalled. In this study it was shown that granulocytes participate actively in the formation of thrombi (3).

PURE CELL SUSPENSIONS

Which methods do we have available to achieve a relatively pure cell suspension for the clinical use of labeled blood cells?

During the past few decades, many different techniques have been developed for the preparation of more or less pure, single cell-type suspensions (4). They are based either on functional or on physical differences of the cells, i.e., adherence, phagocytosis, rosette formation, antibody binding (fluorescent), cytotoxic elimination, immuno affinity, agglutination, as well as charge, density, and size.

I will not go into details of all techniques based on functional properties. In general they have been developed for biochemical or

immunological in vitro studies and are not suitable for in vivo
studies because of loss of sterility, function etc. Moreover,
isolation on the basis of functional properties means selection and
possible changes in cellular metabolism induced by the isolation
technique (5). Thus, maintenance of normal in vivo behavior is
questionable. Separation techniques based on physical properties are
more pertinent to clinical studies. Separations of cells based on
electrophoretic mobility are not very popular because of the need for
special apparatus. Also, only a small number of cells can be
separated, and considerable variations in cell surface charge from a
single cell-type from the same individual donor may occur at different
times (6).

In contrast, the density of the various cellular components of
peripheral blood is less variable. Therefore, this property has been
used extensively to separate blood cells (Table 1). Simple
centrifugation will divide whole blood into a platelet containing
plasma-layer, a so-called buffy-coat consisting mainly of white cells,
and a large erythrocyte-layer at the bottom of the tube. There is,
however, considerable overlap, and hence the recovery of one
particular cell-type in a certain layer is unsatisfactory.

More sophisticated centrifugation methods apply a continuous or
discontinuous density gradient. Many gradient materials have been
used, e.g., albumin, sucrose, stractan, urografin, silicon oil, and
phthalate-esters.

Well-known is the procedure described by Boyum in 1968 (7) for
the isolation of lymphocytes from whole blood or buffy coats. He used
a mixture of a high polymer compound (consisting of sucrose and
epichlorohydrin) which aggregates erythrocytes and a compound with
high density (sodium metrizoate).

Table 1.

Cell Type	Volume (FL)	Diam. (µm)	Count (μl^{-1})	Density (G cm^{-3})
Erythrocyte	90	8	5,000,000	1,092
Thrombocyte	6	3	250,000	1,030
Granulocyte	450	10–15	4,300	1,089
Lymphocyte	210	7–18	2,200	1,070
Monocyte	470	12–20	500	1,069
Plasma				1,027

The mixture, commercially available as a sterile preparation under the name "Lymphoprep"* has a density of 1.077 g/cm³ and is commonly known as ficoll-isopaque. Recently the manfacturer has changed one of the constituents by adding dextran instead of ficoll, resulting in a slightly higher density of 1.082 g/cm³. The preparation has now been called "Lymphopaque".* After centrifugation of diluted, defibrinated or heparinized blood over a layer of the lymphopaque solution, the mononuclear cells(predominantly lymphocytes) float on the interphase and can be harvested easily; the erythrocytes and granulocytes are centrifuged to the bottom of the tube (Fig. 1).

Such isopycnic density centrifugation techniques are widely applied for cell separation. In general, however, extensive washing of the separated cells is necessary to remove the gradient substance. This causes a considerable loss of cells (8) and additional risk for cell damage. Other investigators have reported that ficoll-gradients lead to activation and increase of density of cells due to phagocytosis of the gradient material.

Furthermore, it has been reported several times that the gradient-material affects leukocyte-metabolism (10-14). With this technique granulocytes sediment together with erythrocytes. The latter are removed by a lysis step which might also affect the granulocytes (15-16). Degranulation of platelets has been reported during density-gradient separation (17).

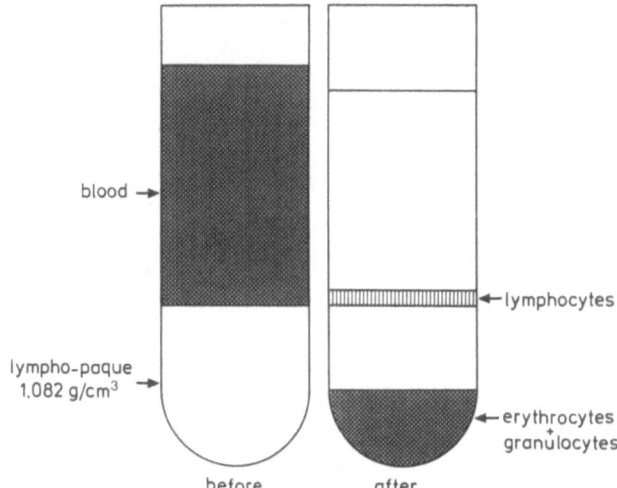

Fig. 1. Cellular partition before and after centrifugation of whole blood over a layer of lymphopaque solution.

*Nyegaard & Co A. S., Oslo, Norway

A relatively new gradient Percoll* has already become popular for cell separation. It consists of dispersed PVP coated silica particles. It has a lower viscosity than ficoll-iosopaque enabling separation at 4°C. There are, however, conflicting reports in the literature about toxicity of Percoll. Wakefield et al (18) demonstrated that Percoll, even when used at 4°C, impairs the adhesive properties of macrophages, but at 37°C there was a considerable ingestion of large numbers of Percoll particles, which were tightly packed within large vacuoles (Fig. 2).

Nathanson et al (19) noted a significant depression of chemotactic and random motility of monocytes separated on a Percoll gradient. On the other hand, Dooley et al (20) and Hjorth et al (21) have shown that neutrophils, isolated on Percoll gradients, were just as active as those harvested by centrifugal elutriation. Last but not least, Saverymuttu et al (22) have recently demonstrated a rapid transit through the lung fields after injection of labeled granulo-

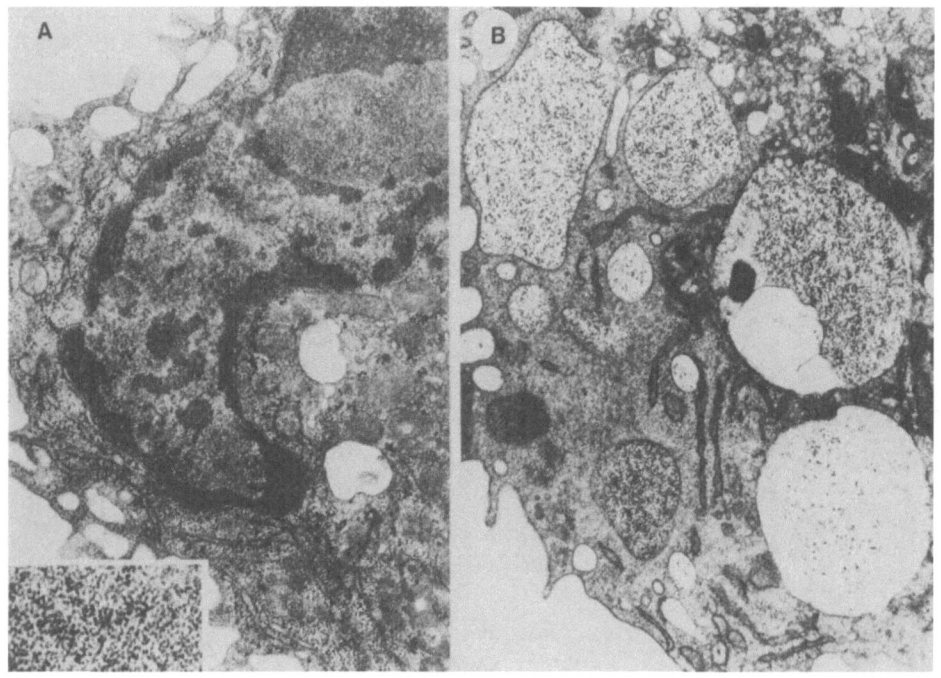

Fig. 2. Electronmicrographs of parts of mouse peritoneal macrophages after incubation with Percoll; (A) after incubation at 4° C (magn. x 22,400); (B) after incubation at 37° C (magn. x 21,600). Inset in (A) is an electronmicrograph of Percoll (magn. x 38,400). (With permission of the Biochemical Society, London, U.K.)

*Pharmacia Fine Chemicals AB, Uppsala, Sweden

cytes isolated on a Percoll-plasma gradient. This was interpreted as
a sign of viability.

Separation methods based on cell size utilize velocity
sedimentation (Fig. 3). Several types of velocity sedimentation have
been described, e.g., specially constructed 1G-velocity sedimentation
apparatus (23) or gelfiltration. These are not suitable for our
purpose. One exception is counterflow centrifugation, or centrifugal
elutriation, which I will discuss later in this chapter.

In general, it is clear that most of the separation techniques
described thus far are used for in vitro studies. In finding optimal
cell isolation methods suitable for labeling and subsequent in vivo
scintigraphic studies, one could look at blood transfusion practice
and especially those methods in use for blood component therapy. The
demands for sterility and viability are the same and since for scintigraphic purposes mostly autologous blood is used, there are no
problems regarding immunological compatibility or donor recruitment.
There are, however, quantitative problems: these techniques are designed for large quantities of blood; for scintigraphic work we use
only 30-50 ml blood. Modifications, enabling the handling of only
50 ml of blood, have not yet been made, as far as I know. For the
same reason, machines like the "cell separator," "celltrifuge," or
"cell processor" cannot be used.

So far, none of the discussed cell separation techniques can be
advocated as being suitable for labeling cells to be used in vivo.
Separation over a density gradient seems to yield the most pure cell
suspensions; however, the consequences of the foreign substances
serving as a density gradient should be studied more extensively
before their routine use for the separation of cells capable of
phagocytosis.

COUNTERFLOW CENTRIFUGATION (ELUTRIATION)

Although the principle of elutriation was practiced during the
years of the gold rush, when the less dense sand grains were washed
out (elutriated), leaving the gold particles (if any!) behind, it was
only 1948 when Lindahl first formally described the principle of
elutriation in his article published in Nature (24). It was not until
1973 that a commercially available elutriator rotor* was constructed.
Since then the use of this technique has increased rapidly.

During centrifugal elutriation the tendency of particles or cells
to sediment in a centrifugal field is offset by a liquid flowing in
the opposite, centripetal, direction (Figs. 4 and 5). When the
centrifugal and counterflow forces are balanced, a wide range of
particle sizes can be concentrated and kept in separate bands in
suspension (Fig. 6).

---

*Beckman Instruments

STOKES LAW : $V = 2 G R^2 \rho_D / 9 \eta K$

V = TERMINAL VELOCITY
G = ACCELERATION OWING TO GRAVITY OR CENTRIFUGAL FORCE
R = PARTICLE <u>RADIUS</u>
ρ_D= DIFFERENCE OF PARTICLE <u>DENSITY</u> FROM THAT OF THE
      SUPPORT MEDIUM
η = VISCOSITY OF THE MEDIUM
K = FACTOR DEPENDING ON THE <u>SHAPE</u> OF THE PARTICLE

Fig. 3. Stoke's Law for cell sedimentation.

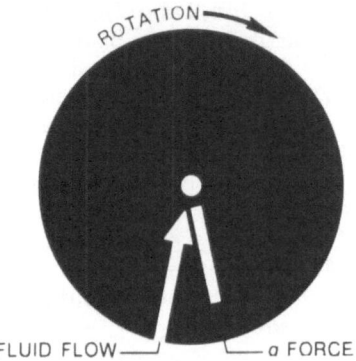

Fig. 4. Principle of counter-flow centrifugal elutriation.

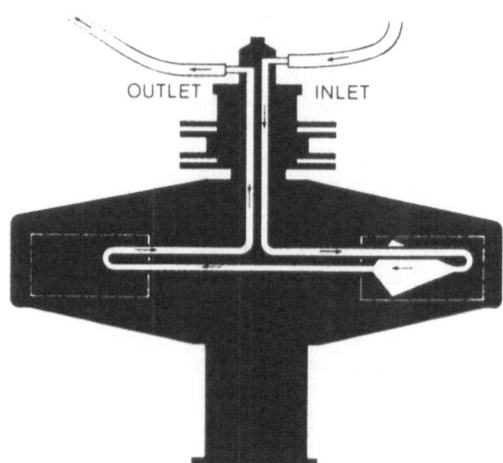

Fig. 5. Cross-section elutriator rotor and flow system.

     The shape of the sedimentation chamber is critical for efficient
separation (Fig. 7). Either by a stepwise decrease of the rotor speed
or by a stepwise increase of the counter-flow rate the various bands
are isolated. Through a built-in stroboscope (Fig. 8), the operator
can view the separation while it occurs in the rotorchamber.

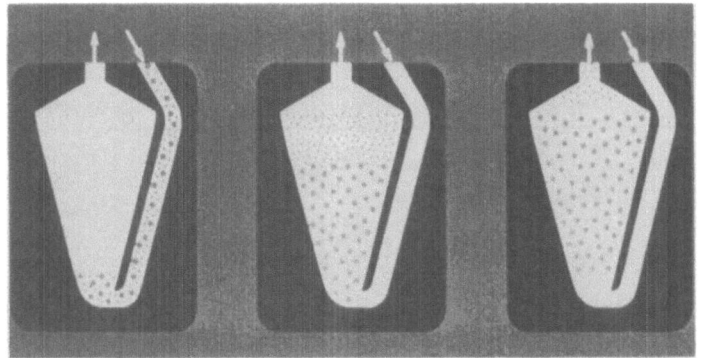

Fig. 6.   Principle of elutriation.

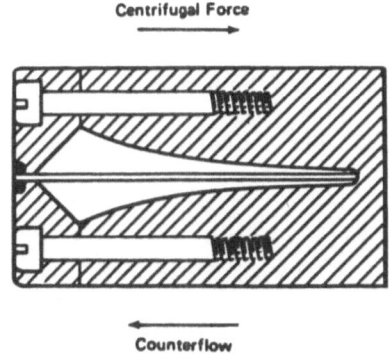

Fig. 7.   Elutriation chamber.

Although the elutriator rotor is rather expensive, the advantages
are numerous. The medium does not have foreign substances which might
activate or damage the cells. Futhermore, low g-forces are employed
and separation is achieved without pelleting and resuspension. When
compared with other methods, the time required for separation is
short, and relatively large numbers of cells (10^8 - 10^9) are separated
in different fractions simultaneously. Since the rotor and
flowsystem are autoclavable and the introduction of the sample, the
separation procedure as well as the collection of various fractions
can be performed in a closed system. Sterility can be maintained at
all the times (Fig. 9).

Thus, apart from having a high cell recovery, the method has been
proven to maintain cell viability and structural integrity.

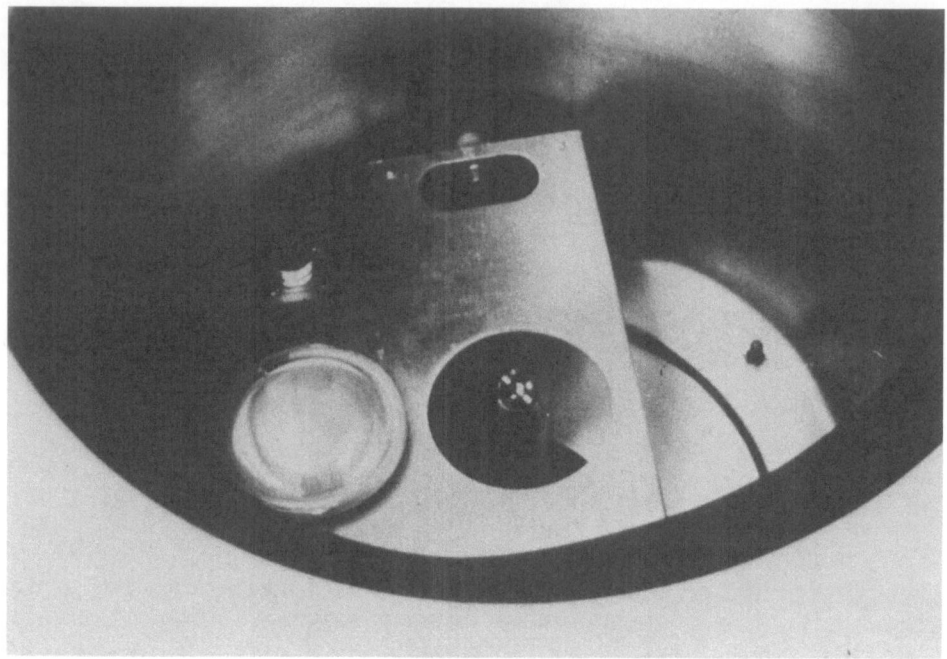

Fig. 8.   Stroboscope built in beneath the rotor (for a better view the
          rotor has been removed).

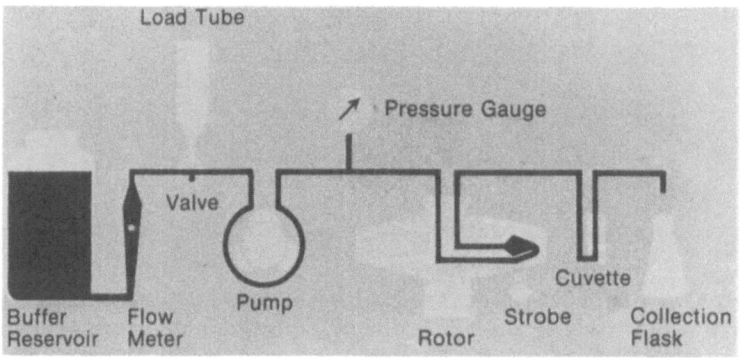

Fig. 9.   Typical setup of elutriation system.

    Nevertheless, several technical and methodological modifications
have been reported during the last few years. A few of them could be
of great interest for our purpose.

    Improved granulocyte separation (with respect to erythrocyte
contamination) has been achieved by performing a crude preseparation
leading to the introduction into the elutriator of a buffy coat or the
supernatant obtained after dextran sedimentation of the red cells
(25).

Granulocyte suspensions prepared acccording to a similar procedure have an overall recovery of 77%, and a purity of 96%. These cells were shown to behave normally in vitro (chemotactic response, stimulated oxygen uptake in the presence of latex beads, bactericidal capacity, and enzymic activity, and furthermore, in the maintenance of their structural integrity (26).

Thrombocytes separated in platelet-rich plasma have been further separated into platelet subpopulations using centrifugal elutriation (27). Recent data have demonstrated that, although the intrinsic functions of these size-dependent platelet subpopulations are similar, the absolute ability to aggregate or to perform the release-reaction correlates closely with their sizes (28).

The separation by counterflow centrifugation of erythrocyte subpopulations according to their cell-volumes is expected to correlate closely with their ages (29).

A stepwise increase of the density of the elutriating medium has made it possible to isolate different human monocyte subpopulations. The monocytes in these fraction have the same size distribution, as judged by electronic sizing, indicating that they differ only in their density (30).

Although 10^8 - 10^9 cells can be processed in the 4.5 ml elutriation chamber, larger amounts of cells are neccessary for granulocyte transfusion. Therefore, larger separation chambers have been designed (Fig. 10), (31,32), or a second standard separation chamber instead of the bypass chamber has been used. However, the orginal rotor is designed for the processing of about 50 ml of blood, and this seems ideal for blood cell labeling purposes.

The reproducibility of the separation was greatly improved by means of a number of modifications that concern both the generation of a constant flow rate by a pulseless hydrostatic or compressed air pressure system instead of a pump and a more accurate electronic rotor speed control, enabling elutriation by variation in the rotor speed at a constant flow rate (33). With these modifications, Figdor et al (34) was able to prepare from whole blood almost pure lymphocytes and a highly purified monocyte suspension with a very high recovery (34).

Coupling the elutriator with continuous flow cytometry optimized the determination of the separation point between 2 cell populations, as compared with the relative "blind" elutriation in fractions, (35) (see Fig. 9).

Although 10^8 - 10^9 cells can be processed in the 4.5 ml chamber, appreciable dilution will take place when the various bands are elutriated and collected. For cell labeling purposes the final suspension obtained this way is too dilute, impairing the labeling

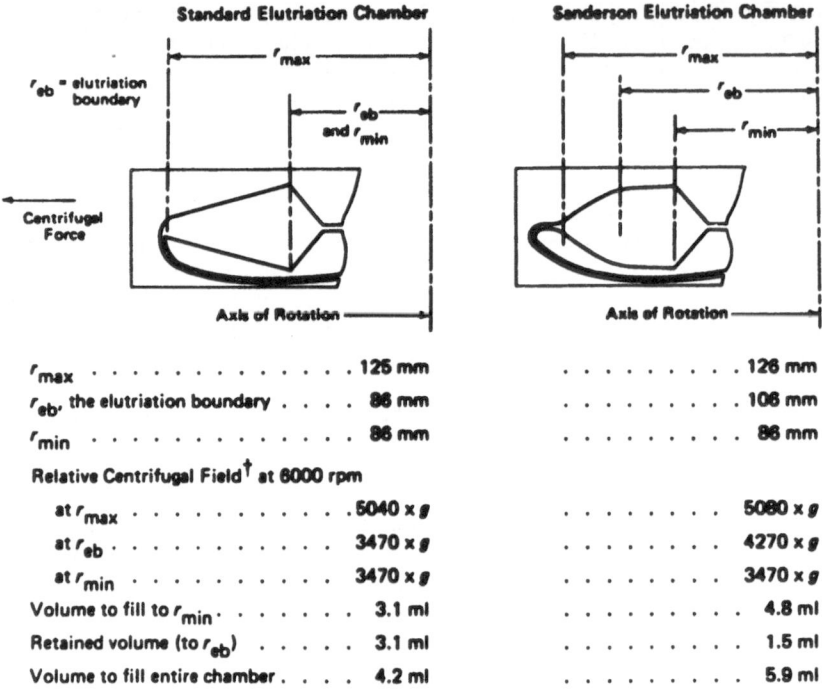

Fig. 10. Elutriation chamber characteristics.

efficiency to an unacceptable degree. In order to prevent this
unwanted dilution during recovery of the separated cells, a simple
modification has been made by installing a septumholder at the top of
the rotor. This enables the operator to recover the content of the
chamber directly (36). It is clear that this method is of value only
for granulocytes, because all smaller cells have to be first
elutriated.

Reversing the direction of flow after the separation has taken
place is another way to harvest the contents of the separation chamber
in a relatively small volume (37).

A very recent development (38) is the use of a second elutriator
to concentrate, on line, the fractions separated by the first
elutriator (Fig. 11). Apart from the rotors, the system consists of a
sample-introduction unit. Sterility is maintained in the closed
system. In a cooperative venture between two centers in Amsterdam,
each having an elutriator, preparation of granulocyte suspensions has
been achieved. The early results are very encouraging: Starting
with 50 ml of blood, concentrations of $10^7 - 10^8$ granulocytes/ml are
obtained with overall recoveries of more than 90%. Furthermore, the
cells have normal chemotactic activity, while labeling efficiencies
with In-111-oxine were in the order of 85-95%. Clinical tests will be
performed very soon.

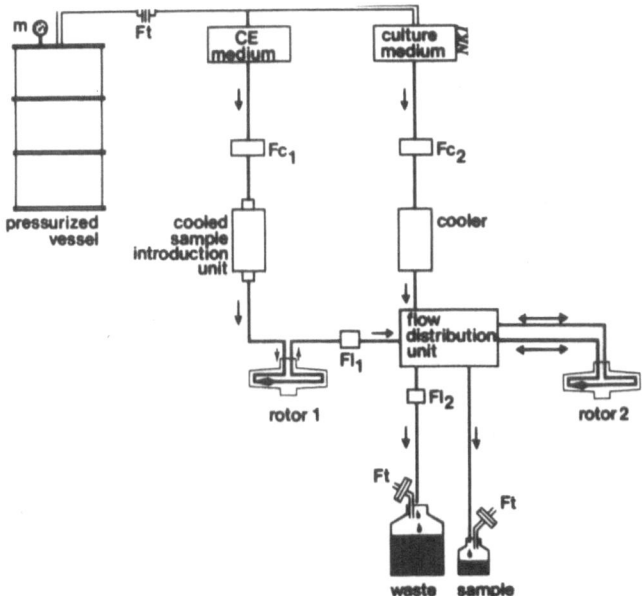

Fig. 11. Schematic representation of the system in which two
          elutriator rotors are coupled in series [see Figdor et al
          (38)].

          FC = flow controller, Fl = flowmeter

          Ft = sterilization filter, m = manometer

          ----> = direction of flow

CONCLUSIONS

1. For scintigraphic purposes, maintenance of relevant cell
    function(s) and viability should be preferred to purity during
    cell purification.

2. The consequences of the use of (foreign) substances, serving as
    gradient for the separation of cells with phagocytosis capability
    need to be studied further before routine use.

3. Centrifugal elutriation is a promising method for the separation
    of pure, viable, and functional cell suspensions.

4. The development of suitable in vitro test(s) predicting the in
    vivo behavior of a cell suspension is needed.

ACKNOWLEDGEMENT

The author thanks Dr. Sussmann, Beckman Instruments International
S.A., Geneva, for his kind permision to use several pictures of the
elutriation techniques.

REFERENCES

1. R. J. Hawker and C. E. Hall, Differential uptake of 111-Indium
    blood platelets and granulocytes, Thromb Hemosta 46:1 (abstr)
    (1981).
2. S. R. Turner, J. A. Tainer, W. S. Lynn, Biogenesis of chemotactic
    molecules by the Arachidonate lipoxygenase system of
    platelets, Nature 257:680 (1975).
3. N. D. Charkes, M. A. Dugan, L. S. Malmud, H. Stern, et al,
    Labeled leucocytes in thrombi, Lancet 2:600 (1974).
4. C. Figdor, Separation of human leucocytes by physical methods,
    Thesis, University of Amsterdam (1982).
5. P. T. Bodel, B. A. Nichols, D. F. Bainton, Appearance of
    peroxidase reactivity within the rough endo-plasmic reticulum
    of blood monocytes after surface adherence, J Exp Med 145:264
    (1977).
6. P. S. Vassar, E. M. Levy, D. E. Brooks, Studies on the
    electrophoretic separability of B and T human lymphocytes,
    Cell Immunol 21:257 (1976).
7. A. Boyum, Isolation of mononuclear cells and granulocytes from
    human blood, Scand J Clin Lab Invest 21 (Suppl 97):77 (1968).
8. W. D. Johnson, B. Bei, Z. A. Cohn, The separation long-term
    cultivation and maturation of the human monocyte, J Exp Med
    146:1613 (1977).
9. F. A. W. Splinter, M. Beudeker, A. van Beek, Changes in cell
    density induced by isopaque, Exp Cell Res 111:245 (1978).
10. C. L. Berger and R. L. Edelson, Comparison of lymphocyte function
    after isolation by Ficoll-Hypaque floation or elutriation, J
    Invest Dermatol 73:231 (1979).
11. J. T. Kurnick, L. Osterberg, M. Stegagno, A. K. Kimura, et al, A
    rapid method for the separation of functional lymphoid cell
    populations of human and animal origin of PVP-Silica (Percoll)
    density gradients, Scand J Immuno 10:563 (1979).
12. J. A. Roth and M. L. Kaeberle, Isolation of neutrophils and
    eosinophils from the peripheral blood of cattle and comparison
    of the functional activities, J Immunol Methods 45:153 (1981).
13. C. Bruch, P. Kovacs, E. Ruber, T. Fliender, Studies on the
    inhibitory effect of granulocytes on human granulopoiesis in
    agar cultures, Exp Hematol 6:337 (1978).
14. H. Pertoft, K. Rubin, L. Kjellen, T. C. Laurent, et al, The
    viability of cells grown or centrifuged in a new density
    gradient medium, Percoll, Exp Cell Res 110:449 (1977).

15. M. L. Thakur, J. P. Lavender, R. M. Arnot, D. J. Silvester, et
     al, Indium-111-labeled autoloques leukocytes in man, J Nucl
     Med 18:1014 (1977).
16. D. C. Dooley and T. Takahaski, The effect of osmotic stress on
     the function of the human granulocyte, Exp Hematol 9:731
     (1981).
17. D. G. Pennington, N. L. Y. Lee, A. E. Roxburgh, J. R. McGready,
     Platelet density and size: the interpretation of
     heterogeneity, Br J Hematol 34:365 (1976).
18. S. St. J. Wakefield, J. S. Gale, M. V. Berridge, T. W. Jordan, et
     al, Is Percoll innocuous to cells? Biochem J 202:795 (1982).
19. S. D. Nathanson, P. L. Zamfirescu, E. I. Drew, S. Wilbur,
     Two-step separation of human peripheral blood monocytes on
     discontinuous density gradients of collodial silic-poly-
     vinylpyrrolidinone, J Immunol Methods 18:225 (1977).
20. D. C. Dooley, J. T. Simpson, H. T. Meryman, Isolation of large
     numbers of fully viable human neutrophils: a preparative
     technique using Percoll Density Gradient Centrifugation, Exp
     Hematol 10: 591 (1982).
21. R. Hjorth, A. Jonsson, P. Vretblad, A rapid method for
     purification of human granulocytes using Percoll. A
     comparison with dextran sedimentation, J Immunol Methods 43:95
     (1981).
22. S. H. Saverymuttu, A. M. Peters, H. J. Danpure, H. J. Reavy, et
     al, Lung transit of Indium-111-labeled granulocytes. Relation
     to labeling techniques Scand J Hematol 30:151 (1983).
23. W. S. Bont and J. E. de Vries, in: "Cell Populations," E. Reid,
     ed., Ellis Horwood, Chicester (1979).
24. P. E. Lindahl, Principle of a counter-streaming centrifuge for
     the separation of particles of different sizes, Nature 161:648
     (1948).
25. T. J. Lionetti, S. M. G. Hunt, P. S. Lin, S. R. Kurtz, et al,
     Preservation of human granulocytes II. Characteristics of
     granulocytes obtained by counter flow centrifugation,
     Transfusion 17:465 (1977).
26. J. F. Jemionek, T. J. Contreras, J. E. French, L. J. Shields,
     Technique for increased granulocyte recovery from human whole
     blood by counter flow centrifugation-elutriation. I. In
     vitro analysis, Transfusion 19:120 (1979).
27. C. B. Thompson, K. A. Eaton, S. M. Princiotta, C. A. Ruskin, et
     al, Size-dependent platelet subpopulation: relationship of
     platelet volume to ultrastructure, enzymatic activity and
     function, Br J Hematol 50:509 (1982).
28. C. B. Thompson, J. A. Jakubowksi, P. G. Quinn, D. Deykin, et al,
     Platelet size as a determinant of platelet function, J Lab
     Clin Med 101:205 (1983).
29. R. J. Sanderson, N. F. Palmer, K. E. Bird, Separation of red
     cells into age groups by counter flow centrifugation, Biophys
     J 15:321 a (1975).

30. C. G. Figdor, W. S. Bont, J. de Roos, et al, Isolation of functionally different human monocytes by counter flow centrifugation elutriation, Blood 60:46 (1982).
31. R. J. Sanderson, K. E. Bird, N. F. Palmer, J. Brenman, Design principles for a counter flow centrifugation cell separation chamber, Anal Biochem 71:615 (1976).
32. J. F. Jemionek, T. J. Contreras, D. N. Stevens, F. W. Bernhards, et al, Use of modified rotor and enlarged separation chamber for isolation of human granulocytes by counter flow centrifugation elutriation, Cryobiology 17:230 (1980).
33. W. L. van Es and W. S. Bont, An improved method for the fractionation of human blood cells by centrifugal elutriation, Anal Biochem 103:295 (1980).
34. C. G. Figdor, W. S. Bont, J. E. de Vries, W. L. van Es, Isolation of large numbers of highly-purified lymphocytes and monocytes with a modified centrifugal elutriation technique, J Immunol Methods 40:275 (1981).
35. P. H. M. de Mulder, J. M. C. Wessels, D. A. Rosenbrand, J. B. J. M. Smeulders, et al, Monocyte purification with counter flow centrifugation monitored by continuous flow cytometry, J Immunol Methods 47:31 (1981).
36. A. D. Nunn and G. Gagne, The recovery, in small volume, of cells from the Beckman JE-6 Elutriator rotor, Transfusion 18:599 (1978).
37. M. D. Persidsky and L. S. Olson, Granulocyte separation by modified centrifugal elutriation system, Proc Soc Exp Biol Med 157:599 (1978).
38. C. G. Figdor, W. L. van Es, J. M. M. Leemans, W. S. Bont: A centrifugal elutriation system to separate small numbers of cells, J Immunol Methods 68:73 (1984).

TECHNIQUES OF CELL LABELING: AN OVERVIEW

M.L. Thakur and S. McKenney

Thomas Jefferson University
11th and Walnut Streets
Philadelphia, PA 19107

INTRODUCTION

Over the past few decades the use of blood cells labeled with
radioactive substances has become increasingly popular. The aims of
investigations have changed from the study of cell origin to the study
of cell kinetics in health and disease and to the use of radiolabeled
cells as a non-invasive diagnostic modality. Changed with the times
are also the techniques of labeling blood cells with radioactive
substances. The importance of techniques lies not only in the
efficacy by which the blood cells are labeled but also in the ease by
which the procedure is performed. The efficacy, the ease of procedure, and the use of nontoxic substances greatly add to the prevention
of cell injury and to the preservation of cell pathophysiologic
function.

Most of the early work was carried out by injecting intravenously
such radioactive compounds as H-3-thymidine(1), C-14 formic acid and
P-32-di-isopropylfluorophosphate(2-4). The pros and cons of the use
of these and more such compounds are discussed elsewhere(5,6). From
the technical point of view, however, a few points must be mentioned.
First of all, most of the radionuclides employed in the early days
were beta emitters. They were aimed at labeling cohort cells of
uniform age before they were released in circulation from bone marrow.
The concept of labeling blood cells in vivo by intravenous injection
was desirable but was only partially successful since the compounds
were nonspecific and labeled other types of cells as well.
Furthermore, the labeling efficiency was generally poor since a major
proportion of the radioactive compound was either excreted or
deposited in other organs in the body.

The work carried out with cohort labels led investigators to the
understanding of cell cycle and kinetics of circulating blood cells.

However, due to the limitations of efficacious in vivo cell labeling
procedures, other radioactive compounds to label circulating blood
cells, in vitro, were investigated. In these procedures, generally,
cells were separated and incubated in vitro with compounds of choice.
Compounds employed incorporated both beta and gamma-emitting
radionuclides. Noteworthy, among the gamma emitters were I-131-sodium
iodide, Cr-51-Na chromate, and Tc-99m-Na pertechnetate.

Iodine-131 was investigated by Morgan et al in 1954 as a
potential surface label for platelet membrane protein(7). Although
only modestly successful, the work paved the way to the acceptance of
the concept of surface cell labeling. The technique was later
evaluated for surface labeling of lymphocytes via I-131 labeled gamma
globulin(8).

Chromium-51 was introduced at the same time by Robertson et
al(9). The radionuclide has a long half-life (27 days) and emits a
gamma photon (320 keV, 7%) that made it easier to detect Cr-51 by
scintillation counter than the detection of previously used beta
emitters. Chromium-51 has, therefore, become a popular agent for cell
labeling and is still being commonly used in the determination of cell
survival. The poor gamma photon emission, however, has prevented its
use in non-invasive applications. Spontaneous elution from labeled
cells in circulation is also considered to be a drawback.

A popular imaging agent available worldwide today at a modest
cost is Tc-99m. The radionuclide is most stable in its oxidation
state 7+ pertechnetate, but in the oxidation state 3^+ and 4^+ it
covalently binds to a number of chelating agents including many bio-
organic molecules. Uchida et al(51) added a small quantity of stannous chloride as a reducing agent to a platelet suspension in isotonic
saline and incubated them for 10 minutes with Tc-99m-Na-pertechnetate.
Although the 6-hour half-life of Tc-99m was not long enough to study
platelet survival of 7-8 days and the results of this application were
not so encouraging, the work led investigators to new approaches of
labeling blood cells with Tc-99m. The major thrust of the work described in this monograph is, however, on In-111 as a tracer for platelets, leukocytes (mixed populations of white cells), and lymphocytes.
This chapter will therefore review In-111 cell labeling techniques and
discuss other important methods described in the literature.

LABELING BLOOD CELLS WITH IN-111:

Preliminary investigations

In 1976, McAfee and Thakur surveyed several radioactive agents to
label human neutrophils in vitro(10,11). Approximately 30 ml of
venous blood was drawn in a syringe containing preservative-free

heparin. The use of such a type of anticoagulant eliminated benzyl
alcohol, a commonly employed preservative, which can affect cell
membrane(12). Three percent (V/V), of 2% (W/V) methyl cellulose was
added to blood, mixed, and the syringe was hung vertically to allow
erythrocytes to sediment rapidly. Within 60 minutes, a large proportion of erythrocytes were sedimented and the plasma remained on the
top. The plasma layer containing leukocytes and platelets was carefully separated, and leukocytes were isolated by centrifugation. They
were then resuspended in plasma or isotonic sodium chloride for incubation with radioactive agents.

The agents consisted of gamma emitting radionuclides such as
In-111, Ga-67, Tc-99m, and I-123. The labeled compounds were classified into particles and soluble agents(10,11). Among the prominent
radioactive particles were In-111-liposomes, Tc-99m-sulfur colloid,
and Tc-99m-labeled heat denatured erythrocytes. These were allowed to
be engulfed by neutrophils. The process was aided by incubation at
37°C or by tumbling the cells at room temperature. The cells were
then separated by centrifugation, washed once with plasma, and radioactivity associated with cells and remaining in the fluid was determined.

It was observed that leukocytes separated by centrifugation
retained up to 40% of the radioactivity added in the particulate form.
However, when plasma containing radioactive particles but no cells
was centrifuged at a similar gravitational force for the same length
of time, 10-15% of the radioactivity was found settled at the bottom
of the test tube. Furthermore, a major proportion of the radioactivity which was thought to have been associated with leukocytes was
eluted from the cells when the cells were further incubated with nonradioactive particles. These results indicated that a) at least some
of the particles were nonspecifically adhered to the cell's surface
and not truly engulfed and b) centrifugation was not a proper way to
eliminate unengulfed particles from labeled cells. Furthermore,
questions were raised as to whether leukocytes or neutrophils, once
allowed to phagocytose a large number particles in vitro, will retain
their ability to function in vivo. The authors concluded, therefore,
that labeling cells with particles was not a preferable technique(11).

Recently, however, there appears to be a renewed interest in
phagocytic labeling of neutrophils. The procedure has led investigators to a third approach of selective neutrophil labeling in whole
blood. The technique is attractive and is discussed later in the
chapter.

Having concluded against the use of particles, McAfee and Thakur
emphasized on soluble agents(10). Among the soluble agents, In-111
labeled 8-hydroxyquinoline (oxine), acetylacetone, and tetraphenyl
porphine were of particular interest. These agents were lipid soluble

and achieved higher labeling efficiency than those which were not. When cells were suspended in isotonic saline, labeling efficiencies were higher than those when cells were suspended in plasma. (Reasons for these and the mechanism of cell labeling are discussed later.)

Among the three agents, acetylacetone was observed to have higher affinity to erythrocytes than to leukocytes. Since leukocytes separated by sedimentation procedures were always contaminated with erythrocytes, In-111-acetylacetone was considered to be unsuitable for leukocyte labeling. The procedure for the preparation of In-111-Tetraphenyl porphine was inefficient and lengthy(13). Oxine was, therefore, the agent of choice.

Technetium-99m oxine could not be prepared as efficiently as In-111-oxine(14). Indium-111 has better physical characteristics (t1/2 - 2.8d, r-173 keV-89%, 247 keV-94%) and higher detectability by the gamma camera than gallium-67 (t1/2 3.2d, r-93 keV-40%, 184 keV-24%, 296 keV-22%, and 388 keV-7%). The half-life of In-111 is also long enough for it to serve as a useful tracer for cells such as platelets and lymphocytes which have longer life-spans than neutrophils. Indium-111-oxine was therefore chosen as the agent of choice for labeling leukocytes.

The Use of Indium-111 Oxine

Preparation of In-111-Oxine. Indium-111 is produced with an alpha or proton particle bombardment of silver or cadmium, respectively, in an accelerator(15) and is available commercially in a no-carrier-added chloride form. When 50 to 100 μg of bidentate oxine dissolved in ethanol is added to trivalent In-111 ions in acetate buffer pH 5 to 6, three molecules of oxine are chelated with one ion of indium. The neutral complex thus formed is extractable in a small volume of solvents such as chloroform or methylene chloride. The extraction step eliminates unwanted salts and purifies the complex from unchelated In-111. The solvent is then evaporated with a gentle stream of nitrogen or air, and the complex is dissolved in 50μl ethanol for use(16). Using polysorbate 80, In-111-oxine has been found to dissolve in aqueous system(17). Water-soluble oxine sulphate also has been used(18).

Mechanism of Cell Labeling

The experimental evidence has led us to conclude that In-111-oxine, when incubated with cells, passes the cell membrane barrier at an ambient temperature(19,20). Although the process is essentially completed within 15 minutes, approximately 70% of the radioactivity is taken up by the cells within the first 5 minutes(16). Indium-111-oxine has lower stability constants (logK=11) than those of In-111-

transferrin (logK=30), and the diffusion process takes place most
efficiently in absence of plasma transferrin and other lipopro-
teins(16,21). Once within the cell, In-111 binds to cytoplasmic
biomolecules of apparent molecular weight 540,000, 82,000, and 3,600
daltons. The free oxine then elutes out of the cell. The results of
the strong association of In-111 with cytoplasmic components are in
agreement with those obtained by the studies with perturbed angular
correlation technique(20) and explain the lack of spontaneous elution
of radioactivity upon washing labeled cells in vitro and the observation of little plasma-bound radioactivity in vivo(16). A large excess
of oxine, however, has been observed to cause elution of radioactivity
from labeled cells(22).

The understanding of this mechanism prompted us to label plate-
lets(23), lymphocytes(24,25) tumor cells(26), as well as bacteria(27)
with In-111-oxine. This ability of In-111-oxine to label several
kinds of cells rendered the agent nonspecific and made cell separation
mandatory. Furthermore, the need for the cells to be suspended in a
nonplasma media caused a concern among investigators, since platelets
labeled in this manner exhibited reduced survival(28,29). These drawbacks have stimulated researchers to develop newer and, it is hoped,
better agents. These will be discussed later.

Indium-111 Leukocytes: Preparations and Problems

The procedure we follow(16) for the preparation of In-111 leukocytes for localization of abscesses has remained practically unchanged. Thirty ml of blood is drawn in a syringe containing preservative-free heparin (6-10 IU/ml blood), and erythrocytes are allowed to
sediment in a laminar flow hood by hanging the syringe supported by a
firmly secured, U-shaped, 19 G needle. Heparin is preferred as an
anticoagulant since it is expected to form platelet aggregates which
are sedimentated along with erythrocytes and thereby minimize platelet
contamination in leukocytes(30). Erythrocytes are allowed to sediment
spontaneously since blood of most patients has a high rate of sedimentation. External sedimenting agents are therefore unnecessary and are
avoided. After sedimentation (60 minutes or less to yield erythrocyte volume approximately 1/2 the initial blood volume),
leukocyte-rich plasma is carefully squirted through the bent needle
into a sterile test tube. This is then centrifuged at 450g for 5
minutes, resulting supernatant is separated and stored aseptically.
The leukocyte button is washed with saline or phosphate buffered
saline (PBS). Many commercially available saline solutions have pH
lower than 6. This should be raised to pH 7 by adding sterile
Sorensen's (0.05M) phosphate buffer pH 7 and autoclaved or filtered
for sterilization. Cells are resuspended in 4 to 5 ml PBS. To the
suspension are then added 500-600 μCi In-111-oxine and incubated at
room temperature for 15 minutes. Two to three ml plasma saved

previously are then added and cells are centrifuged at 450g for 5
minutes. They are then resuspended in 4-5 ml plasma for the measurement of cell-associated radioactivity and for subsequent venous
injections to the subject. The methods of quality control are
discussed previously(27), and the efficacy of clinical results are
evaluated elsewhere in this monograph.

Normally, autologous cells are used. In case of neutropenic
patients, only donar-recipient cross-matched cells should be employed,
since histocompatibility factors play a crucial role in granulocyte
survival in the presence of certain types of antibodies(31) and in
alloimmunized patients(32).

Although the use of In-111 leukocytes has become increasingly
popular, the application of this technique in the studies of in vivo
neutrophil kinetics has largely been hampered by the lack of an
adequate method to separate pure neutrophils from a small volume of
blood(33). The use of most popular Ficoll:Hypaque media for
neutrophil separation required them to be suspended in a hypotonic
medium such as water or ammonium chloride. Neutrophils separated in
this manner suffer from the loss of viability(34). Up to 40% of
neutrophils separated by this technique are taken up by the liver(35).

Recently, we have evaluated four density gradient media for
separation of neutrophils from a small volume of human blood.
Neutrophils were separated by the given procedure, counted, and
evaluated for purity, viability, structural integrity, and phagocytic
ability. The phagocytic ability was evaluated by allowing neutrophils
to engulf In-111-labeled Staphylococcus aureus. Unengulfed bacteria
were lysed by lysostaphin and eliminated by centrifugation. Results
indicated that among the gradients tested, a discontinuous percoll
gradient composed of 55%, 60%, and 65% percoll in saline, produced a
highest number of neutrophils with 100% viability in a shortest period
of time. They, however, were contaminated with 150% erythrocytes(27).
These could be minimized at the expense of lower neutrophil recovery.
The other methods of neutrophil separation are discussed in the
monograph separately by Hardeman and McAfee.

Despite the simplicity of labeling leukocytes with In-111-oxine
in vitro; the heterogenicity in the procedures employed has become
increasingly apparent. Some examples are given in Table 1. The
differences appear in volumes of blood drawn, anticoagulants used,
sedimenting agents added, and gravitational forces employed. We have
observed that use of acid citrate dextrose (ACD) as an anticoagulant
increases platelet contamination in separated leukocytes. A large
number of labeled platelets could increase the blood background and
may adversely affect localization of abscesses.

In our experience, on more than 99% of occasions, erythrocytes in
patient's blood sediment spontaneously and rapidly. The addition of a

Table 1. Examples of Variations in Leukocyte Separation Parameters

Blood Vol.	Anticoagulant	Sedimenting Agent	Centrifugal Force	Ref.
30 ml	6 IU/ml preservative-free heparin	none	450g x 5 min	(35)
40-80ml	10 IU/ml preservative-free heparin	3 ml, 6% Hydroxyethyl starch	?	(70)
50 ml	ACD	3% Hydroxy-ethyl starch	150g x 3 min.	(71)
30 ml	5 ml ACD	2% methyl cellulose	200g x 10 min	(72)
50 ml	5 ml ACD	Polysacchride vortex	350g x 5 min	(73)
80 ml	6.5 ml ACD	Hydroxyethyl starch-0.6 v/v	100g x 5 min	(45)

sedimenting agent does not, therefore, serve any purpose. Low centrifugal forces might decrease the white cell recovery in plasma and thereby increase the number of contaminating erythrocytes. The high centrifugal forces might increase platelet contamination. A uniform procedure should therefore be followed to serve a better compatibility of results from different laboratories.

Indium-111 Platelets: Preparations and Problems

Canine, rabbit, and pig platelets labeled with In-111-oxine in isotonic saline have normal survival of 7-8 days and higher recovery (75% vs 62%) than Cr-51 labeled platelets(23). Human platelets labeled similarly, however, have poor aggregability, low recovery (5.4% + 5.5%), and only 3-4 days survival (28,29). This may be attributed to the loss of sialic acid from human platelets. The loss of sialic acid does not take place from canine, rabbit, or pig plate-lets probably due to the extra acetyl group the animal sialic acid possesses(36). Human platelets labeled with In-111-oxine in plasma have normal survival and excellent aggregability upon stimulation with ADP. However, the labeling procedure is lengthy and produces poor incorporation of radioactivity(28,29,37). The net result has been the

repeated efforts to develop a suitable nonplasma medium in which human
platelets could be labeled with a high labeling efficiency but without
a loss of viability. Some of such methods are summarized in Table 2.

The table provides a testimony for the diversity of the currently
employed procedures for platelet labeling. These differ considerably
in volumes of blood (17 ml-500 ml), volumes of ACD anticoagulant,
centrifugal forces, and media used for platelet suspension. The
smaller the blood volume for labeling the better; but too small a
blood volume may yield fewer platelets and result in inadequate
labeling efficiency. One ml of ACD for every 6 ml of blood has been
found to be the best for optimal anticoagulation and maximal platelet
harvestation(38). Excessively high (>1000g) centrifugal forces or
excessively long (>15 min) centrifugation times progressively
deteriorate platelet aggregation, probably due to the loss of storage
granules(28). Proper care is also important in the step in which
blood is centrifuged for obtaining platelet-rich plasma. In order to
obtain platelet-rich plasma we have used 15 minutes centrifugation at
180g for many years. Inadequate centrifugation may result in cellular
destruction. Erythrocytes are more fragile than platelets. They also

Table 2. Examples of Variations in Platelet Separation Parameters

Blood Vol.	Anticoagulant	Centrifigal Force	Medium	% Aggreg.	Ref.
34 ml	6 ml ACD-pH 4.5	180 x 15 min 1000g x 10 min	MTS	80	(29)
50 ml	10 ml ACD	130g x 20 min 500g x 10 min	PBS	?	(74)
17 ml	1.5 ml ACD (+250ng Prostacyclin)	?	MTS	?	(75)
100 ml	10 ml citrate buffer-pH 5	180g x 25 min 800g x 6 min	PPP	?	(76)
500 ml	75 ml ACD	300g x 15 min 800g x 15 min	0.9% NaCl		(77)
150 ml	20 ml ACD	200g x 10 min 2000g x 20 min	Ringer's citrate	?	(78)
43 ml	7 ml ACD	200g x 15 min 2000g x 10 min	ACD/NaCl	50	(79)

MTS = Modified tyrode's solution, PBS = Phosphate buffered saline,
PPP = Platelet poor plasma, ACD = Acid Citrate Dextrose.

contain ADP. ADP thus released promotes formation of platelet aggregates and prevents homogeneous platelet suspension.

The use of various salt balance solutions, including a mixture of acid citrate dextrose and normal saline, has enabled investigators to achieve 80% to 90% labeling efficiency and preserve a modest (60-80%) platelet aggregability. The variations in these important parameters of platelet labeling procedures have produced equally diverse results of platelet survival and in vivo platelet quantification (Tables 3,4). Although all differences cannot be attributed to any single parameter, a uniform procedure would allow investigators to compare results from one laboratory to another in a more scientific manner than currently possible.

One way to eliminate these problems would be to develop new agents that would allow us to label platelets as well as leukocytes in plasma more efficiently than In-111-oxine. Developed by several investigators are the following agents.

Indium-111-Acetylacetone. This agent, considered undesirable by previous investigators(10), was reinvestigated by the virtue of the

Table 3. Various Methods and In Vivo Survival of In-111 Platelets

Method	Species	Period	Ref.
Saline	Dog	8 days	(23)
Saline	Dog	124.6 $\pm$ 10.5 hrs.	(80)
Plasma	Rabbit	2.8 days	(81)
Ringer citrate dextrose	Dog	7 days	(82)
ACD/saline	Human	7.5 days	(79)
Plasma, plasma/saline dextrose/saline	Human	7-9 days	(83)
Plasma	Human	7 days	(28)
Plasma	Human	9 days	(84)

Table 4. Various Methods and In Vivo Distribution of In-111 Platelets

Method	Species	Liver	Spleen	Ref.
Plasma	Rabbit	40%	14%	(81)
Saline	Dog	20%	59%	(80)
ACD/ saline	Human	10-15%	25-40%	(83)
MTS	Human	12-18%	25-30%	Thakur, Unpublished
Plasma	Human	5%	45%	(28)
Plasma	Human	15.8%	25.9%	(84)

fact that it is soluble in aqueous system but forms a neutral complex
with In-111(39). However, just like In-111-oxine, this agent also
permitted poor incorporation of radioactivity in cells when suspended
in plasma. While some have found it to be useful, others have
observed it to reduce platelet aggregability(40) and too toxic to
cells(41). The toxicity was attributed to the 20 mg acetylacetone,
the minimum required to produce greater than 90% labeling yields.

    Indium-111-Tropolone. Tropolone (2 hydroxy, 2, 4, 6-cyclo-
heptatrienone) was first investigated to form a lipid soluble-Tc-99m
complex to label erythrocytes(42). An Indium-111 complex of tropolone
was evaluated to label platelets and subsequently leukocytes in
plasma(43,44). Like acetylacetone, tropolone is also soluble and
forms a stable complex in aqueous media. Although no stability
constants of In-111-tropolone have been determined, data indicated
that in the presence of plasma, it incorporated a higher percentage of
radioactivity into platelets and leukocytes than did In-111-oxine or
In-111-acetylacetone. However, as the percentage of plasma in saline
or buffer media increased, the percentage of radioactivity incorporated decreased, in a similar number of cells under identical
conditions. Subsequent modifications have succeeded in obtaining
50-80%(45) and an average of 90% (McAfee - this monograph) labeling
efficiency for leukocytes concentrated in plasma.

    While these results are encouraging, labeling platelets in plasma
with In-111-tropolone has not shown to be advantageous, for human
platelets labeled with In-111-tropolone in plasma have been observed
to have 5-6 days survival in normal human volunteers as compared with
6-8 days with those labeled with In-111-oxine in citrated saline(46).

The group has recommended the use of citrated saline as a suspending
medium for labeling platelets with In-111-tropolone(43). Other
investigators have reported that a) they preferred oxine over
tropolone or acetylacetone(47), b) tropolone has no advantage over
oxine(48), c) tropolone has higher affinity for erythrocytes than
leukocytes or platelets(49), and d) for preserving chemotactic ability
of neutrophils oxine technique was preferable to tropolone(50).

Labeling with Tc-99m

Three major investigations have been reported in labeling human
platelets and leukocytes with Tc-99m(51-53). The common concept was
to label cells first with stannous ions so that Tc-99m7+ would be
reduced to Tc-99m4+ and probably bind to membrane proteins. Due to
the virtue of its physical characteristics, worldwide availability,
and low radiation burden, the use of Tc-99m is attractive. However,
it is unlikely that it would provide a good tracer for platelets in
which kinetic studies are to be performed for the entire life-span
(7-8 days) of the cells.

Although initial attempts to label platelets with Tc-99m were
only modestly successful, the work paved the ways for subsequent work
in labeling neutrophils as well as lymphocytes with Tc-99m. In vivo
results of the latter two investigations indicated spontaneous elution
of radioactivity from labeled cells. This was elucidated by only 10%
of circulating activity in association with labeled cells(52) and by
the gamma camera images of a large proportion radioactivity in the
kidneys and bladder(53).

Specific Cell Labeling:

A successful technique will be the one that will label specific
types of cells in whole blood and will eliminate the need for their
isolation from other cell types. The well-characterized phagocytic
ability of neutrophils have, for many years, prompted investigators to
label neutrophils with radioactive particles. As previously
mentioned(10,54), a modest success had been achieved using
Tc-99m-labeled sulfur colloid , but difficulties were encountered.
Nevertheless, recent modifications of those techniques have been more
successful and are discussed below.

Tc-99m-Tin Colloid

Unlike Tc-99m sulfur colloid, technetium tin colloid is soluble
in sodium citrate. Tin colloid containing approximately 25 mCi or
more Tc-99m was therefore incubated in whole blood at 37°C and sodium
citrate was added to dissolve unengulfed particles(55). Blood was

then given back to the subject. Radioactivity was localized in
abscesses, but as much as 70% of the injected radioactivity was taken
up by the liver alone. This is excessively higher than the
radioactivity taken up by the organ when In-111 labeled leukocytes are
injected. Authors had claimed that a major proportion of the 80%
engulfed radioactivity was taken up by macrophages than neutrophils,
but no experimental confirmation was obtained.

Other investigators(56) basically performed similar procedures
but centrifuged the blood and obtained leukocyte-rich plasma after
incubation with radioactive colloid. Only about 20% of the
radioactivity was found in plasma. A major proportion was found in
association with neutrophils. When administered to patients, there
was sufficient activity in abscesses to be imaged by gamma camera, but
a large prorportion was also actively taken up by the bones and
intestines. It appears from these results that this approach is
worthy of further refinement.

In a recent article In-111 ferric hydroxide colloid was employed
for "leukocyte" labeling by phagocytosis(57). Sixty to 80%
radioactivity was cell associated. Unengulfed radioactive colloid was
dissolved with 0.5 ml of 10% (hypertonic) sodium citrate, and free
Indium was then eliminated by centrifugation. This step was essential
since free In-111 would bind to transferrin and if injected would
remain in circulation and increase blood background. Neither in vitro
function nor in vivo distribution of labeled "leukocytes" was
examined. Considering the worldwide availability, modest cost, and
low radiation dose, in my opinion the use of Tc-99m colloid is more
attractive than In-111 colloid for labeling leukocytes to be used for
abscess localization.

Soluble Chemo-attractive Substance

The investigations in the use of soluble chemoattractive substances are based upon the fact that certain chemicals stimulate
neutrophils and interact with specific receptors the cells generate.
The ability of certain synthetic polypeptides to act as chemotactic
agents has been known for many years(58). Among these, N-formyl-
methionyl-leucyl-phenylalanine (FMLP) is most potent. Radioiodinated
FMLP might have reduced biological activity and poor stability due to
a weak carbon-iodine bond strength(59). Covalently bound to a protein
molecule, however, FMLP retains its biological activity(60).
Transferrin has a strong affinity (logk=30) for ionic-In-111.
Transferrin-bound FMLP therefore produced an excellent compound to
label with In-111 and use as a specific neutrophil agent(61). A
modest success (60% labeling efficiency) has been achieved, and
further work is awaited.

Radioiodinated synthetic hexapeptide was also investigated(62).
Approximately 50% labeling efficiency was achieved when isolated

neutrophils were incubated with the agent in a "special" buffer. The uptake, however, was poor when cells were suspended in plasma. Labeling neutrophils specifically in whole blood with these agents with greater than 95% efficiency may be a difficult task, but the approach is noteworthy.

The application of the same principle could be extended to the use of radiolabeled monoclonal antibodies for specific neutrophil labeling in whole blood. A variety of monoclonal antibodies specific for cell surface antigens of blood cells have already been produced, some of which are specific for human granulocytes(63-65). These could be labeled with radioactive iodine or with In-111 via cyclic anhydride of diethylenetriamine-pentaacetic acid (DTPA) covalently bound to the amino group of the protein(66,67). Feasibility studies have already begun in our and Dr. McAfee's laboratory, and encouraging results have been obtained (MLT unpublished, JGM this monograph).

INDIUM-111 MERCAPTORPYRIDINE-N-OXIDE

For the past two years we have been engaged in developing yet another agent, namely, In-111 Mercaptopyridine-N-oxide (Merc). Our interest in Merc stemmed from its ability to form metal chelates and from the persisting need of an agent that would label cells in plasma.

Merc, in its basic, or a sodium salt, form, forms highly colored complexes with the metal ions of elements in the eighth subgroup of the periodic table. The complexes are formed at a wide range of pH and are extractable in chloroform(68). We have observed that with no carrier added, In-111 chloride also forms a complex at a wide range of pH that is quantitatively extractable in chloroform(69). Technetium-99m4+ (reduced with 100 μg SnCl2 in ethanol) also forms a complex with Merc but most efficiently only at pH 4, and only approximately 65% of the added radioactivity is extractable. With ruthenium 103 in acetate buffer pH 7.4, up to 80% of the radioactivity was extractable in chloroform, but incubation of the reaction mixture at 90°C for 15 minutes was necessary. Ruthenium was of interest because of its radionuclide 97, which has physical characteristics (t1/2 2.8d, 215 keV 91%) acceptable for cell labeling.

When In-111-Merc, Tc-99m-Merc, and Ru-103-Merc were incubated at 22°C for 15 minutes. With an equal number of human platelets in plasma (5 x 10⁸/ml), In-111 complex produced best results (57.6% vs. 28.3% and 8.3%, respectively). It is for these reasons the subsequent work was carried out with In-111. Using In-111-Merc, platelets and leukocytes can be labeled with 70-90% labeling efficiency in plasma. However, different quantities of Merc are required for each type of cells. The reasons for this are not clearly understood at the time of this writing, but experiments using S-35-labeled Merc are on the way.

The labeling procedure can be performed in two ways. First, by using preformed In-111-Merc, just like In-111-oxine; and second, by incubating cells in plasma first with Merc and then with In-111. The second method offers a possibility of labeling cells by a kit method. Both methods are briefly described as follows.

Method One

Preparation of In-111-Merc. In this method, In-111-Merc is prepared by adding approximately 500 µCi of In-111 chloride (Medi-Physics, research grade, 50 mCi/ml) and 10 µg Merc (1 mg/ml phosphate buffer) to, a 500 µl 0.9% sodium chloride buffered to pH 7 with 0.05 phosphate buffer. For checking the percentage In-111 incorporated into Merc, a 50 to 100 µl aliquot of this solution (1-2 µg Merc) is diluted to 1 ml with PBS, and the complex is extracted twice with equal volumes of chloroform. Because of a large dilution of 1 to 2 µg Merc, as well as due to cationic impurities in PBS which might bind to Merc and release In-111, a low proportion of In-111 is extracted. Adding 5 to 10 µg Merc before extraction to the above diluted sample increases extraction efficiency to >95%. However, this does not represent the true situation in the original sample and can not be regarded as a true test. We therefore dispense another aliquot of 500 µCi In-111 and carry out extraction. Greater than 95% radioactivity is usually extracted. Indium-111-Merc thus prepared and stored at room temperature is stable for at least a period of two weeks. However, for the extraction test, the Merc concentration must be at least 5 µg/ml.

When In-111-Merc is to be used for labeling neutrophils, the Merc concentration is raised to 20 µg per ml suspension. This quantity of Merc is required for the optimal labeling of leukocytes.

Platelet labeling in plasma. Thirty-four ml venous blood is drawn in 6 ml acid-citrate anticoagulant prepared as described previously(29). Blood is mixed, transferred into two 50 ml sterile plastic tubes in two equal volumes and centrifuged at 180g for 15 minutes. Resultant platelet-rich plasma is carefully separated, combined, and centrifuged at 1000g for 10 minutes. After centrifugation, all but 1.5 ml plasma is separated and platelet button resuspended.

Indium-111-Merc is then added to the platelet suspension and incubated at room temperature for 20 minutes. Greater than 80% radioactivity is incorporated into platelets. Unbound radioactivity is eliminated by centrifugation.

Labeled platelets are then resuspended in platelet-poor plasma obtained from 15 ml blood drawn separately in 1.5 ml 3.8% Sodium citrate as an anticoagulant. Aggregability of labeled platelets is checked using 1 x 10^{-5}M ADP and compared with that of unlabeled platelets (>90%). Platelets are then ready for injecting back to the subject.

Leukocyte labeling in plasma. Leukocyte-rich plasma is obtained from 30 ml venous blood by allowing erythrocytes to sediment spontaneously. It is then separated, centrifuged at 450g for 5 minutes, and all but 1.5 ml plasma is withdrawn. The leukocyte button is resuspended, and In-111-Merc is added for a 15-minute incubation at room temperature. Greater than 80% radioactivity is found in association with leukocytes. The unbound radioactivity is then eliminated and leukocytes are resuspended in fresh plasma for injection.

When In-111 complex containing only 10 μg Merc (as in case of platelets) is added, a much lower labeling efficiency is achieved. Reasons for this are being investigated. Alternatively, leukocytes could be washed once and resuspended in 5 ml PBS and incubated with In-111 complex containing 10 μg Merc. Greater than 95% labeling efficiency is achieved.

Leukocytes labeled by both methods clear rapidly from normal human lungs and localize in abscesses(27). Table 5 compares abscess to tissue ratios in dogs given In-111-Merc- and In-111-oxine-labeled autologous leukocytes and Ga-67-citrate, 24 hours previously. The In-111 radioactivity ratios were consistantly higher than Ga-67 ratios. The abscess to liver and abscess to spleen ratios in dogs receiving In-111-Merc-labeled leukocytes were also higher than in those receiving In-111-oxine-labeled leukocytes. These indicated that a smaller percentage of In-111-Merc-labeled leukocytes was taken by the liver and spleen than that of In-111-oxine-labeled leukocytes. In four dogs given In-111-Merc-labeled leukocytes, the liver received an average of 24.2 percent administered dose. This compared favorably

Table 5. 24 Hour Abscess/Tissue Ratios in Dogs given In-111-Merc- and
          In-111-Oxine-Labeled Leukocytes and Ga-67 Citrate

Abscess To	In-111-Merc	In-111-Oxine	Ga-67 Citrate
Blood	75.2	75.8	11.9
Fat	577.5	946.1	71.1
Muscle	979.6	315.0	38.9
Liver	14.6	4.2	1.8
Spleen	4.2	1.8	2.4

with that of 48.5% In-111-oxine-labeled leukocytes in the same
species(18).

Method Two

For use in this method platelets and leukocytes are separated as
in Method One, with the exception that both platelets and leukocytes
are concentrated in 0.5 ml instead of 1.5 ml plasma.

The concentrated platelet or leukocyte suspension is then
incubated at room temperture with dry 2 and 20 µg Na-Merc,
respectively. (Na-Merc in phosphate buffer solution is added to a
sterile polystyrene tube and evaporated to dryness. A large number of
such test tubes could be prepared and stored for future use. Na-Merc
has higher solubility than Merc).

The Merc-containing cells are then added to In-111 in 0.25 M
acetate or citrate buffer pH 6.0, either in solution or dry. A 15-to-
20-minute incubation provides greater than 80% labeling efficiency for
both types of cells. The dry forms of Na-Merc and In-111 avoid plasma
dilutions that occur in Method One. Above all, this method avoids all
variations that exist in today's cell-labeling procedures and provides
a uniform and easy technique to label cells. Platelet aggregability
and leukocyte phagocytosis remain practically unaltered.

Such procedure carried out using 10 to 50 µg oxine or tropolone
produced 15% and 18.9% labeling efficiencies, respectively.
Furthermore, using 2 µg and 20 µg Na-Merc as chelating agents, Ga-67,
Tc-99m, Tl-201 or I-131 had no advantage over In-111.

We believe that this technique provides a uniform method that
would enable investigators to compare results from different
laboratories. We realize that ultimately the most convenient
technique would be the one that would allow us to label a desired cell
type in whole blood. However, at present, such a selective
cell-labeling method neither exists nor is it on the horizon.
Radiolabeled monoclonal antibodies targeted against specific human
cell antigens may be developed but may require years of work. In the
mean time, dry Merc-In-111 technique may contiue to add to the
usefulness of In-111-labeled blood cells in both diagnostic and
kinetic studies.

REFERENCES

1. J. R. Rubini, E. West-Cott, S. Keller, In vitro DNA-labeling of
    bone marrow and leukemic blood leukocytes with tritiated
    Thymidine-II, ^{3}H Thymidine biochemistry in vitro, J Lab Clin
    Med, 68:566 (1966).

2. T. T. Odell, F. N. Gamble, J. Furth, Life span of naturally labeled platelets of rats, Fed Proc 12:398 (1953).

3. J. A. Cohen and W. G. P. J. Warringa, the fate of P-32-labeled di-isopropylfluorophosphate in the human body and its use as a labeling agent in the study of the turnover of blood plasma and red cells, J Clin Invest 33:459 (1954).

4. D. Grob, J. L. Lilienthal, A. M. Harvey, The administration of di-isopropylfluorophosphate (DFPO) to man, Bull Johns Hopkins Hosp 80:217 (1947).

5. M. L. Thakur and A. Gottschalk, Role of radiopharmaceuticals in nuclear hematology, Radiopharmaceuticals-II The Society of Nuclear Medicine, New York, 341 (1979).

6. M. L. Thakur, Radioisotopic-labeling of platelets: a historical perspective, Semin Thromb Hemostas 9:79 (1983).

7. M. C. Morgan, R. P. Keating, E. H. Reisner, Labeling rabbit platelets with Iodine-131, Proc Soc Exp Biol Med 85:420 (1954).

8. B. Stolc, Iodine metabolism in leukocytes: effects of graded iodide concentrations, Biochem Med 10:293 (1974).

9. J. S. Robertson, W. L. Milne, S. H. Cohn, Labeling and tracing of rat blood platelets with Cr-51, Proc of the 2nd Int Radioisotopic Congress (July 1954) Butterworth Sci. Pub. London, 205 (1954).

10. J. G. McAfee and M. L. Thakur, Survey of radioactive agents for in vitro labeling of phagocytic leukocytes. I. Soluble agents, J Nucl Med 17:480 (1976).

11. J. G. McAfee and M. L. Thakur, Survey of radioactive agents for in vitro labeling of phagocytic leukocytic. II. Soluble agents, J Nucl Med 17:488 (1976).

12. A. W. Segal and A. J. Levi, Factors influencing the entry of dye into neutrophils leukocytes in the nitroblue tetrazolium test, Clin Sci Mol Med 48:201 (1974).

13. A. D. Nunn, The kinetics of incorporation of In-111 into m-tetraphenylporhine, J Radiopharm Chemistry 1-2:291 (1979).

14. Z. D. Grossman, B. W. Wistow, J. G. McAfee, et al, Platelets labeled with oxine complexes of Tc-99m and In-111. Part 2: Localization of experimetnally induced vascular lesions, J Nucl Med 19:488 (1978).

15. M. L. Thakur, Gallium-67 and In-111 radiopharmaceuticals, Int J Appl Radiat Isot 28:183 (1977).

16. M. L. Thakur, R. E. Coleman, M. J. Welch, In-111-Labeled leukocytes for the localization of abscesses: preparation, analysis, tissue distribution, and comparison with Gallium-67 citrate in dogs, J Lab Clin Med 89:217 (1977).

17. Amersham Corporation, In-111-Oxine solution: radiochemical for cell labeling, Code IN 15PA.

18. J. G. McAfee, G. M. Gagne, G. Subramanian, et al, Distribution of leukocytes labeled with In-111-Oxine in dogs with acute inflammatory lesions, J Nucl Med 21:1059 (1980).

19. M. L. Thakur, A. W. Segal, L. Louis, et al, In-111-labeled cellular blood components: mechanism of labeling and

intracellular location in human neutrophils, <u>J Nucl Med</u>
18:1020 (1977).

20. K. J. Hwang, Mode of interaction of (In3+) 8-hydroxyquinoline
    with membrane, <u>J Nucl Med</u> 19:1162 (1978).
21. A. R. Wilkinson, R. J. Hawker, L. M. Hawker, In-111-labeled
    canine platelets, <u>Thromb Res</u> 13:175 (1978).
22. U. Scheffel, M. F. Tsan, P. A. McIntyre, Labeling of human
    platelets with In-111 8-hydroxyquinoline, <u>J Nucl Med</u> 20:524
    (1979).
23. M. L. Thakur, M. J. Welch, J. H. Joist, et al, In-111-labeled
    platelets: studies on preparation and evaluation of in vitro
    and in vivo functions, <u>Thromb Res</u> 9:345 (1976).
24. G. H. Rannie, M. L. Thakur, W. L. Ford, In-111-labelled
    lymphocytes: preparation, evaluation and comparison with Cr-51
    lymphocytes in rats, <u>Clin Exp Immunol</u> 29:509 (1977).
25. J. P. Lavender, J. M. Goldman, M. L. Thakur, et al, Kinetics of
    In-111-Labelled lymphocytes in normal subjects and patients
    with Hodgkins disease, <u>Br Med J</u> 2:797 (1978).
26. J. Ferluga, A. C. Allison, M. L. Thakur, Use of In-111 for
    studies of cytoxicity mediated by lymphocytes or by antibodies
    and compliments, <u>J Clin Lab Immunol</u> 1:339 (1979).
27. M. L. Thakur, C. L. Seifert, M. Madsen, et al, Neutrophil
    labeling: problems and pitfalls, <u>Semin Nucl Med</u>, April (1984).
28. D. A. Goodwin, J. T. Bushberg, P. W. Doherty, et al, In-111-
    labeled autologous platelets for location of vascular thrombi
    in humans, <u>J Nucl Med</u> 19:626 (1978).
29. M. L. Thakur, L. Walsh, H. L. Malech, et al, In-111-labeled human
    platelets: improved method efficacy and evaluation, <u>J Nucl
    Med</u> 22:381 (1981).
30. M. B. Zucker, Effect of heparin on platelet function, <u>Thrombosis
    et Diathesis</u> 33:64 (1974).
31. J. McCullough, B. J. Weiblen, M. E. Clay, et al, Effect of
    leukocyte antibodies on the in vivo fate of In-111-labeled
    granulocytes, <u>Blood</u> 58:164 (1981).
32. J. P. Dutcher, C. A. Schiffer, G. S. Johnston, et al,
    Alloimmunization prevents the migration of transfused
    In-111-labeled granulocytes to sites of infection, <u>Blood</u>
    62:354 (1983).
33. B. J. Weiblen, J. McCullough, L. Forstrom, et al, Kinetics of
    In-111-labeled granulocytes, <u>in</u>: "In-111-Labeled Neutrophils,
    Platelets, and Lymphocytes, M. L. Thakur and A. Gottschalk,
    eds., Triverium, New York (1979).
34. E. Throsby, Cell-specific and common antigens on human
    granulocytes and lymphocytes demonstrated with cytotoxic
    hetero-antibodies, <u>Vox Sang</u> 13:194 (1967).
35. M. L. Thakur, J. P. Lavender, R. N. Arnot, et al, In-111-labeled
    autologous leukocytes in man, <u>J Nucl Med</u> 18:1014 (1977).
36. G. V. R. Born, F. Michal, 5-Hydroxytryptamine Receptors of
    Platelets in Biochemistry and Pharmacology of Platelets,
    Elsevier, p.302 (1975).

37. U. Scheffel, M. F. Tsan, P. A. McIntyre, Labeling of human
     platelets with In-111 8-hydroxyquinoline, J Nucl Med 20:524
     (1979).
38. J. F. Mustard, D. W. Perry, M. G. Ardlie, et al, Preparation of
     suspension of washed platelets from humans, Br J Haematol
     22:193 (1972).
39. H. Sinn and D. J. Silvester, Simplified cell labeling with
     In-111-acetylacetone, Br J Radiol 52:758 (1979).
40. C. J. Mathias, W. A. Heaton, M. J. Welch, et al, Comparison of
     In-111-oxine and In-111-acetylacetone for labeling of cells:
     in vivo and in vitro biological testing, Int J Appl Radiat
     Isot 32:651 (1981).
41. W. T. Goedemans, Simplified cell labeling with In-111-
     acetylacetone and In-111-oxine, Br J Radiol 54:636 (1981).
42. L. A. Spitzangle, C. A. Marino, S. Kasina, Lipophilic chelates of
     Tc-99m: tropolone, J Nucl Med 22:981 (1981).
43. M. K. Dewanjee, S. A. Rao, P. Didisheim, In-111-tropolone, a new
     high affinity platelet label: preparation and evaluation of
     labeling parameters, J Nucl Med 22:981 (1981).
44. H. J. Danpure, S. Osman, F. Brady, The labeling of blood cells in
     plasma with In-111-tropolonate, Br J Radiol 55:247(1982).
45. M. Peters, S. H. Saverymuttu, H. J. Reavy, et al, Imaging of
     inflammation with in-111-tropolonate-labeled leukocytes, J
     Nucl Med 24:39 (1983).
46. J. S. Robertson, M. K. Dewanjee, W. L. Dunn, et al,
     Biodistribution and survival of human platelets labeled in
     buffered media and plasma, J Nucl Med 24:P73 (1983).
47. R. J. Hawker, C. E. Hall, and E. K. Gunson, In-111-tropolone
     versus oxine, J Nucl Med 24:367 (1983).
48. M. R. Hardeman, Tropolone, the favorite ligand for cell
     labeling?, Eur J Nucl Med 7:528 (1982).
49. S. Vallabhajosula, M. L. Greenberg, S. J. Goldsmith, The effect
     of pH on labeling of leukocyte preparation: oxine vs.
     tropolone, J Nucl Med 24:P301 (1983).
50. K. P. Gutner, J. N. Luken, J. A. Clanton, et al, Neutrophil
     labeling with In-111: tropolone vs. oxine, Radiology 149:563
     (1983).
51. T. Uchida, K. Tasunaga, S. Kariyone, et al, Survival and
     sequestration of Cr-51- and Tc-99m-labeled platelets, J Nucl
     Med 15:801 (1974).
52. N. Linhart, B. Bok, M. Meigan, et al, Technetium-99m-labeled
     human leukocytes: in vitro and animal studies, in: "In-111-
     Labeled Neutrophils, Platelets, and Lymphocytes," M. L.
     Thakur, and A. Gottschalk, eds., Trivirum, New York (1979).
53. N. A. Farid, S. M. White, L. L. Heck, et al, Tc-99m-labeled
     leukocytes: preparation and use in identification of abscesses
     and tissue rejection, Radiolog, 148:827 (1983).
54. D. English and B. R. Anderson, Labeling of phagocytes from human
     blood with Tc-99m-sulfur colloid, J Nucl Med 16:5 (1975).

55. H. J. Scroth, E. Oberhausen, R. Berberich, Cell labeling with
     colliodal substances in whole blood, Eur J Nucl Med 6:469
     (1981).
56. F. Lomas, Private Communications
57. M. A. Zimmer and S. M. Spies, Leukocyte labeling with colloidal
     In-111 in whole blood, Int J Appl Radiat Isot 34:1544 (1983).
58. H. Showell, R. J. Freer, S. H. Zigmond, et al, The structure
     activity relationship of synthetic peptides as chemotactic
     factors and inducers of lysosomal enzyme secretion for
     neutrophils, J Exp Med 143:1154 (1976).
59. R. J. Freer, A. R. Day, N. Muthukumarswami, D. Pinon, A. Wu, H.
     J. Showell, E. L. Becker, Formyl peptide chemoattractants: a
     model of the receptor on rabbit neutrophils, Biochemistry
     21:257 (1982).
60. W. A. Marasco, H. J. Showell, R. J. Freer, E. L. Becker,
     Anti-f-Met-Leu-Phe: similarities in fine specificity with the
     formyl peptide chemotaxis receptor of the neutrophil, J
     Immunol 128:956 (1982).
61. S. S. Zogbhi, M. L. Thakur, A. Gottschalk, A potential
     radioactive agent for the selective labeling of human
     neutrophils in vitro, Int J Appl Radiat Isot (in press).
62. R. C. Verma, L. R. Bennett, T. Kawada, et al, Receptor mediated
     selective radiolabeling of neutrophils, J Nucl Med 24:P7
     (1983).
63. H. Zola, P. McNamara, Thomas et al, The preparation and
     properties of monoclonal antibodies against human granulocyte
     membrane antigens, Br J Haematol 48:481 (1981).
64. K. M. Skubitz, Y. Zhen, J. T. August, A human
     granulocyte-specific antigen characterized by use of
     monoclonal antibodies, Blood 61:19 (1983).
65. H. L. Malech, Private Communication.
66. D. J. Hnatowich, W. W. Layne, R. L. Childs, The preparation and
     labeling of DTPA-coupled albumin, Int J Appl Radiat Isot
     33:327 (1982).
67. C. H. Paik, M. A. Ebbert, P. R. Murphy, et al, Factors
     influencing DTPA conjugation with antibodies by cyclic DTPA
     anhydride, J Nucl Med 24:1158 (1983).
68. K. H. Konig, B. Steinberch, G. Schneewieg, et al, Zur
     chromamtographi von metallchelaten fresenius, Z Anal Chem
     297:144 (1979).
69. M. L. Thakur, M. J. Barry, Preparation and evaluation of a new
     In-111 agent for efficient labeling of human platelets in
     plasma, J Lab Com. Radiopharm 19:1410 (1982).
70. S. L. Propst-Proctor, M. F. Dillingham, I. R. McDougall, et al,
     The white blood cell scan in orthopedics, Clin Orthop 168:157
     (1982).
71. J. P. Dutcher, C. A. Schiffer, G. S. Johnston, Rapid migration of
     In-111-labeled granulocytes to sites of infection, N Engl J
     Med 304:586 (1981).

72. M. H. Rovekamp, M. R. Hardeman, J. B. van der Shoot, et al, In-111-labeled leukocyte scintigraphy in the diagnosis of inflammatory disease--first results, Br J Surg 68:150 (1981).

73. G. N. Sfakianakis, W. Al-Sheikh, A. Heal, et al, Comparisons of scintigraphy with In-111 leukocytes and Ga-67 in the diagnosis of occult sepsis, J Nucl Med 23:618 (1982).

74. M. R. Hardeman, E. G. J. Eitjes-van Overbeek, A. J. M. van Velzen, Labeling techniques of granulocytes and platelets with In-111-oxine, Nucl Geneeskunding Bull 4:8 (suppl.) (1982).

75. H. Sinzinger, C. Leithner, R. Hofer, Continuous monitoring of human kidney transplants byautologous-labeled platelets, Nucl Geneeskunding Bull 4:44 (suppl.) (1982).

76. K. H. Laws, J. A. Clanton, V. A. Starnes, et al, Kinetics and imaging of In-111-labeled autologous platelets in experimental myocardial infarction, Circulation 67:110 (1983).

77. A. Heynes, Dup., P. N. Badenhorst, H. Pieters, et al, Preparation of a viable population of In-111-labeled human platelets, Thromb Haemost 42:1473 (1980).

78. J. L. Ritchie, J. R. Stratton, B. Thiele, et al, In-111 platelet imaging for detection of platelet deposition in abdominal aneurysms and prosthetic arterial grafts, Am J Cardiol 47:882 (1981).

79. A. W. Heaton, H. H. Davis, M. J. Welch, et al, In-111: a new radionuclide label for studying human platelet kinetics, Br J Haeamatology 42:613 (1979).

80. M. G. Lotter, P. N. Badenhorst, A. Heyns, Dup. et al, Kinetics, distribution, and sites of destruction of canine blood platelets with In-111-oxine, J Nucl Med 21:36 (1980).

81. Hill-Zoble, U. Scheffel, P. A. McIntyre, et al, In-111-oxine-labeled rabbit platelets: in vivo distribution and site of destruction, Blood 61:149 (1983).

82. A. R. Wilkinson, R. J. Hawker, L. M. Hawker, In-111-labeled canine platelets, Thromb Res 13:175 (1978).

83. I. Klonizakis, A. M. Peters, M. L. Fitzpatrick, et al, Radionuclide distribution following injection of In-111-labeled platelets, Br J Haematol 46:595 (1980).

84. A. Heynes, Dup., M. G. Lotter, P. N. Badenhorst, et al, Kinetics, distribution, and sites of destruction of In-111-labeled human platelets, Br J Haematol 44:269 (1980).

EVALUATION OF ANIMAL MODELS

USING IN-111-LABELED PLATELETS

Carla Z. Mathias and Michael J. Welch

The Edward Mallinckrodt Institute of Radiology
Washington University
St. Louis, MO 63110

Animal models continue to be important for studying many diseases, some of which include vascular and cellular related phenomena. To better elucidate the parameters involved, radiolabeled blood components were employed for determining total blood volume (1), plasma volume (2) cell kinetics (4) and the detection of cellular-mediated disorders such as thrombosis (5) and abscessed tissue (6). As more information about vascular disorders is collected, the roles of platelets and leukocytes are more defined and have been described as participating in the regulation of prostaglandin production and thereby maintaining vascular integrity; the process of joint inflammation; and other subtle, less well-defined roles in atherosclerosis, cerebrovascular disease, and myocardial infarction.

Indium-111 has been used as a complex with several bidentate ligands to radiolabel separated cellular components; the chelates include 8-hydroxyquinoline (oxine), (7-9) acetylacetone (acac), (10, 11) tropolone, (12, 13) and 2-mercaptopyridine-1-oxide (merc) (14). Chromium-51-sodium chromate ($t1/2$ = 27.8d) was used previously to radiolabel platelets; however, In-111-labeled platelets offer many advantages over Cr-51-labeled platelets. These include high labeling efficiency with a high initial in vivo recovery, an appropriate radio-isotope half-life ($t1/2$ = 2.8 days) for measuring most cell survivals, and high energy gamma emissions suitable for scintigraphic detection (173 keV and 247 keV) (15). From an in vivo detection viewpoint, In-111-labeled platelets have been successful for detecting deep vein thrombosis (16, 17) atherosclerotic lesions (18) mural thrombi (19), coronary artery thrombi (20, 21) and platelet accumulation on vascular grafts (22, 23).

Minute platelet accumulation or monolayer adhesion on thrombogenic surfaces is not readily imaged even 24 hours after administration of In-111-labeled platelets (24, 25). Radiolabeled platelets deposited on the lesion are presumably greater than the concentration of radioactivity circulating in the blood. A means for determining the normal blood pool is essential to define areas of platelet deposition by deducting the circulating radioactivity from the area of interest. Red blood cells (RBC) labeled with Tc-99m have been used to scintigraphically demarcate normal blood pool areas (26). This dual isotope technique can effectively enhance the sensitivity and specificity of serial images obtained with the use of In-111-labeled platelets when collected and stored on computer for analysis (27). A noninvasive technique that accurately measures the intravascular platelet deposition is necessary so that quantitative evaluation of scintigraphic images may be performed (28).

Several reactions are observed during the hemostatic process which progress from vasoconstriction, platelet adhesion and aggregation to collagen fibrils (29, 30) to clot reinforcement by fibrin (33), followed by fibrinolysis (34, 35). Numerous factors are produced from these phenomena via the coagulation pathway (36), the vascular surface (i.e., stimulation of prostaglandin production) (37, 38), and from fibrinolysis (stimulation of cAMP) (39). These factors may have an important role in the elucidation of processes that may vary with an increase or a decrease of the concentration of these chemical metabolites. The role of platelets in many disorders has not been fully explained, but the biochemicals produced therein are known to alter platelet behavior and may thereby change the course of disease progression. Other platelet/vascular related diseases (i.e., stroke, hypertension, myocardial infarction, and atherosclerosis) have been described as being the result of a platelet (or platelet-associated factors) mediated process. Modification of these biochemical reactions may, in fact, positively alter the disease progression; and pharmaco-intervention studies are required to define their effectiveness (40).

Known species differences in platelet adhesion (41), platelet response to aggregation stimulating agents (42, 43), effects of platelet inhibitors and other drugs (43), anticoagulant effects (44, 45), platelet size and the involvement at areas of thrombosis (46, 47) or other thrombogenic surfaces (48, 49) should be considered when choosing an appropriate animal model. There is a tremendous variation in platelet counts and size; for example, the volume of a cat platelet is large (12.2 μm^3) compared with man (5.8 μm^3) and rabbit (4.1 μm^3) (50). The circulating platelet count in animals is, in general, higher than that in man (Table 1).

Anticoagulants are essential throughout in vitro platelet manipulation; since heparin and ethylenediaminetetraacetic acid (EDTA) can interfere with platelet function in rabbits, dogs, and cats

Table 1. Species Variation Of Platelet Aggregation Responses √

In Vitro Platelet Aggregation Stimulant

Species	ADP	COLL	TH	Sero-tonin	Adren-alin	Platelet Count	Relative Potency*
							7.0
man	bi	mono	bi	mono	bi	200 - 250 k	6.58
monkey	+bi	mono	mono	mono	+bi	260 k	5.88
rabbit	mono	mono	mono	mono	mono	300 - 400 k	2.32
rat	mono	mono	mono	mono	mono		1.55
cat	bi		bi	bi	bi		
dog	+bi	mono	mono	mono	+bi	320 k	
sheep	mono		mono	mono			1.0

√ data from MacMillan and Sim (61) and Hawkey (66)
* data from Karim and Adarkan (123)
+ biphasic aggregation occasionally
  bi=biphasic aggregation
  mono=monophasic aggregation

(somewhat dependent on the concentration) (51, 52). Sodium citrate
solution (3.8%, pH 7.35) seems more optimal at 1:9 (v:v) concentration
(53). Some species, however, require less sodium citrate (3.0% in
rats) to maintain in vitro platelet aggregation response to adenosine
diphosphate (ADP) (50).

    In vitro platelet aggregation studies were carried out to evaluate
platelet viability before and after radiolabeling (54) (Fig. 1). The
response to various aggregation stimulating agents (ADP, epinephrine,
collagen, arachidonic acid, and thrombin) has been described previously
(55-57); by the turbidometric technique, the formation of aggregates in
a stirred suspension of platelets results in an increase in the
transmission of light through the solution and a positive response
(58-60). The platelet response during aggregation (adherence of
platelets to each other) may be described by a sequence of reactions
that include shape change, adhesion reaction, primary aggregation,
release reaction, and secondary aggregation. The platelets change
from ellipsoids to spiny spheres immediately after contact with an
aggregation stimulus; this is probably not accompanied by any change
in cell volume. The initial reaction is not calcium dependent and is
reversible at very low stimulus levels. The shape change is followed

by an adhesion reaction, the process of platelet adherence to surfaces
other than platelets. Primary phase of platelet aggregation is the
adherence of platelets to each other accompanied by a viscous
metamorphosis and a release reaction. Optimum conditions for
aggregation induction in vitro include pH 6.8 - 8.5, temperature of
37°C, and the appropriate aggregation agent. The release reaction
refers to the biochemical reactions occurring when serotonin,
proteins, amino acids, and adenine nucleotides are extruded rapidly
from platelet alpha granules and dense bodies into extracellular
medium. The release of ADP during the release reaction causes a
second wave of aggregation, which causes further release, which
results in a cumulative effect of total aggregation. The second wave
of aggregation in some cases is not distinguishable, the aggregation
response appears as a monophasic response to the stimulus addition that
represents complete aggregation. There are obvious similarities
between the in vivo platelet aggregation responses between man and
nonhuman primates. Cats are unusual in that they have a biphasic
response to most stimulating agents (61). Dogs have aggregation
response similar to man but require higher concentrations of
stimulating agents (ADP and adrenalin) to obtain biphasic
hyper-response (50, 62). The other relevant species have only
monophasic aggregation responses to most agents.

Plasminogen levels have been correlated to fibrinolytic potentials
(63, 64), and can affect platelet function in vivo. Relative to
humans, pigs have less plasminogen and thereby a lesser fibrinolytic
potential; rats and rabbits have increased (x 2) plasminogen levels,
and dogs have even greater plasminogen levels (x 3). Monkeys have a
plasminogen level only slightly greater than that of humans (the
concentration range includes that of humans, and most likely
approximate the fibrinolytic potential of the human system (65, 68).

Prostaglandin endoperoxides and nonprostanoate derivatives of ara-
chadonic acid may be the chemical initiator for the release reaction
(69, 70). Apparently, aspirin (prostaglandin synthesis inhibitor)
(76) inhibits dense body release induced by ADP, adrenalin, or low
concentrations of collagen but has no effect on alpha granule release
or on dense body release induced by high concentrations of collagen or
thrombin (72). It is important for comparative experiments that the
platelet concentration, pH, temperature, and anticoagulant remain
constant (73). The models discussed evaluate the advantages of
particular animal species to fit the criteria, and the correlation of
human disease is dependent on these data.

ANIMAL MODELS FOR THROMBOSIS

Since 1951, the electrocoagulation method has been utilized to
investigate thrombosis on mesenteric arteries, jugular veins, femoral
veins, carotid arteries, femoral arteries, and coronary arteries (74).

This technique is reproducible in many species of animals and has been
employed to evaluate the efficacy of antithrombotic agents; and
species differences have been reported. However, the question of
vessel size remains unanswered (75). The thrombi produced is the
result of endothelium destruction and exposure of the internal elastic
lamina and media (76); however, some investigators disagree,
suggesting that platelets may adhere to sites of minute endothelial
injury without cell wall breaks (77). Basement membrane (a
collagen-like material) (78), elastin fibers (79), and subendothelial
collagen (80) have been shown to accumulate platelets; this has been
attributed to the destruction of the subendothelium by plasma
proteins, followed by the evolution of thrombin and the resultant
platelet aggregation (81).

The accumulation of platelets at sites of vascular injury has been
confirmed by techniques that utilize mechanical, electrical, and
chemical injury (82, 83). Since the electrical injury technique to
produce thrombi in vivo was so reliable, it was used for radiotracer
modeling (74). Generally, the femoral vein of a dog was utilized to
evaluate thrombus imaging agents such as radioiodinated fibrin (84)
I-123, Tc-99m-urokinase (86), I-123, Tc-99m-streptokinase (87)
I-123-plasminogen (88), and In-111-platelets for deep venous thrombi
(DVT) (89). In-111-platelets and other agents were compared, and the
results demonstrate that platelet deposition in these injured sites
was extensive (90, 91).

To better define the role of platelets and predict a useful treatment for DVT and thromboembolism, the electric injury model was used
to produce DVT in canine femoral veins (92). After clot formation,
In-111-labeled platelets were infused; the accumulation of the
radiolabeled platelets was detected scintigraphically. Maximum
platelet deposition was observed approximately 4 hours after injection
in dogs; however, in monkeys, the clot to blood ratios continued to
increase for about 24 hours. The rapid canine fibrinolytic response
may explain the more rapid platelet accumulation and the more rapid
clot dissolution (92); but primates have a somewhat slower lysis
response, and thereby the thrombus would continue to accumulate
platelets during that period (Fig. 2). The correlation of clot size
and platelet deposition was not linear but was probably related to the
degree of vascular damage. Pulmonary emboli (PE) have been detected
with the use of In-111-platelets (94, 95), but the scintigraphic
localization of PE in canines was difficult after 24 hours when no
evidence of the embolus existed. This model for thrombosis, although
consistent, introduces a complicated interaction to interpret; that
is, the distinction between platelet deposition in propagating thrombi
and platelet interaction on the injured vessel wall. To investigate
the role of platelets in thrombosis, many other mechanisms needed to
be defined, such as the role of platelets in acute arterial thrombosis
on denuded endothelium (96, 97), the white thrombus formation without
vessel injury (98), and the proliferation of atherosclerotic plaques
(99).

Four animal models with potential regions of platelet deposition
have been evaluated using the dual isotope technique with In-111-
labeled platelets and Tc-99m-labeled RBC. These include a group of
nonhuman primates with chronic atherogenic lesions, a canine model for
acute coronary artery thrombosis, an acute arterial endothelial injury
in nonhuman primates, and a canine model to evaluate platelet
deposition on small diameter vascular grafts.

To determine the amount of radiolabeled platelets deposited in the
region of interest, slightly different calculations were performed in
each case. Since each model has intrinsic variations, the technical
suitability of an animal is evaluated for the response to drugs
(including anesthetics), vascular access, and the ability to tolerate
the experiment.

Determination of Platelet Deposition on Atherosclerotic Plaques

Endothelial cell injury has been shown to be the result of hypercholesterolemia (100). Platelets accumulate and aggregate at the
focal desquamation, during which they release their granule
constituents (101). The alpha granule constituents are believed to
cause a smooth muscle cell proliferation and permeability changes in
the vessel wall itself (102). Since the platelet release is
considered an initiating factor in atherogenesis, pharmacologic
inhibition of platelet aggregation may be a reasonable approach to the
prevention of the disease (103).

Animal models for atherosclerosis have been utilized and
techniques to accelerate the onset of plaques have been attempted.

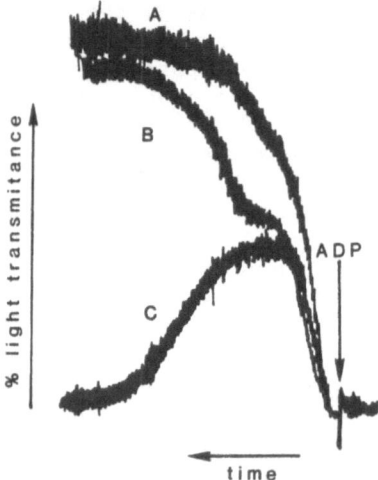

Fig. 1.   Platelet aggregation can be complete (A), biphasic (B), or
reversible (C). Second phase of aggregation is induced by the release
of endogenous stimulating agents contained in platelet granules.

Many animal species have been investigated, including dogs (104), swine (105), pigeons (106), rabbits (107), and nonhuman primates (108-110); the most representative model is, of course, diet-induced atheroma in nonhuman primates. Macaca acatoides that had been on atherogenic diet for 4 or more years were studied. The high cholesterol diet of egg yolks and lard was maintained, and cholesterol and total triglyceride levels were measured. The total cholesterol was elevated to greater than four times as much as baseline. The control animals were fed an identical diet without cholesterol (caiesen substituted). Some documentation of atherosclerotic plaques was angiographically obtained (111), but the role of platelet deposition and perhaps the intervention of the disease progression was not elucidated (112). Autologus In-111-labeled platelets and Tc-99m-RBC were monitored in vivo by scintigraphic detection (113) in the restrained, anesthetized monkeys immediately and 24 hours after administration (Tc-99m-RBC was readministered at 24 hours). Scintigraphic images (200-500K counts) from the 247 keV photopeak of In-111 and from the 140 keV photopeak of Tc-99m were collected. Without computer-assisted processing the localization of platelet deposition was indistinguishable from the vascular blood pool (defined with Tc-99m-RBC); after calculations to estimate the amount of circulating radioactivity were applied, visualization of foci corresponding to the course of the abdominal aorta was possible (Fig. 3).

Because this technique requires the identification of a normal blood pool area (i.e., circulating blood pool but no platelet deposition) within the field of views, the calculations were modified (in atherosclerotic monkeys it would be difficult to define an area with normal blood pool since diffuse platelet adhesion/aggregation should be predominant in most vascular regions). The modification allowed the platelet deposition or percent indium excess (%IE) to be calculated for each horizontal 4-pixel wide row within the vertical limits of the abdominal aorta. After the values were obtained for both Indium-111 and Tc-99m, the In/Tc ratio was determined and the minimum ratio value for each study identified. Even if diffuse platelet deposition in the abdominal aorta exists, the minimum In/Tc ratio will be greater than the true reference ratio. These assumptions, then, limit the sensitivity of this method for detecting platelet deposition. Profiles can be generated with %IE values that represent the platelet deposition along the abdominal aorta, but the mean values for %IE for each 4-pixel row were determined and used for further analysis. Statistical analysis was performed using the Mann Whitney U test (Table 2).

The mean %IE values were generated from the profile information, and it was found that there was no significant difference between the control and diet animals in the immediate studies; however, 24 hours after the radiolabeled platelets were injected the diet animals had significantly greater amounts of platelet deposition in the abdominal aorta compared with the controls.

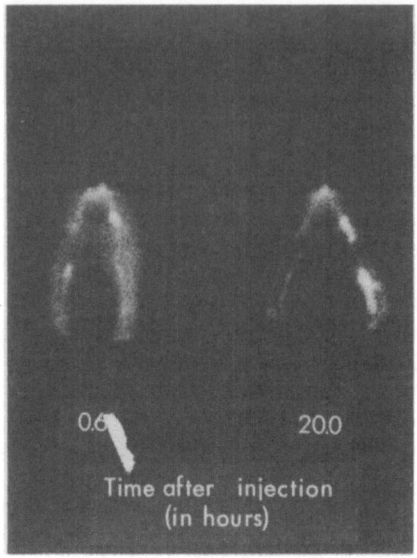

Fig. 2. An electrode-induced deep
vein thrombosis in a nonhuman
primate at 0.6 and 20.0 hours after
injection of In-111-labeled
platelets. Notice the increased
deposition at 20 hours.

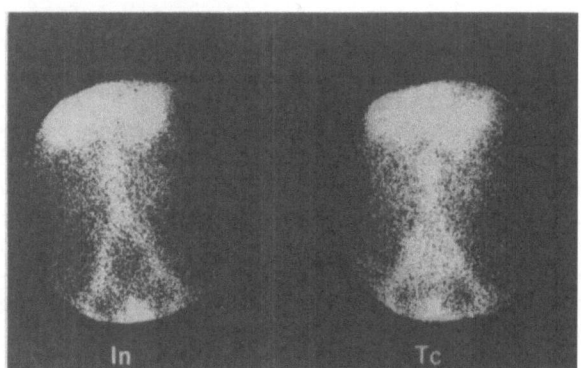

Fig. 3.   Scintigrams obtained using In-111-platelets (left) and Tc-
99m-RBC (right) showing vascular areas with no observable platelet
deposition without the application of blood pool subtraction.

Table 2.   %IE In The Abdominal Aorta 24 Hours After Administration of
In-111-Labeled Platelets

| | Mean % IE | |
	Untreated	Low-Dose ASA
Control diet	15.8 (9.9 – 18.5)	18.7 (14.5 – 22.9)
High cholesterol diet	41.5 (15.4 – 92.5)	16.2 (10.4 – 21.7)

The calculation to determine percent indiums excess (%IE) is:

$$\%IE = \frac{\text{In-111 counts in region of interest} - \text{In-111 counts in blood pool}}{\text{In-111 counts in blood pool}} \times 100$$

(Eq. 5-1)

This same group of animals was studied after administration of low- and high-dose aspirin for 5 consecutive days. Interestingly, either 3mg/kg/day or 30mg/kg/day of aspirin can inhibit platelet deposition on the plaques both immediately and 24 hours after the In-111-labeled platelets were injected. Pathological findings in these animals confirm the scintigraphic findings, reporting diffuse intimal thickening along the course of the abdominal aorta (Table 2).

Canine Model for Acute Coronary Artery Thrombosis

The same dual isotope technique was utilized in the canine model for acute coronary artery thrombosis (114). Because the dog has a rapid fibrinolytic system relative to the human, it probably does not closely mimic the clinical situation; however, nonhuman primates, which have a similar fibrinolytic system, were not justified, since these were terminal experiments. The dog, which has an accelerated fibrinolytic system, may, in fact, provide a means by which the acute infarction may be observed in a relatively short amount of time. Visualization of relatively small thrombi (average 44 ± 28 mg), even in areas with a large blood pool, such as the heart, was possible with the use of the dual isotope technique (20). After transmural myocardial infarction, intracoronary thrombosis occurs (115), and has been identified as a site of rapid fibrinogen turnover (116). Indium-111-labeled platelets can be used for visualization of coronary artery thrombi and interventricular thrombi, but the relatively large blood pool within the region of interest interferes with the immediate observation of radiolabeled platelet localization (117).

The coronary artery thrombus was induced by placing a small copper wire coil (5-7.5 mm in length) in the left anterior descending coronary artery. Five to fifteen minutes after removing the guide wire from the coil, a thrombus occluded the artery (118). Indium-111-labeled platelets and Tc-99m-labeled RBC were prepared and injected either just prior to clot induction or 1 hour after the clot formation was complete as determined by arteriography. Alternating sequential images were collected, digitized, and stored on computer for later processing. In this model, a reference region was chosen as the origin of the right carotid artery. A subtraction algorithm was applied, corrected by the reference ratio for the appropriate pair of images, to obtain representative %IE images. The images obtained from either In-111 or Tc-99m scintigrams were unremarkable; however, the

Table 3. Evaluation of Coronary Artery Thrombus Dissolution

Time (h) after In-111-Platelets Injected	n	% IE	Clot Blood Ratio	Clot Weight(mg)
1	5	20.7 ± 12.8	27.5 ± 21.7	32.9 ± 24.8
6	2	17.2 ± 1.2	25.1 ± 2.3	27.6 ± 10.1
1 SK	3	10.1 ± 9.5	9.1 ± 1.9	5.1 ± 4.2

%IE image, even in the immediate situation, was suitable to define an abnormal focal accumulation corresponding to the location of the coil as arteriographically determined.

The %IE values in the region of the clot were determined and the in vitro clot/blood ratios were calculated (Table 3). The clot/blood ratio and the %IE in either case (radiolabeled cells administered either before or after clot formation) were similar; the difference between the values obtained when radiolabeled platelets were injected either before or after clot formation were not statistically significant. A linear correlation exists of the in vitro counts in the clot and the in vivo %IE in the region of the clot (Fig. 4).

The efficacy of thrombolytic therapy, such as plasminogen inhibitors, streptokinase, and urokinase, can be evaluated using this model. Radiolabeled cells were administered either before or after clot formation confirmed by arteriography and processed scintigrams. Image pairs were collected for 3 hours after injection of the radiolabeled cells, when the intracoronary infusion of the drug was begun. Thrombolysis occurred approximately 45 minutes after the beginning of the therapy using streptokinase (4000 U/min), as also demonstrated by reperfusion dysrhythmias in EKG tracings. More scintigraphic images were collected and stored for later processing, the dog underwent a repeat arteriogram for confirmation and was then sacrificed. The excised heart was examined, and specimens including the reperfused tissue, the copper wire coil, the artery surrounding the coil, the infarcted tissue, and the blood were obtained for in vitro counting.

Thrombolysis with streptokinase was successful as demonstrated by the reduced clot weight and the %IE values. It should be noted that even though the vessel is reperfused, there is some residual clot strands remaining, making the in vitro values slightly high.

Acute Arterial Endothelial Injury in Nonhuman Primates

Several approaches to evaluate platelet-endothelium interaction after an acute endothelial injury have been investigated (110, 119). It has been reported that after a single injury in arteries of normal rabbits, a monolayer of platelets accumulates on the injured surface, which may lead to more severe thrombosis or dissipate (107). The rate of plasmin formation and of fibrinolysis is, of course, related to this phenomenon. Prostacyclin (PGI_2), a potent smooth muscle relaxant and inhibitor of platelet function, is a potentially effective agent for the treatment of thrombosis (120). The inhibition of platelet accumulation in areas of vascular damage and exposed subendothelium has been demonstrated by several investigators. The dose response of PGI_2 varies within species, and some secondary negative effects have been documented (121). Many species of animals have been used to examine the effects of PGI_2 on experimentally induced vascular lesions (120, 122, 123). Scintigraphic results suggest platelet accumulation in areas of vascular damage, but in vitro confirmation by counting excised vascular specimens has thus far been required (124, 125). Nonhuman primates have a similar coagulation system and in vitro platelet aggregation responses to human (in vitro platelet aggregation inhibition and aggregation reversal is observed in both human and non-human primates) (123), and these experiments would be lengthy, so the animal chosen for this experimental protocol has to tolerate the anesthesia necessary. Nonhuman primates were thereby chosen for their attributes and their similarities to humans. Technetium-99m-labeled RBC were infused 24 hours after the administration of In-111-labeled autologous platelets, and the image collection was carried out. Sequential 1-minute images for the 247 keV photopeak of In-111 and the 140 keV photopeak of Tc-99m were collected, digitized, and stored as a 64 x 64 matrix on a computer. The endothelial damage induced by an inflated balloon catheter produced a rapid marked platelet deposition (Fig. 5). Without blood pool subtraction, significant differences between the undamaged and damaged artery were not detected. From the stored images, time-activity curves before and after vessel damage were generated and then followed during platelet inhibitory drug administration. Platelet deposition on exposed endothelium and the reversal of the mural platelet thrombi can be induced with PGI_2 (126-128).

Platelet Deposition on Small Diameter Vascular Grafts

Small diameter prosthetic grafts (4 mm-I.D.) are used for arterial reconstruction when suitable autologous vessel is not available (48). The low patency rate of these grafts in the early post-operative period is a major limitation (129). Since PTFE (the most commonly used graft biomaterial) has individual fibers protruding into the intravascular space with fiber diameter similar to collagen fibrils (130), platelets (known to adhere to exposed subendothelium that contains collagen fibrils) might be expected to accumulate on these

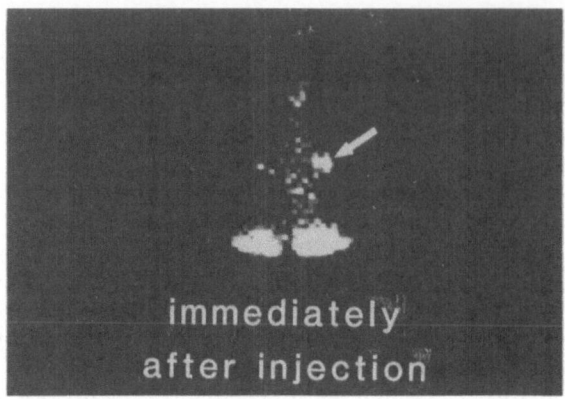

Fig. 4.   A processed image of a coronary artery thrombus representing
In-111-platelet deposition at the location of the coil as determined
angiographically.

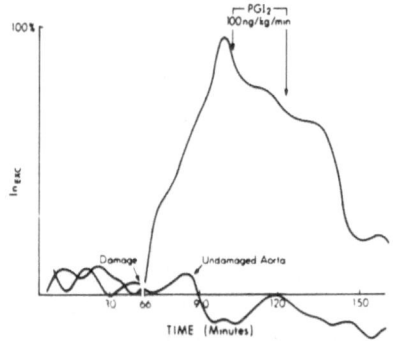

Fig. 5.   Time-activity curve representing platelet deposition on a
balloon catheter damaged aorta and on undamaged aorta. After damage,
an infusion of PGI2 (100ng/kg/min) for 20 minutes reverses the acute
accumulation on the damaged region

biomaterials as part of the hemostatic process (131). The use of
platelet inhibitors in the presence of various biomaterial grafts has
been investigated, but the results may not accurately reflect the
platelet function in vivo since evaluation is based on graft patency.
Indium-111-labeled platelets have been used to evaluate abnormal
platelet deposition on arterial grafts (22, 23, 132-134). This dual
isotope imaging technique may also be useful to elucidate the
platelet-vessel interaction, the production of pseudointima (135), and
the mechanism of platelet inhibitory drug on vascular grafts (136).

     In a canine model, small-diameter grafts of either PTFE or dacron
knit were implanted in the carotid or femoral artery. Since the
canine has a more active fibrinolytic system, the measurements at
approximately 3 months are equivalent to 2 years in humans. The
canine model, then, lends itself to studies to evaluate anti-platelet

therapy to reduce platelet deposition and increase graft patency. As the blood flow was restored to the implanted grafts, the In-111 platelets were injected and thereby exposed to the graft material in vivo. The dogs were scanned 2 hours and 24 hours after injection. Technetium-99m-labeled RBC were administered just prior to scanning. The images (100-200 k counts) were collected and stored for processing. Followup scans were carried out 1 or 2 weeks, 3 or 4 weeks, 1 month, 2 months, or 3 months after operative implantation.

The various graft materials were compared for patency and the correlation of platelet deposition (%IE). The difference between the grafted autologous vein and the dacron velour grafts was statistically significant ($p=0.01$ and $p=0.05$, respectively), but no significant difference between PTFE and vein was observed. The explanted patent grafts were counted in vitro and correlated with the in vivo %IE values; a linear relation ship was found.

In conclusion, choosing an appropriate animal model should be an important criterion for experimental design. The models described are only a few relative to the many described in the literature. The dual isotope subtraction technique allows quantitative evaluation of scintigrams and, it is hoped, better elucidates platelet interaction in various phases of thrombosis and atherosclerosis and pharmacointerventions of these processes.

ACKNOWLEDGMENT

This work was supported in part by NIH Grants HL14147 and NS 06833.

REFERENCES
1. R. Wennesland, E. Brown, J. Hopper, Red cell, plasma, and blood volume in healthy men measured by radiochromium (Cr-51) cell tagging and hematocrit: influence of age, somatotype, and habits of physical activity on the variance of regression of volumes to height and weight combined, J Clin Invest 38:1065 (1959).
2. P. J. Hurley, Red cell and plasma volumes in normal adults, J Nucl Med 16:46 (1974).
3. W. A. Heaton, H. H. Davis, M. J. Welch, C. J. Mathias, J. H. Joist, L. A. Sherman, B. A. Siegel, Indium-111: new radionuclide label for studying human platelet kinetics, Br J Haematol 42:613 (1979).
4. A. Heyns, M. G. Lotter, P. N. Badenhorst, O. R. vanReenen, H. Pieters, P. C. Minnaar, F. P. Retief, Kinetics, distribution, and sites of destruction of 111-Indium-labeled human platelets, Br J. Haematol 44:269 (1980).
5. D. A. Goodwin, J. T. Bushberg, B. U. Doherty, M. J. Lipton, F. K. Conley, C. J. Diamenti, C. F. Meares, Indium-111-labeled

autologous platelets for location of vascular thrombi in humans, J Nucl Med 19:626 (1978).

6. M. L. Thakur, R. E. Coleman, M. J. Welch, Indium-111-labeled leukocytes for the localization of abcesses: preparation, analysis, tissue distribution and comparison with Gallium-67-citrate in dogs, J Lab Clin Med 89:217 (1977).

7. U. Scheffel, P. A. McIntyre, B. Evatt, J. A. Dvornisky, T. K. Natarajan, D. R. Bolling, E. A. Murphy, Evaluation of Indium-111 as a new high-photon yield gamma-emitting "physiological" platelet label, Johns Hopkins Med J 140:285 (1977).

8. M. L. Thakur, M. J. Welch, H. J. Joist, R. E. Coleman, Indium-111-labeled platelets: studies on preparation and evaluation of in vitro and in vivo functions, Thromb Res 9:345 (1976).

9. J. G. McAfee and M. L. Thakur, Survey of radioactive agents for in vitro labeling of phagocytic leukocytes, J Nucl Med 17:480 (1976).

10. C. J. Mathias, W. A. Heaton, M. J. Welch, P. G. Douglas, J. D. Kelly, Comparison of In-111-oxine and In-111-acetylacetone for the labeling of cells: in vivo and in vitro biological testing, Int J Appl Radiat Isot 32:651 (1981).

11. H. Sinn and D. J. Silvester, Simplified cell label with Indium-111-acetylacetone, Br J Radiol 52:758 (1979).

12. M. K. Dewanjee, S. A. Rao, P. Didisheim, Indium-111-tropolone, new high-affinity platelet label: preparation and evaluation of labeling parameters, J Nucl Med 22:981 (1981).

13. H. J. Danpure, S. Osman, F. Brady, The labeling of blood cells in plasma with In-111-tropolonate, Br J Radiol 55:247 (1982).

14. M. L. Thakur and M. J. Barry, Preparation and evaluation of a new Indium-111 agent for efficient labeling of human platelets in plasma, J Lab Cmpds Radiopharm 19:1410 (1982).

15. G. V. R. Born, Jr., Proceedings of the British Institute of Radiology. Cell labeling with gamma-emitting radionuclides for in vivo study, Brit J Radiol 53:922 (1980).

16. R. P. Grimley, E. Rafimi, R. J. Hawker, Z. Drole, Imaging of 111-In-labelled platelets - a new method for the diagnosis of deep vein thrombosis, Br J Surg 68:714 (1981).

17. A. French, J. K. Hussey, F. W. Smith, P. O. Dendy, B. Bennett, A. S. Douglas, Diagnosis of deep vein thrombosis using autologous Indium-111-labeled platelets, Br Med J 282:1020 (1981).

18. H. H. Davis, B. A. Siegel, J. H. Joist, W. A. Heaton, C. J. Mathias, L. A. Sherman, M. J. Welch, Scintigraphic detection of atherosclerotic lesions and venous thrombi in man with Indium-111-labeled autologous platelets, Lancet 1:1185 (1978).

19. H. H. Davis and M. J. Welch, Radioisotopic detection of arterial thrombi, in: "Venous and Arterial Thrombosis: Pathogenesis, Diagnosis, Prevention, and Therapy," J. H. Joist and L. A. Sherman, eds., p 295, Grune & Stratton, New York (1979).

20. S. R. Bergmann, R. A. Lerch, C. J. Mathias, B. E. Sobel, M. J. Welch, Non-invasive detection of coronary thrombi with In-111-platelets, J Nucl Med 24:130 (1983).
21. M. D. Ezekowitz, E. O. Smith, A. C. Cox, F. B. Taylor, Failure of aspirin to prevent incorporation of In-111-labeled platelets into cardiac thrombi in man, Lancet 440 (1981).
22. B. T. Allen, C. J. Mathias, M. J. Welch, R. E. Clark, Platelet deposition on vascular grafts: the accuracy of in vivo quantitation and the significance of in vivo platelet reactivity, circulation (in press).
23. J. Megerman, J. T. Christenson, K. C. Hanel, H. W. Strauss, W. M. Abott, Imaging vascular grafts in vivo with Indium-111-labeled platelets, Ann Surg 198:178 (1983).
24. A. L. Riba, M. L. Thakur, A. Gottschalk, B. L. Zaret, Imaging experimental coronary artery thrombosis with In-111-platelets, Circulation 60:767 (1979).
25. W. J. Powers, B. A. Seigel, H. H. Davis, C. J. Mathias, H. B. Clark, M. J. Welch, Evaluation of cerebrovascular disease with Indium-111-platelet scintigraphy, Neurology 32:939 (1982).
26. K. D. Schwartz and M. Kruger, Improvement in labeling erythrocytes with Tc-99m-pertechnetate, J Nucl Med 12:323 (1977).
27. W. J. Powers, C. J. Mathias, K. T. Hopkins, B. A. Siegel M. J. Welch, Dual radiotracer technique for improved scintigraphic detection of thrombi, in: "Nuclear Medicine and Biology," vol 1., C. Raynaud, ed., p 1163, Pergamon Press, Paris (1982).
28. C. B. Sutherland, M. E. King, S. J. Peerless, W. C. Begina, G. W. Brown, M. J. Chamberlain, Platelet interaction within giant intracranial aneurysms, J Neurosurg 56:53 (1982).
29. P. N. Walsh, Collagen-platelet interaction in coagulation, hemostases and thrombosis, in "Platelets and Thrombosis," D. C. B. Mills and F. I. Pareti, eds., p 125, Academic Press, New York (1977).
30. G. A. Jamieson, Interaction of platelets and collagen, in: "Platelets: Production, Function, Transfusion, and Storage," M. G. Baldini and S. Ebbe, eds., p 171, Grune & Stratton, New York (1974).
31. J. F. Mustard, S. Moore, M. A. Packham, R. L. Kinlough-Rathbone, Platelets, thrombosis and atherosclerosis, Prog Biochem Pharmacol 14:312 (1977).
32. S. Niewiarowski, E. Regoeczi, C. J. Stewart, A. Senyi, J. F. Mustard, Platelet interaction with polymerizing fibrin, J Clin Invest 51:685 (1972).
33. S. Niewiarowski and C. J. Steward, Interaction of blood cells with fibrinogen and polymerizing fibrin, in: "Platelets: Multidisciplinary Approach," G. Gaetano and S. Garottini, eds., p 131, Raven Press, New York (1978).
34. J. Vermylen, Physiology of haemostasis, in: "Platelets: Multidisciplinary Approach," G. Gaetano and S. Garottini, eds., p 3, Raven Press, New York (1978).

35. U. Hedner and I. M. Nilsson, The role of fibrinolysis, clinics in haematology 10:327 (1981).

36. R. G. MacFarlane, Haemostasis, in: "Human Blood Coagulation, Haemostasis and Thrombosis," R. Briggs, ed., p 543, Blackwell Scientific Publications, London (1972).

37. S. Moncada and J. R. Vane, Arachidonic acid metabolites and the interactions between platelets and blood vessel walls, New Engl J Med 300:1142 (1979).

38. C. Tsao, Vascular substances that modulate blood-to-vessel interactions, Artery 5:246 (1979).

39. R. M. Nalbandian and R. L. Henry, Platelet-endothelial cell interactions, Seminars in Thrombosis and Haemostasis 5:87 (1978).

40. S. Sherry, The role of the platelet in thrombosis, in: "Platelets and Thrombosis," D. C. B. Mills and F. I. Pareti, eds., p 111, Academic Press, New York (1977).

41. D. C. B. Mills, Platelet aggregation and platelet nucleotide concentration in various species, in: "Haemostasis in Man and Other Animals," R. G. Macfarlane, ed., p 99, Academic Press, London (1970).

42. Z. Sinakos and J. P. Caen, Platelet aggregation in mammalians (human, rat, rabbit, guinea-pig, horse, dog) a comparative study, Thromb Diath Haemorrh 17:99 (1967).

43. V. P. Addonizio, EL. H. Edmunds, R. W. Colman, The function of monkey (M.Mulatta) platelets compared to platelets of pig, sheep, and man, J Lab clin Med 91:989 (1978).

44. M. A. Packham and J. F. Mustard, Clinical pharmacology of platelets, Blood 50:555 (1977).

45. J. R. Vane and S. Moncada, The anti-thrombotic effects of prostacyclin, Acta Med Scand (Suppl) 642:11 (1980).

46. M. M. Guest, B. M. Daly, A. G. Ware, W. H. Seegers, A study of antifibrinolyisin activity in the plasma of various animal species, J Clin Invest 27:785 (1948).

47. C. M. Hawkey, The relationship between blood coagulation and thrombosis and atherosclerosis in man, monkeys and carnivores, Thrombosis et Diathesis Haemorrhagica 31:103 (1974).

48. H. V. Roohk, J. Pick, R. Hill, E. Hung, R. H. Bartlett, Kinetics of fibrinogen and platelet adherence to biomaterials, Trans Am Soc Artif Intern Organs 22:1 (1976).

49. A. I. Schafer and R. I. Handen, The role of platelets in thrombotic and vascular disease, Prog Cardiovas Dis 22:31 (1979).

50. W. J. Dodds, Platelet function in animals: Species specificities, in: "Platelets: A Multidisciplinary Approach," G. Gaetano and S. Garattini, eds., p 45, Raven Press, New York (1976).

51. M. C. Scrutton and C. M. Egan, Divalent cation requirements for aggregation of human blood platelets and the role of the anti-coagulant, Thromb Res 14:713 (1979).

52. R. H. Aster, Factors affecting the kinetics of isotopically-
     labeled platelets, in: "Platelet Kinetics," J. M. Paulus,
     ed., p 5, North Holland Publishing Co., Amsterdam (1971).
53. M. Kien, F. A. Belamarich, D. Shepro, Effect of adenosine-
     related compounds on thrombocyte and platelet aggregation, Am
     J Physiol 221:604 (1971).
54. M. J. Welch and C. J. Mathias, Platelet viability following
     Indium-111-oxine labeling in electrolyte solutions, in:
     "Indium-111-Labeled Neutrophils, Platelets, and Lymphocytes,"
     M. L. Thakur and A. Gottschalk, eds., p 93, Trivirum, New
     York (1980).
55. G. V. R. Born, Aggregation of haemostatic cells as an example of
     specialized cell function, in: "Platelet Aggregation and
     Drugs," L. Caprino and E. C. Ross, eds., p 1, Academic Press,
     New York (1974).
56. J. F. Mustard and M. A. Packham, Factors influencing platelet
     function: adhesion, release, and aggregation, Pharmacol Rev
     22:97 (1970).
57. H. J. Weiss, Platelet physiology and abnormalities of platelet
     function, New Engl J Med 293:531, (1975).
58. H. Holmsen, Are platelet shape, change, aggreagation, and
     release reaction tangeable manifestations of one basic
     platelet function, in: "Platelets: Production, Function,
     Transfusion, and Storage," M. Baldini and S. Ebbe, eds.,
     p 207, Grune & Stratton, New York (1974).
59. P. R. Roper-Drewinko, B. Drewinko, G. Corrigni, D. Johnston, K.
     P. McCredie, E. J. Freireuk, Standardization of platelet
     function test, Am J Hematol 11:767 (1981).
60. H. J. Weiss, Pathophysiology and detection of clinically-
     significant platelet dysfunction, in: "Platelets:
     Production, Function, Transfusion and Storage," M. G. Baldini
     and S. Ebbe, eds., p 253, Grune & Stratton, New York (1974).
61. D. C. MacMillan and A. K. Sim, A comparative study of platelet
     aggregation in man and laboratory animals, Thromb Diath
     Haemorrh 24:385 (1970).
62. R. H. Harris, R. Nichols, J. W. Schmeling, P. W. Ramwell,
     Thromboxane A2 and the endoperoxides mediate canine platelet
     activation, Thromb Res 23:521 (1981).
63. J. L. Wautier and J. P. Coen, Pharmacology of platelet
     suppressive agents, Sem Thromb Hemostasis 5:293 (1979).
64. D. Ogston and B. Bennett, Surface-mediated reactions in the
     formation of thrombin, plasmin and Kallikrein, Br Med Bull
     34:107 (1978).
65. M. J. Gallimore, M. V. Nulkar, J. T. B. Shaw, A comparative
     study of the inhibitors of fibrinolysis in human, dog, and
     rabbit blood, Thrombpsos et Diathesis Haemorrhagica 14:145
     (1965).
66. C. M. Hawkey, Fibrinolysis in animals, in: "The Haemostatic
     Mechanism in Man and Other Animals," R. G. MacFarlane, ed.,
     p 143, Academic Press, London (1970).

67. R. G. Mason and M. S. Read, Some species differences in
     fibrinolysis and blood coagulation, J Biomed Mater Res 5:121
     (1971).
68. R. F. Doolittle, J. L. Omcley, D. M. Surgenor, Species
     differences in the interaction of thrombin and fibrinogen, J
     Biol Chem 237:3123 (1962).
69. J. F. Smith, A. W. Sedar, C. M. Ingerman, M. J. Silver,
     Prostaglandin endoperoxides: platelet shape change,
     aggregation, and the release reaction, in: "Platelets and
     Thrombosis," D. C. B. Mills and F. I. Pareti, eds., p 83,
     Academic Press, New York (1977).
70. M. Hamberg, B. Svensson, B. Samuelson, Thromboxanes new group of
     biologically-active compounds derived from prostaglandin
     endoperoxides, Proc Natl Acad Sci USA 72:2994 (1975).
71. G. J. Roth, N. Stanford, J. W. Jacobs, P. W. Majerus,
     Acetylation of prostaglandin synthetase by aspirin
     purification and properties of acetylated proteins from sheep
     vesicular gland, Proc Natl Acad Sci USA 72:3073 (1975).
72. J. B. Smith and A. L. Willis, Formation and release of
     prostaglandins by platelets in response to thrombin, Br J
     Pharmacol 40:545 (1971).
73. S. S. Tang, M. M. Frozinovic, The effects of pCO2 pH on platelet
     shape, change and aggregation for human and rabbit
     platelet-rich plasma, Thromb Res 10:135 (1977).
74. C. R. Cowan and F. C. Monkhouse, Studies on electrically-induced
     thrombosis and related phenomena, Cand J Physiol Pharmacol
     44:881 (1966).
75. P. Didisheim, Animal models useful in the study of thrombosis
     and antithrombotic agents, in: "Progress in Hemostasis and
     Thrombosis," vol. 1, T. H. Spaet, ed., p 165, Grune &
     Stratton, New York (1976).
76. H. R. Baumgartner and C. Haudenschild, Adhesion of platelets to
     subendothelium, Ann NY Acad Sci 201:22 (1972).
77. S. A. Johnson, Formation of thrombi on injured endothelium in
     mesenteric arterioles in guinea pig, Thrombosis et Diathesis
     Haemorrhagica (Suppl) 28:65 (1968).
78. N. A. Kefalides and R. J. Wingler, The chemistry of glomerular
     basement membrane and its relation to collagen, Biochemistry
     5:702 (1966).
79. R. Ross and P. Bornstein, The elastic fiber, I. The separation
     and partial characterization of its macromolecular
     compoenents, J Cell Biol 40:366 (1969).
80. H. R. Baumgartner, Platelet interaction with collagen fibrils in
     flowing blood, Thromb Haemost 37:1 (1977).
81. G. Majno and G. E. Palade, Studies on inflammation. I. The
     effect of histamine and serotonin on vascular permeability.
     An electron microscopic study, J Biophys Biochem Cytol 11:571
     (1961).
82. M. I. Barnhart and S-T. Chen, Vessel wall models for studying
     interaction capabilities with blood platelets, Semin Thromb
     Hemostas 5:112 (1978).

83. Z. D. Grossman, B. W. Wistow, J. G. McAfee, G. Subramanian, F. D. Thomas, R. W. Henderson, R. F. Rohner, M. L. Roskopf, Tc-99m-oxine and In-111-oxine-labeled platelets Part II Localization of experimentally induced vascular lesion, J Nucl Med 19:488 (1978).
84. J. F. Harwig, S. S. Harwig, J. O. Eichling, R. E. Colemen, M. J. Welch, I-123-Labeled soluble fibrin: preparation and comparison with other thrombus imaging agents, Int J Appl Radiat Isot 28:157 (1976).
85. M. J. Welch and K. A. Krohn, Critical review of radiolabeled fibrinogen: its preparation and use, in: "Radiopharmaceuticals," G. Subramanian, B. A. Rhodes, J. F. Cooper, V. J. Sodd, eds., p 493, Soc Nucl Med, New York (1975).
86. B. A. Rhodes, W. R. Bill, L. S. Malmud, M. E. Siegel, H. N. Wagner, Labeling and testing of urokinase and streptokinase: new tracers for the detection of thromoemboli, in: "Radiopharmaceuticals and Labeled Compounds," vol 2, Vienna, IAEA, p 163 (1973).
87. B. R. Persson and V. Kemper, Labeling and testing of 99m-Tc-streptokinase for the diagnosis of deep vein thrombosis, J Nucl Med 16:474 (1977).
88. S. S. L. Harwig, J. F. Harwig, L. A. Sherman, R. E. Coleman, M. J. Welch, Radioiodinated plasminogen: an imaging agent for pre-existing thrombi, J Nucl Med 18:42 (1977).
89. M. J. Welch, C. J. Mathias, B. A. Siegel, Clinical experience with Indium-111-labeled platelets, in: "Indium-111-Labeled Neutrophils, Platelets, and Lymphocytes," M. L. Thakur and A. Gottschalk, eds., p 171, Trivirum, New York (1980).
90. A. M. White and S. Heptinstall, Contribution of platelets to thrombus formation, Br Med Bull 34:123 (1978).
91. L. C. Knight, J. L. Primeau, B. A. Siegel, M. J. Welch, Comparison of In-111-labeled platelets and iodinated fibrinogen for the detection of deep vein thrombosis, J Nucl Med 19:891 (1978).
92. S. I. Schwartz, Effects of electric environment on thrombosis, Clin Neurosurg 10:291 (1962).
93. G. H. R. Rao, G. J. Johnson, R. K. Reddy, J. F. White, Rapid return of cyclo-oxygenase-active platelets in dogs after a single oral dose of aspirin, Prostaglandins 22:761 (1981).
94. K. M. Moser, M. Grusan, E. E. Bartimo, In vivo and postmortem dissolution rates of pulmonary emboli and venous thrombi in the dog, Circulation 48:170 (1973).
95. G. McIllmoyle, H. H. Davis, M. J. Welch, J. L. Primeau, L. A. Sherman, B. A. Siegel, Scintigraphic diagnosis of experimental pulmonary embolism with In-111-labeled platelets, J Nucl Med 18:910 (1977).
96. R. S. Cotran, Ultrastructural studies of endothelial injury in the micro circulation with special reference to thrombosis, in: "Thrombosis," S. Sherry, K. M. Brinkhous, E. Genton, J. M. Stengle, eds., p 437, National Academy of Science, Washington D.C. (1969).

97. B. C. Sheppard and J. E. French, Platelet adhesion in the rabbit
     abdominal aorta following the removal of the endothelium: a
     scanning and transmission electron microscopical study, Proc
     R Soc Lond (Biol) 176:427 (1971).
98. L. A. Harker, R. Ross, J. A. Glomset, The role of endothelial
     cell injury and platelet response in atherogenesis, Thromb
     Haemost 39:312 (1978).
99. W. H. Welch, The structure of white thrombi, Trans Path Soc
     Philadelphia 13:281 (1887).
100. R. Ross and L. Harker, Hyperlipidemia and atherosclerosis.
     Chronic hyperlipidemia initiates and maintains lesions by
     endothelial cell desquamation and lipid accumulation,
     193:1094 (1976).
101. R. Ross, The arterial wall and atherosclerosis, Ann Rev Med 30:1
     (1979).
102. B. C. Bullock, N. D. M. Lehner, T. B. Clarkson, M. A. Feldner,
     W. D. Wagner, H. B. Lofland, Comparative primate
     atherosclerosis. I. Tissue cholesterol concentration and
     pathologic anatomy, Exp Mol Pathol 22:151 (1975).
103. D. Steinberg, Research related to underlying mechanisms in
     atherosclerosis, Circulation 60:1559 (1979).
104. H. Malmros and N. H. Sternby, Induction of atherosclerosis in
     dogs by a thiouracil-free semisynthetic diet containing
     cholesterol and hydrogenated coconut oil, Prog Biochem
     Pharmacol 4:482 (1968).
105. S. C. Nam, W. M. Lee, J. Garmalyck, K. T. Lee, W. A. Thomas,
     Rapid production of advanced atherosclerosis in swine by a
     combination of endothelial injury and cholesterol feeding,
     Exp Med Pathol 28:369 (1973).
106. J. C. Lewis, V. Fuster, B. A. Kottke, Spontaneous endothelial
     cell injury in the intimal cushions of atherosclerotic
     pigeons, Prog Biochem Pharmacol 14:220 (1977).
107. H. M. Groves, R. L. Kinlough-Rathobone, M. Richardson, S. Moore,
     J. F. Mustard, Platelet interaction with damaged rabbit
     aorta, J Clin Lab Invest 40:194 (1979).
108. R. W. Wissler and D. Vesselinovitch, Atherosclerosis in nonhuman
     primates, Adv Vet Sci Comp Med 21:351 (1977).
109. R. Pick, P. J. Johnson, G. Glick, Deleterious effects of
     hypertension on the development of aortic and coronary
     atherosclerosis in stumptail macaques (Macaca speciosa) on an
     atherogenic diet, Circ Res 35:472 (1974).
110. M. B. Stemerman and R. Ross, Experimental arteriosclerosis. I.
     Fibrous plaque formation in primates, an electron microscope
     study, J Exp Med 136:769 (1972).
111. A. J. Honour, R. D. Carter, J. I. Mann, The effects of changes
     in the diet on lipid levels and platelet thrombus formation
     in living blood vessels, Br J Exp Path 50:390 (1978).
112. W. D. Wagner and T. B. Clarkson, Comparative primate
     atherosclerosis. II. A biochemical study of lipids,
     calcium and collagen in atherotic arteries, Expe Mol Pathol
     23:96 (1975).

113. W. J. Powers, C. J. Mathias, M. J. Welch, L. A. Sherman, B. A. Siegel, T. B. Clarkson, Scintigraphic detection of platelet deposition in atherosclerotic macaques: a new model for investigation of antithrombotic drugs, Thromb Res 25:137 (1982).

114. S. R. Bergmann, C. J. Mathias, B. E. Sobel, M. J. Welch, Evaluation of thrombolytic therapy in coronary artery thrombosis: scintigraphic detection with the use of In-111-labeled platelets, in: "Nuclear Medicine and Biology," vol.1, C. Raynaud, ed., p 65, Pergamon Press, Paris (1982).

115. M. A. Dewood, J. Spores, R. Notske, L. T. Mouser, K. Burroughs, M. S. Galden, H. T. Lang, Prevalence of total coronary occlusion during the early hours of transmural myocardial infarction, New Engl J Med 303:897 (1980).

116. L. R. Erhardt, T. Lundman, H. Millstedt, Incorporation of I-125-labeled fibrinogen into coronary arterial thrombi on acute myocardial infarction in man, Lancet 1:387 (1973).

117. M. D. Ezekowitz, R. D. Burrow, P. W. Heath, T. Streitz, E. O. Smelk, D. E. Parker, Diagnostic accuracy of Indium-111 platelet scintigraphy in identifying left ventricular thrombi, Am J Cardiol 51:1712 (1983).

118. R. Kordenat and K. Kizidi, Experimental intracoronary thrombosis and selective in situ lysis by catheter technique, Am J Cardiol 30 (1972).

119. C. A. Ramirez, M. B. Stemerman, K. A. Isaacson, C. K. Colton, KI. A. Smith, R. S. Lees, Morphological and morphometric characterization of platelet adhesion to the exposed subendothelium of the rabbit thoracic aorta in vivo, Microvasc Res 21:320 (1981).

120. R. H. Bourgain, The inhibitory effect of PGI2 (Prostacyclin) on white platelet arterial thrombus formation, Haemostasis 8:117 (1979).

121. D. M. Creasy, M. Fallenfant, D. A. James, A. D. Dayan, Preliminary toxicity testing of prostacyclin, in: "Prostacyclin," J. R. Vane and S. Bergstrom, eds., p 385, Raven Press, NY (1979).

122. H. Sinzinger, P. Clopath, K. Silberbauer, The effect of ballooning onminipeg aortic prostacyclin formation - a time course, Artery 1:23 (1980).

123. S. M. M. Karim, P. G. Adarkan, Some pharmacological studies with prostacyclin in baboon and man, in: "Prostacyclin," J. R. Vane and S. Berstrom, eds., p 419, Raven Press, NY (1979).

124. M. R. Buchanan, E. Dejana, M. Gent, J. F. Mustard, J. Hirsch, Enhanced platelet accumulation onto injured carotid arteries in rabbits after aspirin treatment, J Clin Invest 67:503 (1981).

125. S. Jaeger and H. Berntsen, Deposition of human labeled platelets on damaged rabbit aorta before and after ingestion of acetylsalicylic acid, Haemostasis 8:99 (1979).

126. B. Adelman, M. B. Slemmerman, D. Mennell, R. E. Hardin, The
      interaction of platelets with aortic subendothelium:
      inhibition of adhesion and secretion by Prostaglandin I-2,
      Blood 58:198 (1981).
127. H. J. Weiss and V. T. Turitto, Prostacyclin (Prostaglandin I-2,
      PGI2) inhibits platelet adhesion and thrombus formation on
      subendothelium, Blood 53:244 (1979).
128. R. J. Gryglewski, R. Korbut, A. Ocetkiewicz, Reversal of
      platelet aggregation by prostacyclin. Pharmacol Res Commun
      10:185 (1978).
129. M. Goldman, H. C. Norcott, R. J. Hawker, Z. Drolc, C. N.
      McCollum, Platelet accumulation on mature dacron grafts in
      man, Br J Surg 69 (Suppl) S38 (1982).
130. S. Berger, E. W. Salzman, E. W. Merrill, P. S. Wong, The
      reaction of platelets with prosthetic surfaces, in:
      "Platelets: Production, Function, Transfusion, and Storage,"
      M. G. Baldini and S. Ebbe, eds., p 299, Grune & Stratton, New
      York (1974).
131. C. D. Forbes and C. R. Prentice, Thrombus formation and
      artificial surfaces, Br Med Bull 34:201 (1978).
132. V. Fuster, M. K. Dewanjee, M. P. Kaye, M. Josa, M. P. Metke, J.
      H. Chesebro, Noninvasive radioisotopic technique for
      detection of platelet deposition in coronary artery bypass
      grafts in dogs and its reduction with platelet inhibitors,
      Circulation 60:1508 (1979).
133. J. T. Christensen, J. Megerman, K. C. Hanel, G. J. L'Italien, H.
      W. Strauss, W. M. Abbott, The effect of blood flow rates on
      platelets deposition in PTFE arterial bypass grafts, Trans Am
      Soc Artif Intern Organs 27:188 (1981).
134. J. L. Ritchie, J. R. Stratton, B. Thiele, G. W. Hamilton, L. N.
      Warrick, T. W. Huang, L. A. Harker, In-111 platelet imaging
      for detection of platelet deposition in abdominal aneurysms
      and prosthetic arterial grafts, Am J CardioL 47:882 (1981).
135. G. P. Clagett, M. Robinowitz, Y. Maddox, J. M. Langloss, P. W.
      Ramwell, The antithrombotic nature of vascular prosthetic
      pseudointima, Surgery 91:87 (1982).
136. P. Hagen, S. Wang, E. M. Mikat, D. B. Hackel, Antiplatelet
      therapy reduces aortic intimal hyperplasia distal to small
      diameter vascular prosthesis (PTFE) in nonhuman primates, Ann
      Surg 195:328 (1982).

PLATELET KINETICS

A. M. Peters

Medical Division
Glaxo Group Research Ltd
Ware, Hertfordshire
SG12 7EH

Department of Diagnostic Radiology
Hammersmith Hospital
London
W12

INTRODUCTION

The subject of platelet kinetics can be conveniently considered
under three separate headings: platelet distribution, analysis of
platelet survival data, and platelet destruction or consumption; the
latter is obviously closely related to platelet life-span.

In some regions of the body, notably the spleen, platelets are
concentrated with respect to plasma, and a platelet "pool" can be said
to exist. Other regions of platelet pooling may be in lung and liver.
As with erythrocytes, therefore, one can visualize a range of platelet
"haematocrits" throughout the body, with the mean body hematocrit
being the ratio of total platelet volume to total plasma volume.
Abnormal platelet pooling can be considered to exist in splenomegaly
and also may exist in some diseased organs which normally do not show
pooling.

Measurement of platelet life span is an important component of
platelet kinetic studies. The pathological causes of reduced platelet
life-span can best be considered under platelet destruction, although
the formulation of models which take into account possible modes of
destruction is necessary before mathematical analysis of platelet
survival data is possible. This subheading of platelet kinetics is
primarily concerned with such mathematical analysis.

Platelet destruction, and its impact on platelet life-span, is, perhaps, the most difficult area of platelet kinetics to study. Destruction can be physiological or pathological. The former may be confined to the reticuloendothelial system (RES), or it may include a component of intravascular "consumption," which is the result of basal ongoing inapparent hemostasis. Pathological destruction may also be confined to the RES or result from intravascular platelet deposition. Some forms of pathological destruction, such as in idiopathic thrombo-cytopenic purpura (ITP), are confined to the RES. The latter may also be responsible for uptake of platelets which have undergone various vascular insults. Intravascular deposition may be temporary, in which case it merges with platelet distribution abnormalities, or it may be irreversible and lead to microthrombus formation. It can be appreciated from this introduction, therfore, that the subject of platelet kinetics is a complex one and that there is considerable overlap between the different aspects, created for the sake of descriptive convenience.

PLATELET DISTRIBUTION

About 30 minutes after the injection of radiolabeled platelets in the normal human subject, only about two thirds of the dose (the recovery) can be accounted for in circulating blood(1,2). Since the introduction of In-111 as a cell-labeling radionuclide, it has been possible to confirm that most of the other third enters the spleen(3-6). Furthermore, by dynamic imaging with the gamma camera immediately following reinjection, it has been shown that the activity in the spleen is rapidly and freely exchangeable with the activity in the circulating blood, as in a closed, well-mixed, two-compartmental system(7). In other words, platelets pool in the spleen in numbers that depend on the balance of input (dependent on splenic blood flow - SBF) and output (dependent on the mean intrasplenic platelet residence time - t). In splenomegaly the splenic platelet pool is expanded and the recovery falls to as low as 10%(8,9). In contrast, the recovery in asplenic subjects approaches 100%(8,9).

The presence of nonsplenic platelet pools has been suggested by Freedman et al(10) who showed an increase in platelet count following vigorous exercise in asplenic subjects, and Vilen et al(11) who showed an increase following epinephrine infusion. The lung and liver are likely candidates for such extrasplenic platelet pools. In the dog, for instance, an excess of Cr-51 activity relative to Tc-99m activity has been demonstrated in lung slices prepared following the injection of Cr-51-labeled platelets and Tc-99m-labeled red cells, suggesting a lung transit time of platelets about twice that of red cells(12). Furthermore, in man, epinephrine infusion results in an increase in the arterial platelet count, which precedes the increase in the peripheral venous count(13). The liver certainly appears to sequester platelets during hypothermia(14) and following protamine sulphate infusion(15), but whether it normally pools platelets is less clear.

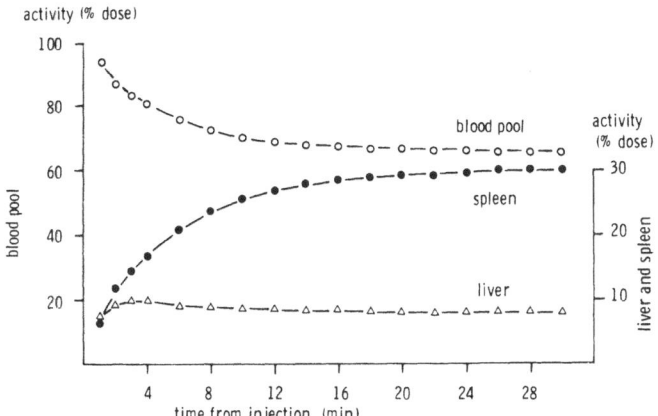

Fig. 1 Time activity
curves recorded over
the cardiac blood
pool, spleen and
liver following
injection of 111-In-
oxine labeled autolo-
gous platelets in a
normal subject. Note
2 ordinate scales.

Following injection of In-111-labeled platelets in normal subjects, activity enters the spleen and to a lesser extent the liver
(Fig 1). Uptake in the latter is completed before 5 minutes and
varies between about 5 and 15% of the dose(3-6). Following completion
of liver uptake, the splenic and cardiac blood pool time activity
curves mirror each other and approach stable values over the same
monoexponential time course of about 0.15 per minute(13,7). The percentage of the dose present in the spleen at the steady state is about
30%(3-6). The interpretation of the splenic uptake time course seems
clear, i.e., uptake reflects dynamic equilibration between blood
platelets and splenic pool platelets, and this has been confirmed by
directly injecting radiolabeled platelets into the splenic artery and
monitoring peripheral venous activity(7) (Fig. 2). In effect, this
reverses the compartments, with the result that the peripheral venous
time activity curve becomes an uptake curve resembling, and with a
similar time constant to, the splenic uptake curve recorded following
I.V. injection.

The interpretation of liver activity following In-111-platelet
injection is not so clear. Possible explanations are 1) uptake of
nonviable platelets damaged during labeling and 2) pooling of platelets within the liver. Neither of these, however, exactly fit the
kinetic data. The curve of uptake of damaged platelets would be
expected to resemble that of uptake of sulphur colloid, i.e., a
maximum (say 95%) completion time (determined by liver blood flow) of
no less than 10 minutes. The uptake completion time frequently
observed of much less than this suggests pooling (Fig. 3). However,
pooling would require that the liver and blood pool time activity
curves parallel each other after completion of liver uptake, but this
often is not the case (see Figure 3). It seems probable that because
it often shows a slow fall at a time when blood pool activity is
constant, liver activity represents uptake of "activated" platelets

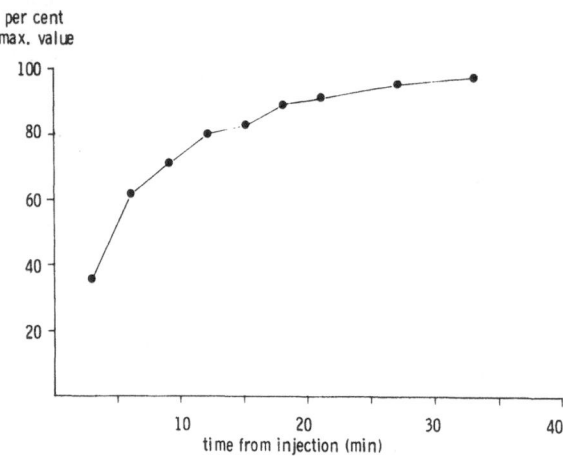

Fig. 2. Periphenal venous time activity curve recorded following injection of 111-In-oxine labeled platelets directly into the splenic artery in a subject undergoing splanchnic angiography.

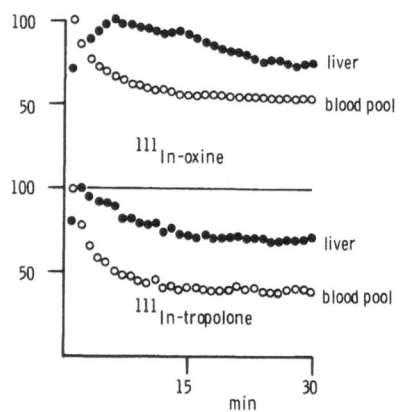

Fig. 3 Time activity curves recorded over the cardiac blood pool and liver following injection of 111-In-oxine labeled platelets (upper) and 111-In-tropolone labeled platelets (lower). Ordinate: per unit of maximum count rate. The lower recovery seen in the lower panel was due to splenomegaly.

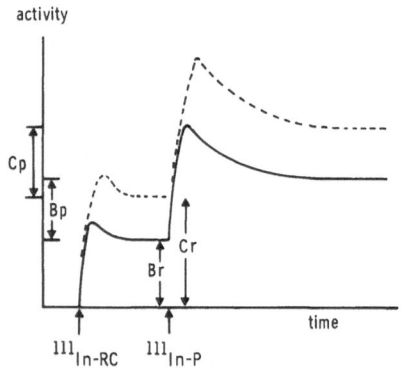

Fig. 4 Calcuation of platelet (p) transit time as a factor of red cell (r) transit time. Discontinuous line - gamma camera signal (C) Continuous line - peripheral blood signal (B) Ratio of transit time, tp/tr is equal to:

$$\frac{C_p}{C_r} \times \frac{B_r}{B_p}$$

which undergo a reversible, artifactual pooling. This is clearly seen
with In-111-labeled granulocytes in the liver and lung(16) and has
been described(17) for platelets labeled with Cr-51 by techniques
using EDTA in which platelets have been manipulated without prior
inhibition by acidification or prostaglandin exposure. Of course
elements of physiological pooling and/or irreversible uptake of
damaged platelets may be superimposed on artefactural pooling, as may
irreversible uptake of free (plasma bound) indium(18).

It is of interest that platelets labeled entirely in plasma using
indium tropolone seem to undergo less liver concentration as
manifested by less prominent liver images and, from a kinetic
viewpoint, earlier completion of uptake and more subsequent
parallelism with blood pool activity (see Figure 3). This suggests
that, following plasma labeling, liver activity predominantly reflects
physiological pooling. In an attempt to quantitate this pooling, we
have compared the hepatic signal following injection of
In-111-platelets with the signal produced by In-111-labeled red cells
given prior to the platelets (Fig. 4). By comparing these signals
with the simultaneous corresponding blood activities, platelet transit
time through the liver was estimated to be about 1.5 that of red
cells (unpublished). The same approach can be applied to the lung,
the activity in which parallels blood activity and through which
platelet transit time appears to be about 1.2 that of red cell transit
time (unpublished) (1.2). This figure is somewhat lower than that
suggested by Martin et al(12) for Cr-51-labeled platelets (about 2)
and our own group for In-111-labeled granulocytes (about 2.5)(19).
However, quantitative scanning data suggest that if extra-hepato-
splenic platelet pools exist, they must be small, because soon after
the injection of In-111-platelets, all the activity can be accounted
for in the spleen, liver, and circulating blood(3-6).

The kinetic data recorded dynamically following In-111-platelet
injection have been applied to the measurement of SBF and mean intrasplenic platelet transit time (t)(7,20-22). These two parameters,
being the input and output functions of the splenic platelet pool, are
the sole determinants of the capacity of the latter and can be
calculated from the splenic and blood pool time activity curves,
provided liver uptake is minimal and completed early relative to
completion of splenic uptake. SBF is given in units of percent total
blood volume (TBV) per minute assuming that no platelet pools exist
other than in the spleen. (If they do exist, then the apparent
distribution of platelets will be greater than TBV and so SBF
underestimated. Thus, if liver and lung platelet transit times were
1.5 those of red cells and their combined blood volumes taken as 15%
each of TBV, then the effective extrasplenic distribution volume of
platelets would be 115% of extrasplenic TBV, and SBF underestimated by
13%.) SBF by this technique agrees favorably with more invasive
techniques(23) and, as might be expected, increases with increasing
spleen size(20). An interesting group of patients with immune complex

related diseases, such as systemic lupus erythematosus and essential
mixed cryoglobulinemia, has been found to have markedly increased SBF
but no significant splenomegaly(24). We have also been able, with
In-111-labeled platelets, to measure splenic extraction ratio of
modified erythrocytes(25) and to investigate its response, and that of
SBF, to therapeutic manoeuvres such as plasma exchange and high-dose
methyl-prednisolone therapy (unpublished).

Intrasplenic platelet transit time is a parameter that until
recently had not been investigated. Unlike SBF, it appears not to be
related to spleen size but is dependent on splenic perfusion (i.e.,
SBF per unit spleen volume)(22). In other words, it is rapid in
patients without splenomegaly but who have elevated SBF, and it is
prolonged in patients, who have normal or low SBF associated with mild
to moderate splenomegaly, such as those with cardiac disease(20).
Polycythemia secondary to congenital heart disease was found to be
associated with very prolonged values of t, but the latter was normal
to rapid in primary polycythemia(20). The relationships betweeen
splenic perfusion and t and between SBF and spleen size result in a
tendency for the splenic platelet pool capacity to remain constant in
any subject, in the face of changes in SBF, and to be proportional to
spleen size from subject to subject.

Why the spleen concentrates platelets in such large numbers is
entirely obscure, and, apart from the relationship between t and
splenic perfusion - which is not a surprising one - the factors
controlling platelet transit through the spleen are also unknown.
Macrophage function may play a role, as suggested by the scanning
electromicroscopic observations of Weiss(26) that many platelets can
be seen adhering to splenic macrophages. Platelet age may also be
important, as suggested by the studies of Schulman et al(27), who,
using nonisotopic techniques, concluded that young platelets are
preferentially "sequestered" in the spleen. Furthermore, Freedman et
al(10), by showing that the thrombocythemic response to exercise was
greater with respect to the megathrombocytes and that these responses
were reduced in asplenic subjects, also suggested that young platelets
are preferentially retained in the spleen. Conceptually it can be
regarded, therefore, that young and/or large platelets have a long
intrasplenic transit time. The relationship between platelet size and
age, however, remains unclear(28,29).

The abnormal kinetics of platelet distribution can be subgrouped
into 1) abnormal pooling at sites (e.g., spleen) where physiological
pooling otherwise occurs and 2) abnormal concentration at sites where
physiological pooling is not expected. Thus, with regard to the
first, can abnormalities within the spleen lead to enhanced or
decreased platelet "trapping", or can abnormalities within the
platelet result in enhanced splenic pooling or pooling elsewhere?
With regard to the second, are platelets reversibly concentrated
within, say, the rejecting renal allograft (as distinct from

irreversible uptake), or are platelets taken up on atheromatous
plaques in dynamic equilibrium with circulating platelets?

1) Apart from increased t in cardiac disorders(21) and decreased t
in some of the rheumatological conditions(20), changes which would be
anticipated from abnormalities of splenic perfusion, no abnormalities
of platelet transit time have been identified in any specific condition. Of course, increased platelet pooling is seen in splenomegaly,
but this is the result of increased platelet input into the spleen
(i.e., increased SBF) with no alteration in t(7,20). No platelet
abnormalities have been described to result in abnormal splenic
pooling, though it might be suggested that the low platelet recovery
observed by some workers in idiopathic thrombocytopenic purpura
(ITP)(30) may be due to increased splenic pooling. Because the mean
age of platelets in ITP is reduced, the evidence of Schulman et al
that young platelets are preferentially retained in the spleen would
suggest that the splenic platelet pool in ITP should be enlarged as a
result of an increased platelet t. However, the fact that homologous
and autologous labeled platelets have been found to have similar
recoveries in ITP(31) argues against this. The percentage responses,

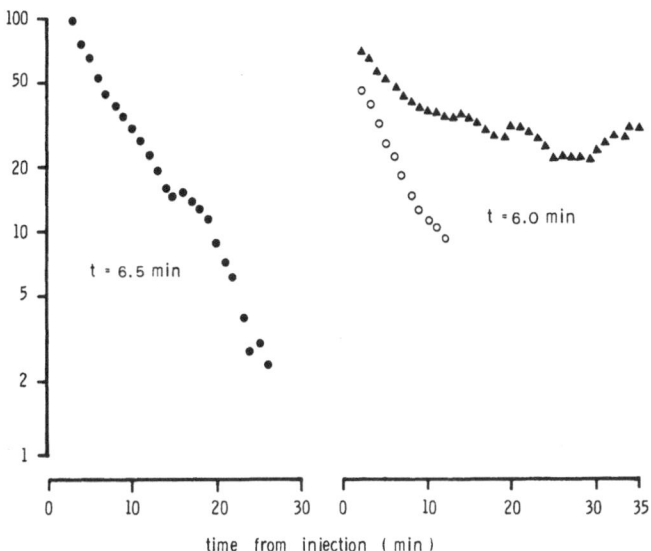

Fig. 5 Splenic platelet "washout" curves derived from deconvolution
        analysis(33) of the blood pool and splenic time activity
        curves. Ordinate: per cent of initial activity; t – mean
        platelet transit time. Left: normal; right: patient with
        idiopathic thrombocytopaenic purpura. The open circles,
        derived by subtraction of the asymptote from the curve
        (triangles), represent washout of "un-extracted" platelets
        with t of 6.0 min.

furthermore, of the peripheral platelet count to epinephrine infusion
are similar in normal patients and in patients with ITP(32).

   Interpretation of platelet recovery in severe ITP is complicated
by the greatly increased rate of platelet destruction which becomes
superimposed on splenic pooling during the early kinetics following
radiolabeled platelet injection. We have attempted to resolve these
two components of the early platelet loss from the circulation by the
technique of deconvolution analysis(24,33) applied to the simultaneously recorded time activity curves over spleen and blood pool. This
computerized technique predicts the splenic time activity curve that
would be obtained if labeled platelets were injected as a bolus into
the splenic artery and did not recirculate. Mean platelet transit
time is obtainable from the slope of the curve (Fig. 5). In subjects
with very short platelet survival, the curve approaches an asymptote
(see Figure 5), the height of which is an indication of the splenic
platelet extraction fraction (unpublished). A comparison between this
approach and compartmental analysis for the estimation of t has been
reported in dogs(34) and in subjects without markedly reduced platelet
survival(24). Platelet transit time in ITP was variable but
essentially normal, although we have been unable to test autologous
platelets in severely thrombocytopenic patients, in whom one might
expect a decreased t. At any rate, it would appear that
antibody-coated platelets are not significantly delayed in their
transit through the spleen.

   2)The kinetics of platelets taken up by diseased organs or
tissues are not well understood, especially in chronic conditions. In
extensive atheromatous disease, for example, platelet survival is
known to be reduced, but whether this is because of "consumption" of
platelets by the atheromatous plaque or because of damage inflicted on
the platelet by interaction with the plaque (an increased rate of
"hits," as envisaged in the multiple-hit model - see below) with
subsequent, normal, removal by the reticuloendothelial system has not
precisely been resolved. It can be shown theoretically that substantial abnormal consumption of platelets is required to produce readily
measurable reductions in platelet survival(21). Any difficulty experienced in visualizing In-111 uptake in conditions in which platelet
survival is clearly reduced would suggest RE destruction of platelets
rather than "peripheral" consumption. Another general point to bear
in mind is that thrombotic occlusion would be anticipated if mobilization and dispersion of platelet material was not achieved pari passu
with platelet uptake. Thus, the fact that prosthetic arterial grafts
continue to be scintigraphically visible for many years,(35,36) and
remain patent suggests that platelets are in dynamic equilibrium
between the graft surface and blood. The role of platelet aggregation
and disaggregation in these kinetics also has to be considered.
Obviously, a clinically significant platelet-fibrin embolus represents
a form of platelet output from a platelet-active focus, but ongoing
clinically silent microscopic aggregate output is also likely to

occur. Techniques for measuring platelet aggregates in vivo have been developed(37), and the presence of such aggregates in a variety of diseases associated with platelet abnormalities has been descri- bed(38). The presence of arteriovenous platelet count. gradients in vascular beds, such as the myocardium, may be related to the discharge of aggregates into the venous blood. Thus Mehta and Mehta(39) described a mean fall in count of 233 to 174 x 10^9 per 1 in 6 subjects with coronary artery disease. This gradient (i.e., extraction ratio of 25%) is not possible on the basis of platelet consumption in the coronary bed because, with a coronary blood flow of 5% TBV per minute it would be expected to produce a platelet life span of less than 90 minutes. A reverse gradient in platelet count has been described across the pulmonary bed(40) and interpretated to indicate platelet production from megakaryocytes in the lung(28). However, such gradients would not be measurable with normal platelet survivals of 9 days, and it may be that the lung normally disaggregates cell aggregates.

In acute events, In-111-labeled platelet uptake is presumably related to thrombotic phenomena and of a less subtle nature, at least from the kinetic standpoint. Thus, taking the rejecting renal allograft as an example, most workers have described a ratio of counts over the allograft to counts over a reference point (based on blood pool) which increases with time(41). This is consistent with irreversible platelet uptake and can be contrasted with chronic rejection in which the ratio apparently remains stable but higher than that given by labeled red cells (Sinzinger, personal communication). The reversible element of uptake, superimposed on the irreversible element predominating in acute situations, but possibly predominating in chronic situations, could generally be expressed as platelet transit time through the organ.

PLATELET LIFE-SPAN

The graph of platelet survival in the normal subject, determined by random In-111-labeling, appears to be linear. With Cr-51 the survival has in some hands been linear and in others curvilinear. In general, as platelet life-span becomes more severely reduced in disease, the survival deviates further from linearity. There are countless reports in the literature on platelet life-span in various disorders, and no attempt will be made here to comprehensively review them. Instead, a general approach to platelet survival data will be outlined, with comments on the various analytical techniques that have been applied.

The purpose of acquiring platelet survival data is twofold: first to determine mean platelet life-span (MPLS), and second to obtain information on the mode of platelet destruction and the severity of its abnormal component in a given case. The main task

when analyzing platelet survival data is the detection of an
appropriate model which adequately describes the platelet destruction.
This presents no problem in the normal subject, in whom the platelet
survival profile appears to be linear, indicating that destruction is
predominantly an ordered process in which platelets are removed from
the circulation when they have reached a particular age. When
platelet life-span is severely reduced, as in idiopathic thrombocytopenic purpura, the survival profile is almost monoexponential,
indicating that destruction is predominantly a random process. In
these two cases the mode of destruction seems clear and the MPLS
easily obtained from the familiar equations describing linear and
exponential functions, namely,

$$N(t) \quad = \quad N(o) - bt$$
$$\text{and} \quad N(t) \quad = \quad N(o)e^{-kt}$$

respectively, where $N(t)$ is the number of labeled platelets circulating at time t, $N(o)$ is the number at time zero, and b and k the
appropriate constants describing the rate of destruction, i.e.,

$$MPLS \quad = \quad N(o)/b$$
$$\text{and} \quad MPLS \quad = \quad 1/k.$$

Most abnormal platelet survival profiles are neither linear nor
exponential, and this has led to difficulty 1) in calculating MPLS and
2) in comparing MPLS between patients with differently shaped survival
profiles. Two techniques are commonly used to analyze these
curvilinear profiles: the gamma function, based on the so-called
multiple-hit model(42), and the Mills-Dornhorst equation(43,44).

The multiple-hit model states that a platelet is removed from the
circulation after it has sustained a required number of hits. The
latter might be thought of as "insults" to the platelet resulting in
accumulating damage which ultimately leaves the platelet nonviable.
The equation describing $N(t)$ is complicated but depends on the number
(n) of hits required and the reciprocal (a) of the waiting time
between hits. The advantage of the gamma function is its flexibility,
i.e., it can be applied to any curvilinear profile intermediate
between monoexponential and linear. It provides not only a measure of
MPLS, but also the "biological" information given by n and a, essentially n gives information about the platelet; thus an abnormal platelet
will require fewer hits for destruction than a normal platelet. On
the other hand, a gives information about the platelet "milieu"; thus
a would be expected to increase in vascular disease in which the
insult rate is presumably accelerated. It should be pointed out that
when $n = 1$, the survival is exponential with rate constant equal to a,
and when $n = \infty$, the survival is linear with a slope proportional to a,
since $MPLS = n/a$.

The Mills-Dornhorst equation

$$N(t) = N(o) \frac{e^{-kt} - e^{-kT}}{1 - e^{-kT}}, \text{ from which MPLS} = \frac{1 - e^{-kT}}{k}$$

is based on the model in which platelet destruction occurs by simultaneous random (i.e., exponential) and ordered (i.e., linear) processes, where T is the "deterministic" life-span, i.e., the life-span attained by cells escaping random destruction. The random component is minimal in the normal subject in which the survival profile is essentially linear and increases in relative magnitude as the survival profile departs from linearity. This approach, like the gamma function, is also flexible and, in addition to providing the value of MPLS, gives quantitative information about the two simultaneous components of destruction, in particular the magnitude of random destruction (k), presumed to be the abnormal component. This latter component may itself be a composite of a number of random destruction processes, corresponding to different compartments, each with its own rate constant. The sum of these rate constants will then be equal to the overall rate constant k.

Of course, any particular model is not validated simply because a curve will fit it but requires independent validation before it can be considered appropriate for platelet destruction. Thus the "best" model, multiple-hit or random superimposed on ordered, cannot be identified since independent evidence does not appear to exist to validate either.

Two further simple techniques have been suggested to obtain the MPLS from curvilinear survival profiles. One of these is to calculate the weighted harmonic mean of exponential and linear fits(45). The other gives MPLS by forward extrapolation of the initial portion of the survival curve to the time axis, which, for any curve between exponential and linear, is cut at the MPLS. This technique has some appeal, not only because of its simplicity but also because it is potentially quite accurate. Thus curves based on the mulitple hit model are, in fact, almost linear between 100% (time zero) and 50% for all n values greater than 2 (see Figure 3 of Murphy and Francis42). Unfortunately, however, many survival profiles commence with a plateau (see below) resulting in uncertainty in the identification of the initial slope.

Trowbridge and Martin(46) have recently put forward another approach to analyze platelet survival data. Their technique, like the Mills-Dornhorst model, is based on the principle that platelet destruction is both ordered and random. A mathematical argument is developed which shows that the ratio (c) of the numbers of platelets destroyed by the two respective processes is related to and can be readily calculated from the ratio of the t 1/2 and t_s, where the t 1/2

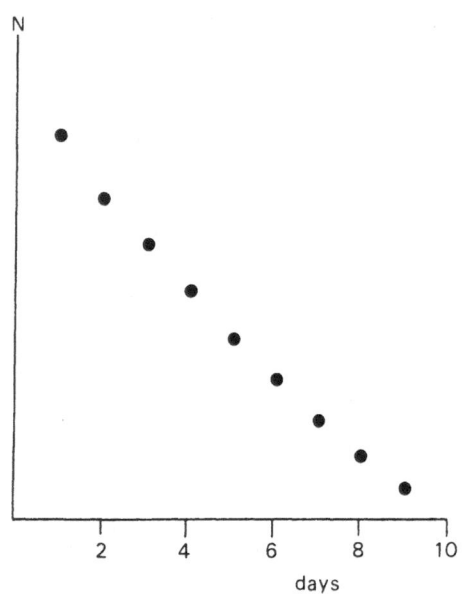

N

| 2 | 4 | 6 | 8 | 10 |

days

Fig. 6 Theoretical platelet survival curve based on the Mills-Dornhorst (M-D) equation such that C (Trowbridge and Martin[46]) is equal to unity. MPLS by linear fit is 2-3 days longer than by fitting the M-D equation. Nonetheless, note the high correlation coefficient given by the linear fit to the non-linear data.

ORDERED: RANDOM   = 1
MPLS (M-D)   = 7.2 DAYS
DESTRUCTION RATE PER DAY   = 6.9% BY SENESCENCE
                             6.9% BY RANDOM
LINEAR FIT: R   = 0.995
MPLS   = 9.5 DAYS

is the time taken for the labeled platelet activity to fall to 50% of its initial activity (after equilibration in the splenic pool) and t_s is the time to disappearance of the labeled platelets.

$$\frac{t_s}{t\ 1/2} = \frac{\log_e(1 + 1/c)}{\log_e[(2c + 2)/(2c + 1)]}$$

The ordered (k_1) and random (k_2) components are obtainable from

$$t_s \cdot k_2 = \log_e(1 + 1/c)$$

and

$$k_1 = k_2 c$$

This technique has the attraction that it requires only two points. However, to be reliable, these must be obtained from the survival curve fitted to the m experimental data points by an (m-1)th degree polynomial.

These authors emphasize the point that even when c is equal to unity the survival curve looks almost linear (Fig. 6). Similarly, inserting increasing values of k in the Mills-Dornhorst equation demonstrates that only when k becomes substantial does the survival curve begin to visibly depart from linearity(21). These considerations indicate that for reasonably accurate estimations of the random destruction component, and pathological interpretations based on it, it is essential to have good-quality platelet survival data. Examples in the literature of attempts to quantitate the random component are the papers of Dassin et al(47) who measured platelet survival in diabetes mellitus, Hanson et al(48) who calculated that 7300 platelets per ul per day are randomly consumed in the maintenance of vascular integrity (see below), and Harker et al(49) who correlated random destruction with the presence of arteriovenous cannulae in baboons.

On a more practical note, most workers find that, at least in normal subjects, many of their platelet survival profiles have a plateau of up to 24 or even 48 hours, followed by a sustained closely linear portion of 5-6 days and finally a tail of about 2 days. The explanation for the plateau is not clear but may be due to 1) net shifts of platelets from the liver and/or spleen to the circulation (i.e., a progressive reduction in intrasplenic platelet transit time following reinjection) or 2) a failure of older platelets to become labeled. Indirect evidence in favor of the first is that the specific activity of In-111-labeled dense platelets increased the day following labeling in intact, but not in splenectomised, subjects(50). The tail can to some extent be accounted for by the multiple-hit and Mills-Dornhorst models described above. Alternatively it may be a manifestation of the preferential pooling of young platelets, discussed above, which would result in a relative deficiency of young platelets in the blood sample drawn for random labeling. The slope of the survival curve would then fall as these young platelets, in their turn, reached senescence at the end of the study. Another possible relative failure to label young platelets might result from the relationship between platelet age and size. Thus, big, young

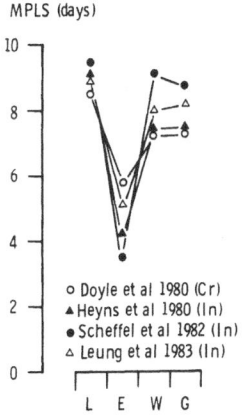

Fig. 7 Comparison of MPLS based on 4 analytical techniques: linear (L), exponential (E), weighted mean (W) and gamma function (G). The subjects of Leung et al. were patients with various disorders; the subjects of the other 3 groups were normal.

platelets may be lost in the first centrifugation for the harvesting
of platelet rich plasma. Finally, an artifactual tail may result from
failure to account for plasma isotope, or residual labeled
erythrocytes, although the Birmingham group have suggested that plasma
In-111 activity is itself an artifact resulting from postsampling
manipulations(51). Whatever their cause, these "blemishes" on
platelet survival profiles will obviously give rise to problems of
ranging severity in the application of the various techniques of
platelet survival data analysis, described above.

Figure 7 illustrates comparative values of MPLS obtained by three
groups of authors who compared linear, exponential, weighted mean and
gamma function fits in normal subjects(6,52,54), and one group who
compared them in a large heterogenous series of patients(53).

PLATELET DESTRUCTION

The accuracy with which radioisotopic methods are able to quantitatively define the distribution of platelet destruction depends on
the physical and biological properties of the radioisotope itself.
For instance, if the radioisotope elutes readily from an organ within
which platelets are destroyed(55,61) or binds nonspecifically to the
foreign material of a prosthetic graft, then misleading information
will result from the use of that radioisotope to quantitate platelet
destruction or "consumption." The precise biochemical form of the
radioisotope when presented to a tissue is important. Thus does the
bone marrow image seen toward the end of a platelet kinetic study
reflect platelet destruction by bone marrow or uptake of plasma-bound
indium in a manner analogous to uptake of transferrin-bound iron?
Attention has recently been drawn to another interesting example along
these lines by Gunson et al(56) who showed that In-111 released from
extravasated labeled platelets was taken up in the kidney, whereas
others note that I.V.-injected plasma In-111 is not(18). These
kinetic aspects of radiolabeled cells have received little attention
but are clearly of importance.

Platelet destruction can be subgrouped into normal and abnormal.
The latter can be the result of appropriate destruction of abnormal
platelets (as in ITP) or destruction by abnormal tissues of normal
platelets.

Normal Destruction

On the basis of Cr-51 labeling it seemed clear that platelets are
normally destroyed in the liver and spleen when they become
senescent(57). Because of the superior gamma emission of In-111, and
therefore the availability of quantitative imaging, the introduction
of In-111 as a cell label provided the opportunity of much more

accurate identification and quantitation of platelet destruction.
This data is largely in agreement with chromium data, especially with
regard to splenic destruction. Areas of controversy, however, are the
quantitative roles of 1) liver destruction, 2) bone marrow
destruction, and 3) physiological widespread peripheral vascular
consumption as a defense against hemorrhage.

There is general agreement that the spleen destroys 30–50% of
platelets(3–6). It is of interest that following In-111-platelet
injection, the total activity in the spleen alters little between
completion of platelet equilibration in the splenic pool at about 30
minutes and completion of platelet life-span some 9 days
later(3–6,57,58). This indicates that the fraction of platelets
destroyed in the spleen parallels the fraction pooling there.

Estimates of liver destruction have been more varied than those
of the spleen. Klonizakis et al(4), for instance, concluded that the
liver played a minor role in platelet destruction, whereas Heyns et
al(3) put the figure at about 40%. Liver activity is somewhat more
difficult to relate to destruction because of the uncertain
contribution of activity from nonviable platelets that may be produced
as a result of labeling and may immediately undergo irreversible
hepatic uptake. The apparently reduced scope for this particular
artifact afforded by tropolone labeling in plasma may help to clarify
the role of the liver.

Quantitation of total bone marrow activity is, with present techniques, impossible because of its diffuse distribution although
quantitation of the chest wall marrow has been reported(3,59).
Because a bone marrow image becomes increasingly obvious throughout
the course of platelet survival studies, Klonizakis et al (4)
speculated that the difference between the total activity that could
be accounted for in spleen, liver, and circulating blood and the dose
of injected activity represented bone marrow uptake of In-111. This
difference was essentially zero at 1 hour following injection and
increased linearly thereafter. This group went on to argue that since
In-111 activity deposited in the spleen undergoes a very slow rate of
elution(55), the In-111 activity present in bone marrow represented
In-111-labeled platelet uptake rather than nonspecific uptake of
plasma In-111. Other workers have since demonstrated in animal
studies that significant amounts of In-111 can be recovered in a
variety of tissues including muscle and skin(60,61), arguing against
bone marrow as an important site for platelet destruction, although
there does appear to be considerable species differences(62). It
would appear that in human studies, quantitation of total bone marrow
uptake will require comparison of In-111 activities (directly sampled
or externally monitored) with activities of other isotopes, such as
those of iron, which at the time of sampling can be assumed to be
confined to bone marrow. Figure 8 compares the quantitative organ
data recorded by four groups following In-111 platelet injection.

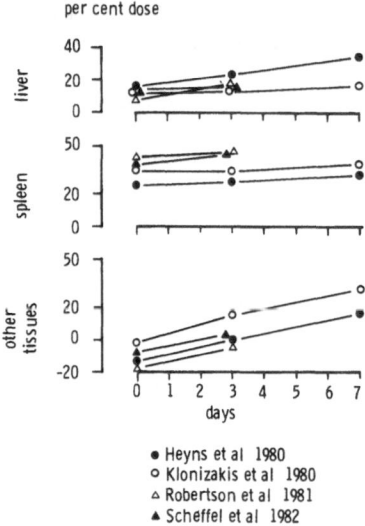

Fig. 8   Distribution of In-111 measured by quantitative scanning in
         normal subjects by 4 groups. "Other tissues" derived by sub
         tracting total of liver, spleen and blood activities from
         100%. Abscissa: time from injection

     Since thrombocytopenic patients suffer episodes of spontaneous
bleeding when their platelet counts fall to levels of 5 x 10^9 per 1 or
lower, it has been suggested that some platelets are consumed in the
physiological defense of hemorrhage in small blood vessels throughout
the body. The scintigraphic evidence for this is scanty, although, as
it presumably represents a rather small proportion of total platelet
destruction and also occurs on a diffuse basis, it would not be
expected to be readily apparent. In a series of patients with
aplastic anemia but without antiplatelet antibodies, Hanson et al(48),
using Cr-51, observed reduced homologous platelet life-span and
recorded survival curves from which it was apparent that, following
the application of the Mills-Dornhorst equation (see above), the
percentage of platelets undergoing random destruction was greatly
increased. They suggested that utilization of platelets for defense
against hemorrhage proceeded quantitatively normally in
thrombocytopenia, implying that fewer platelets survived to
senescence, having been "pressed" into service for hemostasis.
Repetition of this work with In-111 would be useful since greatly
reduced deposition of activity in the RES would be anticipated and
quantifiable.

Abnormal Destruction

     Appropriate destruction of abnormal platelets is exemplified by
the platelet kinetics of ITP. Platelet life-span in this condition is

reduced and may be as short as a few hours. Many investigators have
studied platelet kinetics in ITP, not only in an attempt to improve
our understanding of the disease but also to define those patients who
will benefit from splenectomy. It is generally regarded that an
indication for the latter is scintigraphic evidence of splenic
destruction of labeled platelets. This evidence has assumed various
forms in different centers, being regarded as positive if splenic
counts progressively increased over the duration of platelet
survival(63), if the splenic to hepatic count ratio was elevated(64),
if the splenic to precordial count rate was elevated(65), if total
splenic counts were increased(58), if splenic counts corrected for
pooled platelet activity were increased(66), if the absolute quantity
of activity (In-111) in the spleen was increased(67), and, more
recently, if the ratio of destruction to pooling within the spleen was
increased above unity (submitted for publication). This last approach
is based on the observation that in a wide range of pathologies
(excluding those associated with antiplatelet antibodies or evidence
of "peripheral" consumption of platelets) the counts over the spleen
failed to alter very much from the earliest establishment of splenic
equilibration to the completion of platelet life span, irrespective of
spleen size or platelet life-span. This parallelism can be explained
on the basis of splenic pooling, which determines the total time spent
by a platelet within the splenic pool and therefore the level of
probability of a platelet being in the spleen, as opposed to another
organ of the RES, at the time at which the platelet is due for
destruction. In other words, the pathological significance of splenic
platelet uptake should be viewed against the effective volume of the
splenic platelet pool. The presence of antiplatelet antibodies may
then redirect platelet destruction towards or away from the spleen,
depending on their nature and/or titer, with the result that the
splenic destruction to pooling ratio may be greater or less than
unity. Other factors unrelated to splenic function would also be
expected to modify the ratio. Thus consumption of platelets in a
thrombus should decrease the ratio, whereas liver disease might be
expected to increase it. It is too early to say whether or not this
approach has any usefulness. It may well be that none of them do,
since, as Aster and Keene(63) have suggested, the predominance of
splenic or hepatic destruction may change throughout the course of a
patient's disease, depending primarily on antibody titer, which, when
high, favors hepatic destruction. With In-111-labeling, it is also
apparent that the bone marrow may be involved in accelerated platelet
destruction in ITP(67). This probably occurred in at least one of our
studies and was manifested as 1) very short platelet life-span 2) a
bone marrow image, 3) low hepatic activity, and 4) a splenic
destruction to pooling ratio of about 0.5 (normal about 1.0).

Reduced life-span of normal, or at least unsensitized, platelets
is seen in a wide variety of conditions, principally of the
cardiovascular system, and has been interpreted as the result of
platelet consumption. Such uptake my occur in large vessels, as on

atheromatous plaques(68), or in small vessels, as in the rejecting
renal allograft(59,69-71). The kinetics of such uptake are ill
understood and probably complex. The extent of its reversibility is
unknown and, as discussed above, is likely to involve macroscopic
embolization of platelet-fibrin deposits, expulsion of microscopic
platelet-aggregates, and exchange of individual platelets. The effect
of platelet-vessel interaction on platelet life expectancy is also
superimposed on this dynamic system.

Methods of detecting platelet destruction at these non-RES sites
have included 1) imaging following injection of In-111-labeled
platelets(68) 2) imaging of In-111-labeled platelets following
subtraction techniques based on Tc-99m-labeled red cells(72,73), and
3) comparison of In-111 platelet counts over suspected foci with
In-111 counts from reference regions, which, it is assumed, contain
only blood pool counts (36,74). This latter approach is, of course, a
form of blood pool subtraction. The first approach is only applicable
to relatively active lesions such as acute thrombosis with rapid
platelet "turnover." The other two have been applied to more chronic
lesions. Thus the Amsterdam(72), St Louis(75), and Bloemfontein(73)
groups use Tc-99m red cell subtraction to visualize arterial disease.
Comparison with reference regions has been extensively used to detect
uptake in prosthetic arterial grafts, such as described by Goldman et
al(36,74), who introduced the "thrombogenicity index," and Stratton et
al(35), who referred counts over the graft to peripheral blood counts.
An analogous approach to detect In-111 platelet uptake in the diseased
arteries of familial hypercholesterolemia was described by Pieters et
al(73), who compared the disappearance slope of radioactivity over the
abdominal aorta and lower limbs with that of peripheral blood. The
relevance of all this to "platelet kinetics" resides in the kinetic
nature of the uptake, such as the existance of reversibility and the
rate of its equilbration. Thus, the use of reference regions is
appropriate for the detection of irreversible uptake but would miss a
reversible component unless this reached equilibrium very slowly or
counting was started very early. On the other hand, Tc-99m red cell
subtraction will encompass both forms of uptake but fail to quantitate
them separately. It should also be borne in mind that since the
normal rate of platelet destruction is about 10% per day, relatively
large quantities of In-111, which should be easily detected by the
gamma camera (unless it is diffuse), need to be deposited in abnormal
foci in order to have much impact on the platelet survival curve(21).
Inability to detect such deposits with the gamma camera in
circumstances in which platelet life-span is clearly reduced suggests
that RES destruction of "insulted" platelets may be an important
component. The reverse emphasizes this point. Thus Goldman et al(36)
and Davis et al(69) have both described visible In-111 uptake (in
prosthetic grafts and carotid disease, respectively) associated with
normal platelet life-span. In other words, the use of the platelet
survival curve alone as an indicator of peripheral platelet
consumption is not only unreliable but also insensitive.

ACKNOWLEDGMENTS

My colleagues Drs. S. H. Saverymuttu and J. P. Lavender are gratefully acknowledged for invaluable help, which takes many forms. The
manuscript was expertly typed by Miss B. M. E. Shambrook.

REFERENCES

1. R. H. Aster and H. H. Jandl, Platelet sequestration in man: II
    immunological and clinical studies, J Clin Invest 43:856-869
    (1964).
2. L. A. Harker, The kinetics of platelet production and destruction
    in man, Clin Haematol 6:671 (1977).
3. A. duP. Heyns, M. G. Lotter, P. N. Badenhorst, O. R. Van Reenen, H.
    Pieters, P. C. Minnaar and F. P. Retief, Kinetics, distribution
    and sites of destruction of Indium-111 labeled human platelets,
    Br J Haematol 44:269 (1980).
4. I. Klonizakis, A. M. Peters, M. L. Fitzpatrick, M. J. Kensett, S.
    M. Lewis, J. P. Lavender, Radionuclide distribution following
    injection of 111-indium labeled platelets, Br J Haematol
    46:595 (1980).
5. J. S. Robertson, M. K. Dewanjee, M. L. Brown, V. Fuster, J. J.
    Chesebro, Distribution and dosimetry of In-111-labeled
    platelets, Radiology 140:169 (1981).
6. U. Scheffel, M. F. Tsan, T. G. Mitchell, E. E. Camargo, H. Braine,
    M. D. Ezekowitz, E. L. Nickoloff, R. Hill-Zobel, E. Murphy and
    P. A. McIntyre, Human platelets labeled with In-111-8 hydro-
    xyquinoline: kinetics, distribution and estimates of radiation
    dose, J Nucl Med 23:149 (1982).
7. A. M. Peters, I. Klonizakis, J. P. Lavender, S. M. Lewis, Use of
    Indium-111 labeled platelets to measure spleen function, Br J
    Haematol 46:587 (1980).
8. R. H. Aster, Pooling of platelets in the spleen: role in the patho-
    genesis of "hypersplenic" thrombocytopenia, J Clin Invest 45:645
    (1966).
9. L. A. Harker and C. A. Finch, Thrombokinetics in man, J Clin Invest
    48:963 (1969).
10. M. Freedman, N. Altzuler and S. Karpatkin, Presence of a non-
    splenic platelet pool, Blood 50:419 (1977).
11. L. Vilen, K. Freden, J. Kutti, Presence of a nonsplenic platelet
    pool in man, Scand J Haematol 24:137 (1980).
12. B. A. Martin, R. Dahlby, I. Nicholls, J. C. Hogg, Platelet
    sequestration in the lung with hemorrhagic shock and re-infusion
    in dogs, J Appl Physiol 50:1306 (1981).
13. H. R. Bierman, K. H. Kelly, F. L. Cordes, R. L. Byron, J. A.
    Polhemus, S. Rapoport, The release of leukocytes and platelets
    from the pulmonary circulation by epinephrine, Blood 7:683
    (1952).
14. E. A. Hessel, G. Schmer, D. H. Dillard, Liver sequestration of
    platelets during hypothermia, J Surg Res 28:23 (1980).

15. A. duP. Heyns, M.G. Lotter, P.N. Badenhorst, H. Kotze, F.C.
     Killian, C. Herbst, O.R. Van Reenen and P.C. Minnaar, Kinetics
     and in vivo redistribution of Indium-111-labeled human platelets
     after intravenous protamine sulphate, Thromb Haemost 44:65
     (1980).

16. S. H. Saverymuttu, A. M. Peters, H. J. Danpure, H. J. Reavy, S.
     Osman, J. P. Lavender, Lung transit of Indium-111 labeled
     granulocytes: relationship to labelling techniques, Scand J
     Haematol 30:151 (1983).

17. R. H. Aster and J.H. Jandl, Platelet sequestration in man. I.
     Methods, J Clin Invest 43:843 (1964).

18. F. Vieras, Splenic uptake of In-111, J Nucl Med 16:1205 (1975).

19. A. M. Peters, S. H. Saverymuttu, H. J. Reavy, R. N. Bell, J. P.
     Lavender, The kinetics of human autologous granulocytes labeled
     in plasma with Indium-111 tropolonate, submitted for publication.

20. A. M. Peters and J. P. Lavender, Factors controlling the
     intrasplenic transit of platelets, Eur J Clin Invest 12:191
     (1982).

21. A. M. Peters, A. Rozkovec, R. N. Bell, K. A. Hallidie-Smith, J. F.
     Goodwin, J. P. Lavender, Platelets kinetics in congenital heart
     disease, Cardiovasc Dis 16:391 (1982).

22. A. M. Peters and J. P. Lavender, Platelet kinetics with Indium-111
     platelets: comparison with chromium-51 platelets, Sem Thromb
     Hemost 9:100 (1983).

23. R. Williams, R. E. Condon, H. S. Williams, L. M. Blendis, L. Kreel,
     Splenic blood flow in cirrhosis and portal hypertension, Clin
     Sci 34:441 (1968).

24. A. M. Peters, M. J. Walport, R. N. Bell, J. P. Lavender, The
     measurement of splenic blood flow and intrasplenic platelet
     transit time in man using autologous Indium labeled platelets: a
     comparison of compartmental analysis with alternative analytical
     techniques, J Nucl Med, (in press).

25. A. M. Peters, M. J. Walport, K. B. Elkon, H. J. Reavy, P. P.
     Ferjencik, J. P. Lavender, G. R. V. Hughes, The comparative
     blood clearance kinetics of modified radiolabeled erythrocytes,
     Clin Sci (in press).

26. L. Weiss, A scanning electron microscopic study of the spleen,
     Blood 43:665 (1974).

27. N. R. Schulman, S. P. Watkins, S. B. Itscoitz and A. B. Students,
     Evidence that the spleen retains the youngest and hemostatically
     most effective platelets, Trans Assoc Am Phys 81:302 (1968).

28. J. F. Martin, D. N. Slater, E. A. Trowbridge, Abnormal intrapul-
     monary platelet production: a possible cause of vascular and
     lung disease, Lancet 1:793 (1983).

29. C. B. Thompson, K. A. Eaton, S. M. Princiotta, C. A. Ruskin, C. R.
     Valeri, Size dependent platelet subpopulations: relationship to
     platelet volume to ultrastructure, enzymatic activity and
     function, Br J Haematol 50:509 (1982).

30. I. Branehog, Platelet kinetics in idiopathic thrombocytopaenic
     purpura before and at different times after splenectomy, Br J
     Haematol 29:413 (1975).

31. L. A. Harker, Thrombokinetics in idiopathic thrombocytopaenic purpura, Br J Haematol 19:95 (1970).

32. I. Branehog, A. Weinfeld, B. Roos, The exchangeable splenic platelet pool studied with epinephrine infusion in idiopathic thrombocytopaenic purpura and in patients with splenomegaly, Br J Haematol 25:239 (1973).

33. D. L. Williams, Improvement in quantitative data analysis by numerical deconvolution techniques, J Nucl Med 20:568 (1979).

34. A. M. Peters, R. T. Mathie, M. J. Walport, H. J. Reavy, R. N. Bell, J. P. Lavender, Measurement of splenic blood flow in the anaesthetised dog using electromagnetic flowmetry and indium labeled platelets, Cardiovasc Res, (in press).

35. J. R. Stratton, B. L. Thiele, J. L. Ritchie, Platelet deposition on dacron aortic bifuriation grafts in man: quantitation with Indium-111 platelet imaging, Circulation 66:1287 (1982).

36. M. Goldman, H. C. Norcott, R. J. Hawker, Z. Drolc, C. N. McCollum, Platelet accumulation on mature dacron grafts in man, Br J Surg 69:S38 (1982).

37. K. K. Wu and J. C. Hoak, A new method for the quantitative detection of platelet aggregates in patients with arterial insufficiency, Lancet 11:924.

38. Y-C Chen and K. K. Wu, A comparison of methods for the study of platelet hyperfunction in thromboembolic disorders, Br J Haematol 46:263 (1980).

39. J. Mehta, P. Mehta, C. J. Pepine, Platelet aggregation in aortic and coronary venous blood in patients with and without coronary disease, Circulation 58:881 (1978).

40. A. Kallinikos-Maniakas, Megakaryocytes and platelets in central venous and arterial blood, Acta Haematol 42:330 (1969).

41. S. T. Chandler, J. A. C. Buckels, Z. Drolc, R. J. Hawker, A. D. Barnes, C. N. McCollum, A bedside technique for the diagnosis of acute rejection in renal transplants using In-111-platelets, in: "Proceedings of the Third World Congress of Nuclear Medicine and Radiology", C. Raynaud ed., 1575-1578, Perganon Press, Paris (1982).

42. E. A. Murphy and E. A. Francis, The estimation of blood platelet survival II. The multiple hit model, Thromb Diath Haemorrh 25:53 (1971).

43. J. N. Mills, The life span of the erythrocyte, J Physiol 105:16P (1946).

44. A. C. Dornhorst, The interpretation of red cell survival curves, Blood 6:1284 (1961).

45. International Committee for Standardization in Haematology, Recommended methods for radioisotope platelet survival studies, Blood 50:1137 (1977).

46. E. A. Trowbridge and J. F. Martin, A biological approach to the platelet survival curve with criticism of previous interpretations, Phys Med Biol, (in press).

47. E. Dassin, Y. Najean, O. Poirier, P. Passa and D. Bensoussan, In vivo platelet kinetics in 31 diabetic patients: Correlation with the degree of vascular impairment, Thromb Haemost 40:83 (1978).

48. S. R. Hanson, S. J. Slichter, L. A. Harker, The platelet require-
     ment for normal haemostasis in man, Thromb Haemost 50:189
     (1983).
49. L. A. Harker, S. R. Hanson, T. R. Kirkman, Experimental arterial
     thrombembolism in baboons, J Clin Invest 64:559 (1979).
50. H. K. Watson and C. A. Ludlam, Survival of In-111 platelet
     subpopulations of varying density in normal and post-
     splenectomised subjects, Thromb Haemost 50:213 (1983).
51. R. J. Hawker, C. E. Hall, M. Goldman, C. N. McCollum, In-111 loss
     from platelets by in-vitro and ex-vivo manipulation, in: P. Cox
     ed., Mastinus and Nijhoff, Netherlands (in press).
52. D. J. Doyle, C. N. Chesterman, J. F. Cade, J. R. McGready, G. L.
     Rennie, F. J. Morgan, Plasma concentration of platelet specific
     proteins correlated with platelet survival, Blood 55:82 (1980).
53. J. Y. Leung, J. Dykes, M. Goldman, R. J. Hawker, C. N. McCollum,
     How should platelet survival be measured? Thromb Haemost 50:217
     (1983).
54. A. duP. Heyns, M. G. Lotter, P. N. Badenhorst, O. R. Van Reenen, H.
     Pieters, P. C. Minnaar, Kinetics and fate of Indium-111-oxine
     labeled platelets in asplenic subjects, Thromb Haemost
     44:100 (1980).
55. A. M. Peters, I. Klonizakis, J. P. Lavender, S. M. Lewis, Elution
     of Indium-111 from reticuloendothelial cells, J Clin Pathol
     35:507 (1982).
56. B. K. Gunson, C. E. Hall, R. J. Hawker, W. A. Jurewicz,
     Extravasation of Indium-111 labeled platelets, Thromb Hemost
     50:15 (1983).
57. R. H. Aster, Studies of the fate of platelets in rats and man,
     Blood 34:117 (1969).
58. C. A. Ries and D. C. Price, Cr-51 platelet kinetics in thrombo-
     cytopaenia: correlation between splenic sequestration of
     platelets and response to splenectomy, Ann Intern Med 80:702
     (1974).
59. A. duP. Heyns, M. G. Lotter, H. Pieters, F. H. Pauw, P. N.
     Badenhorst, P. Wessels, P. C. Minnaar, A quantitative study of
     Indium-111-oxine platelet kinetics in acute and chronic renal
     transplant rejection, Clin Nephrol 18:174 (1982).
60. A. duP. Heyns, M. G. Lotter, H. F. Kotze, H. Pieters, P. Wessels,
     Quantification of in vivo distribution of platelets labeled with
     Indium-111 oxine, J Nucl Med 23:943 (1982).
61. R. J. Hawker, C. E. Hall, B. K. Gunson, S. T. Chandler, J. D.
     Kelly, Distribution of 114m-In labeled platelets: implications
     to radiation dosimetry and mechanism of platelet destruction,
     Thromb Haemost, 50:38 (1983).
62. U. Scheffel, R. Hill-Zobel, M. F. Tsan, Quantification of in vivo
     distribution of platelets labeled with indium-111-oxine, J Nucl
     Med 23:944 (1982).
63. R. H. Aster and W. R. Keene, Sites of platelet destruction in
     idiopathic thrombocytopenic purpura, Br J Haematol 16:61 (1969).

64. P. J. Toghill, S. Green, R. Ferguson, Platelet dynamics in chronic liver disease with special reference to the role of the spleen, J Clin Pathol 30:367 (1977).

65. Y. Najean and N. Ardaillou, The sequestration sites of platelets in idiopathic thrombocytopenic purpura: its correlation with the results of splenectomy, Br J Haematol 21:153 (1971).

66. M. Kotilainen, Platelet kinetics in normal subjects and in haematological disorders, with special reference to thrombocytopenia and the role of the spleen, Scand J Haematol, supplement 5 (1969).

67. A. duP. Heyns, M. G. Lotter, P. N. Badenhorst, F. deKock, H. Pieters, C. Herbst, O. R. Van Reenen, H. Kotze, P. C. Minnaar, Kinetics and sites of destruction of 111-indium oxine labeled platelets in idiopathic thrombocytopenic purpura: a quantitative study, Am J Haematol 12:167 (1982).

68. H. H. Davis, W. A. Heaton, B. A. Siegel, C. J. Mathias, J. H. Joist, L. A. Sherman, M. J. Welch, Scintigraphic detection of arteriosclerotic lesions and venous thrombi in man by Indium-111 labeled autologous platelets, Lancet 1:1185 (1978).

69. A. Fenech, A. Nicholls, F. W. Smith, Indium labeled platelets in the diagnosis of renal transplant rejection: preliminary findings, Br J Radiol 54:325 (1981).

70. N. Smith, S. Chandler, R. J. Hawker, L. M. Hawker, A. D. Barnes, Indium labeled autologous platelets as diagnostic aid after renal transplantation, Lancet II:1241 (1979).

71. C. Leithner, H. Sinzinger, M. Schwarz, Treatment of chronic kidney transplant rejection with prostacyclin; reduction of platelet deposition in the transplant, prolongation of platelet survival and improvement of transplant function, Prostaglandins 22:783 (1981).

72. M. D. Trip, M. R. Hardeman, E. A. Van Royen, C. A. Viser and J. Vreeken, Accumulation of In-111 labeled thrombocytes in left ventricular and aortic aneurysms, Thromb Haemost 50:334 (1983).

73. H. Pieters, A. duP. Heyns, P. Wessels, M. G. Lotter, M. Locke, P.N. Badenhorst, Kinetics and Indium-111 oxinate labeled platelets in familial hypercholesterolaemia, Thromb Haemost 50:137 (1983).

74. M. Goldman, H. C. Norcott, R. J. Hawker, C. Hail, Z. Drolc and C. N. McCollum, Femoropopliteal by-pass grafts - an isotope technique allowing in vivo comparison of thrombogenicity, Br J Surg 69:380 (1982).

75. W. J. Powers and B. A. Siegel, Thrombus imaging with Indium-111 platelets, Sem Thromb Haemost 9:115 (1983).

IN VIVO QUANTIFICATION AND PLATELET SURVIVAL STUDIES:

WHAT IS THE BEST APPROACH?

A. du P. Heyns

M.R.C. Blood Platelet Research Unit
University of the Orange Free State
P.O. Box 339(G2)
Bloemfontein 9300
South Africa

INTRODUCTION

Although the role of blood platelets in health and disease may be assessed with in vitro methods such as platelet aggregometry, these have limited value and have been found wanting in the clinical situation, except in a few well-defined diseases. Recently investigators have relied increasingly on in vivo measurements of platelet behaviour. For many years the determination of platelet survival time was regarded as the best determinant of in vivo platelet behavior. It was also accepted that this was best done with sodium-dichromate as platelet label (1). In an important systematic investigation of alternative gamma-emitting radionuclides, Thakur(2) recommended In-111 as a suitable and preferable platelet label. The discovery of oxine and other chelates, has now established the superiority of In-111-complexes over Cr-51 as a platelet label.

The major advantages of In-111 are, first, that only a small volume of blood is necessary for adequate platelet studies, and, second that the modern technology for external imaging of in vivo platelet distribution may be exploited. The scintillation camera is readily accessible and is an excellent imaging device widely used for static organ visualization. However, the scintillation camera with a dedicated computer with dynamic image analysis capability has opened exciting new possibilities for the in vivo study of the behavior of platelets and other cellular blood elements. These techniques have already been successfully applied in studies of normal platelet kinetics and in the detection of platelet deposition in several disease states(3,4).

In this paper I will discuss briefly the requirements of an ideal
platelet label and focus on the measurement of platelet survival and
platelet imaging. I will outline the different techniques that are
available and, although it is obviously a complex and time-consuming
technique, recommend a relatively simple approach to the study of
platelet kinetics. I will also point out some of the important
pitfalls which may affect techniques and the interpretation of data
and introduce some new, still experimental, approaches to the study of
in vivo platelet behavior.

IDEAL PLATELET LABEL

There is as yet no ideal platelet label. The minimum
requirements are: a low rate of elution; no significant reutilization
of the label after destruction of the labeled cell; acceptable
physical characteristics; low radiation dose and general availability
if the label is a radionuclide; and ease of the labeling procedure
with no damage to the platelet product.

Chromium-51-sodium chromate has been the label most frequently
used in platelet studies, but its long physical half-life and only 7%
abundance of its 320 keV gamma photon makes it far from ideal. Also,
although a number of nonisotope methods have been used for the study
of platelet survival, these have been of limited value. In-111
complexed to oxine, or other chelates such as Merc (Mercapto pyridine-
N-oxide) is lipid soluble, has a high labeling efficiency for platelets, acceptable radiation dose for human platelet kinetic studies,
and has abundant gamma photons (90% and 94% for the 173 and 247 keV
energies respectively). The high flux of radiations are easily colli-
mated and may thus be efficiently detected by routinely used scintillation cameras. Indium-111 is thus now widely regarded as the platelet label of choice.

Cell separation techniques have already been discussed at this
meeting, but I would like to again emphasize a few important facets.
Platelets are usually isolated by differential centrifugation. This
procedure is acceptable, and by the modification of washing the red
cell layer repeatedly, it is possible to harvest more then 90% of the
platelet population. This is especially important for studies with
autologous platelets in the face of severe thrombocytopenia(5). It is
important to do the final washing step for the removal of free and
plasma-bound In-111 with plasma, and to resuspend labeled platelets in
plasma for about 1 hour before reinjection into the recipient. This
minimizes the "collection injury." I am not sure that results of in
vitro platelet aggregation tests reflect platelet viability, and our
group prefers to assess this by the recovery rate of labeled platelets
in the circulation and by the shape of the disappearance curve of
labeled platelets from the peripheral blood. In normal subjects this
curve should fit a linear function best.

THE COLLECTION INJURY

    In normal subjects, the platelet survival curve may be divided
into three, usually distinct, phases. Initially, platelet activity in
the blood may reach a plateau, but on the first day after reinjection
of labeled platelets the radionuclide levels in the blood
characteristically remain unstable and variable. Thereafter, blood
radioactivity decreases as a linear function for about 8 days.
"Tailing" of the curve, reflected by the remaining 5% of the
radioactivity present on day 8, then becomes evident.

    The early, unstable, phase seen on the first day after
reinjection of platelets may be ascribed to the "collection injury," a
term referring to the damage caused by the isolation and labeling of
the platelets. This varies considerably in intensity from subject to
subject and may reflect the gentleness of the labeling procedure and
the skill of the technician. We have observed an unexplained phenomenon in a patient with systemic lupus erythematosus studied with
autologous In-111-oxine labeled platelets. Platelets accumulated in
the lungs immediately after reinjection of labeled platelets. This
accumulation was transient, and the following day these platelets had
been mobilized into the circulation. Platelet survival was not
affected as shown by results of a study repeated a week later. This
seems to be a variant of a severe "collection injury," and demonstrates the unique platelet de-aggregating ability of the lung.

    There is, however, an almost inevitable collection injury seen in
all the patients we have studied with dynamic in vivo platelet kinetic
studies (3). There is a transient accumulation of platelets in the
liver evident shortly after reinjection of labeled platelets. This
hepatic radioactivity reaches a maximum at about 15 minutes and then
decreases to reach equilibrium 60 - 90 minutes after reinjection.

    The collection injury must clearly influence the calculation of
platelet recovery in the circulation if this is determined by
measurement of the radioactivity of a 5 minute blood sample.
Estimates of platelet survival, which include the points of blood
radioactivity of the first 60 minutes, must also be affected. The
suggestion that platelet survival may be accurately determined by
extrapolation of the radioactivity of only the first few data points
must also be regarded with caution, and results must be carefully
interpreted.

    Recently, Badenhorst et al(6) investigated the effect of the
platelet collection injury in an animal model. The survival, in vivo
distribution after equilibrium, and sites of destruction of
In-111-labeled canine platelets were compared with those of
In-111-labeled platelets of a matched dog previously injected with the
same In-111 platelet preparation. The matched dog thus served as a
biological "filter" for In-111 platelets with a collection injury.

The life-spans of the "filtered" and "unfiltered" platelets were
similar, although the shape of the survival curve of the filtered
platelets was slightly more linear than that of the unfiltered
platelets. The splenic pool size and sites of sequestration of
labeled platelets were also similar in the pairs of dogs. It was
concluded that the isolation and labeling procedures used in this
study did not appreciably injure platelets and that the collection
injury is largely reversible.

Schmidt has also demonstrated that labeled rabbit platelets are
not altered markedly by blood collection if assessed by transmission
electron microscopy and morphometric analysis. Furthermore, these
platelets were functionally effective and shortened the prolonged
bleeding time of thrombocytopenic rabbits.

These studies all indicate that the collection injury associated
with platelet labeling is minimal, and thus findings and results of in
vivo kinetic studies may be interpreted with some confidence. I would
like to emphasize that all in vivo platelet studies should always be
performed with adequate and rigid controls. Results are not to be
relied upon if the platelet recovery in the circulation of the
recipient, the in vitro aggregation response of labeled platelets as
measured in the aggregometer, and the shape of the platelet survival
curve are not reported and taken into account. Obviously, a
scintillation camera whole-body image demonstrating normal in vivo
platelet distribution is a refined and elegant quality-control
technique.

ANALYSIS OF PLATELET SURVIVAL CURVES

Although several other methods with different radionuclides and
also nonisotopic methods have been used to estimate platelet survival
and turnover, In this paper I will concentrate on our data with
In-111-oxine as a platelet label.

It should always be kept in mind that the determination of
platelet survival is in all instances dependent on the ratio of the
Y-intercept.

MATHEMATICAL MODELS FOR THE ANALYSIS OF PLATELET SURVIVAL CURVES

Linear Function

This is appropriate only for the calculation of platelet survival
in normal subjects and reflects an age-dependent destruction and
removal of platelets from the circulation.(1)

Exponential Function

This is appropriate only when platelets are removed from the circulation in a random fashion, independent of age.(1)

Weighted Mean of the Linear and Exponential Functions

This is a mathematical procedure whereby the relative weights of the linear and exponential survival times are determined by the sum of the squares of the deviations of data points from the fitted functions. This method has the advantage of simplicity, is not committed to a mathematical model, but gives no information on the shape of the curve (1).

Multiple Hit (Gamma Function) Model

This is probably the most widely used of the various mathematical models. It is based on physiological assumptions, some of which may not be valid. The mean platelet survival is calculated as a computed number of "hits" multiplied by the waiting time between these hits. A "hit" is defined as an insult imposed on the platelet by its environment and may vary in intensity. The number of hits may also give some indication of the shape of the platelet survival curve. Thus, a single hit is found in a simple exponential curve, whereas a linear survival is reflected by an infinite number of hits. If one estimates the number of hits, rather than consider this number as fixed, one may use this parameter as an indication of the shape of the survival curve (7).

Dornhorst Model

In this model it is assumed that platelets have a determined life-span in the circulation and that they are at risk of destruction by external mechanisms. Platelet survival is calculated from these parameters. The model may also conveniently be extended to give the mean age of the circulating platelets and for calculation of the production rate of platelets(8).

Alpha-order Model

This model has been employed in pharmacokinetic studies. It is appropriate for estimation of platelet survival time in the range from a linear to an exponential function of curve fitting. The formula is derived from a general compartment, non-linear differential model(9):

$$dC/dt = kC^\alpha$$

$$\text{where } C = \text{concentration}$$
$$t = \text{time}$$
$$k = \text{rate parameter } (k < 0)$$
$$\alpha = \text{order parameter } (k \geq 0)$$

The shape of the survival curve may be conveniently estimated by calculating alpha in the above equation. Zero order ($\alpha = 0$), and first order ($\alpha = 1$) kinetics are particular cases.

Other Models

Several other mathematical models have been used for the calculation of platelet survival. The polynomial model was extensively investigated by Paulus,[10] and Meuleman in 1980[11] proposed another model. We have also investigated these models in a manner similar to that which I will present, and analysis of our data indicated that the results of these mathematical manipulations also corresponds closely to that of the multiple hit, Dornhorst, and alpha-order models.

Results of Curve Fitting with Mathematical Models

A study was performed to assess and compare results obtained with several mathematical models commonly used for the estimation of platelet survival in different clinical situations. No systematic studies to compare the efficacy of these models have been performed to my knowledge. The patients were divided into three groups: those with platelet survival within the normal range (180-240h); those with an intermediate platelet life-span (120-180h); and patients with a platelet life-span less than 120h. We investigated 55 patients, and the three groups comprised 19, 13, and 23 patients respectively. Platelet survival was estimated with six mathematical models: linear function, exponential function, weighted mean function, multiple hit model, Dornhorst model, and alpha-order model. The mathematical models were fitted to the survival curve data by at least squares nonlinear Newton iteration method. Iteration was terminated when fractional changes in regression constants were less than 0.0001. The calculations were performed on a computer program written for a Medical Data system A2 imaging system. Curve fitting was with those data points lying between the stage after recovery from the collection injury, and when blood sample In-111-radioactivity was less than 10% that of the maximum curve value. The goodness of fit of the different models to the data was evaluated by calculating the standard deviation of the data points around the fitted line as a percentage of the regression curve Y-intercept. One-way variance analysis was performed to determine if the differences between the goodness of fit and survival times for each of the models investigated were statistically significant.

The estimated platelet survival with the different models is illustrated in Table I.

It is clear that the one-way analysis of variance demonstrated no significant differences in the goodness of fit of the weighted mean,

Table 1. Estimated Platelet Survival (hours $\pm$ SD)

	Linear	Exponential	Weighted Mean	Multiple Hit	Dornhorst	Alpha-Order
			MODEL			
Reference group Variance analysis	234 % 21 S	130 % 25 S	216 % 19 NS	223 % 16 NS	211 % 27 NS	220 % 26 NS
All patients Variance analysis	191 % 55 S	104 % 36 S	150 % 59 NS	155 % 67 NS	147 % 67 NS	142 % 70 NS

S = Statistically significant   NS = Statistically not significant

multiple hit, Dornhorst, and alpha-order functions. The simple
expedience of fitting the data to either a linear or an exponential
function was, however, inadequate.

The standard deviation of the data points around the fitted line
was also not significantly different for the weighted mean, multiple
hit, Dornhorst, and alpha-order models, as is illustrated in Table 2.

We conclude that there is no statistical significant difference
in platelet survival if calculated with the above models. Since the
models were evaluated over a wide spectrum of platelet survival times,
the results suggested that any of these models would be appropriate
for the analysis of platelet survival curve data. We thus recommend
that if computer facilities are not available, the weighted mean
method may be quite suitable. Those with facilities for sophisticated
computer analysis may use either the multiple hit, Dornhorst,
alpha-order, polynomial, or Meuleman models.

I must again stress that since the survival time is calculated by
a ratio of the Y-intercept and the initial slope of the regression
curve, frequent and accurate measurements must be obtained during the
early part of the study. It is also apt at this stage to point out
that technical factors such as accurate radioactivity counts of blood
samples with correction for geometry, accurate sample volume
determinations, adequate blood sample technique, and appropriate
spacing of time intervals between the blood samples will certainly
have a far greater influence on estimates of platelet survival than
the choice of the mathematical model for analysis of data and curve
fitting. Also keep in mind that determination of platelet survival in
patients with extremely rapid clearance of platelets from the
circulation, such as is seen in immune thrombocytopenic purpura, is
complicated by the fact that platelet survival is in many instances
shorter than the time taken for platelet redistribution in the various
body compartments to reach a steady state.

Platelet Survival and Shape of the Survival Curve

Since there are difficulties associated with the analysis of the
platelet survival curve and determination of platelet life-span, we
have also attempted to gain additional information from this data by
analyzing the shape of the platelet survival curve. It must, however,
be emphasized that this is still an experimental approach that has not
been validated.

It should be recognized that platelet survival is to some extent
dependent on the shape of the clearance curve. It thus seems
appropriate to analyze the shape of the curve. We have evaluated the
curve shape as follows. Firstly, by the number of hits generated in
the multiple hit model. Platelet removal by a single hit presents

Table 2. Standard Deviation (SD) of Data Points around the Fitted Line (n = 55)

	Linear	Exponential	MODEL Weighted Mean	Multiple Hit	Dornhorst	Alpha-Order
Mean SD	6, 0	4, 3	3, 6	3, 1	3, 7	3, 4
Variance Analysis	S	S	NS	NS	NS	NS

S = Statistically significant   NS = Statistically not significant

random exponential destruction (7). As platelet removal becomes more
and more age-dependent and more linear, the number of hits increases
and eventually approaches infinity. Secondly, the alpha value of the
alpha-order model also represents the shape of the curve(9). In this
model when $\alpha = 0$ a linear, and if $\alpha = 1$, an exponential survival curve
is obtained. The values between $\alpha = 0$ and $\alpha = 1$ will thus reflect
varying degrees of curvilinearity. Thirdly, a simple mathematical
analysis of the shape of the curve has been proposed by Lötter. This
so-called variance curve shape factor has the advantage that it is
independent of the mathematical model used for the analysis of
platelet survival.

The variance curve shape factor is calculated as follows:

$$\text{Variance curve shape factor} = \frac{Sp - Sr}{(Sp-Sr) + (Se-Sr)}$$

where Sr = sum of the squares of the deviations of
the data points from a model fitted
regression curve

Sp = sum of the squares of the deviations of
the data points from a linear fitted
regression curve

Se = sum of the squares of the deviations of
the data points from a exponential
fitted regression curve

The value of the variance curve shape factor varies between 0 for
fully linear and 1 for exponential platelet survival. We have tested
the statistical differences between the variance shape factor and
platelet survival estimated with each of the mathematical models by
one way analysis of variance. The variance shape factor was also
correlated with the number of hits, the alpha value, and platelet
survival.

The variance shape factor correlated well with the alpha value of
the alpha-order model, and the results indicated that these two
parameters adequately reflect the shape of the survival curve. The
variance shape factor and the number of hits of the multiple hit model
were, however, not linearly related. The variance shape factor and
the alpha value correlated well with platelet survival in the range of
120h to 260h. If platelet survival is less than 120h, the survival
curve is always exponential, as was reflected by a variance shape
factor and alpha value close to unity.

These results suggest that the determination of the variance
curve shape factor and alpha of the alpha-order model (and possibly
even the number of hits of the multiple hit model) warrants further

investigation in different clinical situations. This may be an
appropriate and relatively sensitive method to obtain additional
information on platelet consumption in various disease states before
and after therapeutic intervention with antiplatelet drugs.

PLATELET IMAGING AND QUANTIFICATION OF

IN VIVO DISTRIBUTION OF PLATELETS

     The physical properties of In-111 allow external imaging of its
in vivo distribution and areas of localization. This image is
detected by an imaging system with a recording device. The two most
important classes of instruments used for this purpose are the
rectilinear scanner and the scintillation camera. The rectilinear
scanner has only rarely been used in In-111 platelet studies, and its
use will therefore not be discussed.

     Imaging is now mostly performed with the Anger scintillation
camera. Briefly, this device consists of a large sodium iodide
crystal viewed by an array of photomultiplier tubes. The gamma
photons emitted from the object to be imaged are directed with a
collimator placed at the face of the detector. A collimator typically
consists of multiple parallel channels (or holes), separated by lead
septa. Thus, only those photons parallel in direction to the long
axes of the collimator channels will reach the detector. The gamma
photon interacts with the sodium iodide crystal, and visible light is
emitted. A portion of this light is collected and converted into an
electronic signal by the photomultiplier tube. The signals from all
the phototubes are summed and the resulting signal analyzed by a
pulse-height spectrometer. The quantity of light emitted by the
crystal is proportional to the energy transferred to the crystal by
the gamma photon, and it is, therefore, also possible to derive
quantitative information with the scintillation camera. This is
typically achieved with a dedicated computer interfaced to the camera.
This computer allows one to store the acquired images, and the data
may subsequently be manipulated so as to obtain an ideal image and to
extract the maximum useful information.

     Several limitations of the in vivo quantification of
In-111-labeled platelets with a scintillation camera and a
computer-assisted image analysis system should be clearly recognized
at the outset. The scintillation camera does not provide an image
that is a perfect representation of the object being imaged. The
resolution capability of the instrument is limited, and with the best
available scintillation camera the intrinsic spatial resolution is 2
to 4 mm. The collimator naturally also leads to further loss of
information, and this may halve the resolution capability of the total
system. Also, since radioactive decay occurs at random, there is
inherent statistical variability of all the measurements. This tends

to blur the distinction between two regions with different
radioactivities. It may thus be quite difficult, or even impossible,
to delineate a region of interest perfectly with no personal bias.
Computer programs for selection of regions of interest may be an
improvement. It should also be recognized that a scintillation camera
reflects a three-dimensional object as a two-dimensional image.
Tissue overlying or underlying a region of interest are inevitably
superimposed, and the contrast between the region of interest and its
surrounds is markedly reduced.

A major problem is that tissue is not transparent to photons, and
quantification of organ In-111 radioactivity must take the influence
of attentuation and geometry into account. This is illustrated when
only an anterior image is acquired: The radioactivity of In-111
platelets in the spleen, which is posteriorly situated, may be
underestimated by as much as 15%. Similarly, a posterior image will
significantly underestimate hepatic radioactivity.

These technical problems may be overcome by various experimental
approaches, but one should recognize that despite these correction
factors it is prudent to consider conventional scintigraphy as a
semiquantitative method. I would like to emphasize that all
investigators employing such quantitative in vivo scintigraphic
methods should validate their methodology by independent in vitro
measurements (12). Unfortunately, this has not been done often in
studies with In-111-labeled platelets.

The geometrical mean method that we recommend for in vivo
quantification of In-111 platelet distribution is practical and
relatively simple. A whole body image is acquired anteriorly and
posteriorly. Radiation within organs, or regions of interest, is also
determined by anterior and posterior measurements of radioactivity and
corrected for attentuation. This method measures radioactivity in the
organ as if the organ were situated in the anterior-posterior midplane
of the body. Geometrical mean organ radioactivity is expressed as a
percentage of geometrical mean whole body activity, thus overcoming
problems inherent in measuring absolute radioactivity of the organ.
Although this method may be subject to error, experimental validation
has indicated that organ radioactivity may be quantitated with an
accuracy better than 6%(12). The method is somewhat time consuming
but is easy to perform and reproduciable.

Since the method is described in detail elsewhere(13), I will
limit myself to a discussion of some relevant important facets. The
preferred gamma camera should have a large field of view, a crystal
thickness of 9.5 or 12.7 mm, and multiple spectrometers to permit
simultaneous detection of both the 173 and 247 keV gamma photons. The
use of dual pulse-height analysers will give nearly double the
counting rate obtained with a single spectrometer set for only one
photopeak.

The choice of the collimator is also important. We recommend the medium energy collimator optimized for imaging Gallium-67. This commercially available collimator is an acceptable compromise between high-sensitivity and high-resolution capability.

Several types of dedicated computer systems for dynamic image analysis are commercially available. The key component of the computer hardware is the central processing unit. This unit is unlikely to be a limiting factor since the dynamic imaging rate of platelet kinetic studies is not high. Data are fed into the computer and are quickly stored in a fast memory. The data are then more permanently stored: The single, removable magnetic disc is best from the point of view of cost, capacity, and speed. The data are displayed in TV video format. The system has the capability to expand 64 x 64 or 128 x 128 images into larger matrix sizes. A zoom facility is very useful. The image is normally displayed in shades of gray, but a color facility which may be purchased as an optional extra is very useful.

The utility programes of the computer software can handle general data acquisition, display of images, region of interest selection, and routine data analysis. Ideally, the raw data should be stored on the magnetic disk of the computer system. The computer permits subtraction of background activity, smoothing of the image, and contrast enhancement. Groups of images may be generated, replayed at various gray shade levels or in color, and the optimum method of display selected. The diagnostic capabilities of the system are thus greatly improved. The image displayed may be direct or retrospective and may be done with the scintillation camera or the computer. Several different options are available as standard or accessory equipment in commercially available systems. We use the persistence oscilliscope unit of the gamma camera for interactive quality control and proper positioning of the patient. We prefer to routinely record the image on an X-ray film. This is done with an optional gamma camera accessory. The quality of the image is high, and lesions are easier to distinguish than on Polaroid film. Retrospective image recording with the computer may be done on X-ray film, in black and white or color, on polaroid film or in 35 mm format.

Static imaging is performed by positioning the patient with the collimator face as near as possible to, and in the correct position relative to, the region of interest. One then accumulates counts over the area. Although 500000 acquisition counts are necessary for an image of excellent quality, this is not practical with In-111-labeled platelets. Counts are therefore accumulated for a fixed time, usually 15 minutes. This time may be increased, and it should be recognized that the minimum counts for an acceptable image are 20000 to 50000. It is not always possible to obtain an ideal image, and practical considerations will limit the sensitivity of the procedure. These static images must generally be obtained at several times after the

injection of In-lll-labeled platelets. Only rarely are lesions
visualized immediately, or within a few hours after reinjection of
labeled platelets. Most lesions are not clearly visible until 24
hours or later. This should be taken into account in the planning of
the protocol for the investigation of thrombotic lesions.

In vivo quantification of In-lll-labeled platelets requires a
scintillation camera with digital imaging processing capabilities. We
divide the procedure into equilibrium and a platelet survival phases.
The equilibrium phase is the dynamic measurement of the accumulation
of radioactivity within selected organs during the first 90 minutes
after reinjection of the In-lll-platelets. The platelet survival
phase is the serial measurement of platelet distribution until the end
of the platelet life-span in the circulation.

Images of only a single projection (usually the anterior view)
are acquired in the equilibrium phase, and the geometrical mean method
for calculation of platelet distribution must therefore be indirectly
applied. The data are normalized by using the geometrical mean method
estimate of In-lll platelet distribution measured on day 1 of the
platelet survival phase. Data are generally represented as time-
radioactivity curves of organs and regions of interest. Since
radioactivity counts are expressed per image element per second, the
curve represents count density related to time. These curves are
stored on magnetic disk and are later normalized when the platelet
survival phase data become available.

In the platelet survival phase the whole body image is usually
acquired with a scintillation camera and bed with a scanning facility.
Regions of interest are selected by extraction from the whole body
image. Organ radioactivity is calculated by the geometrical mean
method and expressed as a percentage of the corrected whole body
radioactivity.

The mean equilibrium, and end of platelet life-span In-lll
radioactivity of the whole body and a few important organs as
determined in normal subjects, is given in Table 3.

The interpretation of static gamma camera images is similar to
that of radiographs. It is important to interpret the image with full
knowledge of relevant clinical and other data, and it may be necessary
to correlate the image with radiographic or angiographic features. A
positive scintigram is a discreet area of increased radioactivity
clearly greater than that of the background blood pool. The activity
of this area must increase with time if compared with a remote
background blood pool such as the ileofemoral vessels. Images
displayed in black and white on Polaroid film allow defects with a 20%
increase of radioactivity to be easily seen; those recorded in
transmission format on X-ray films are somewhat easier to interpret.

Table 3. Equilibrium and Final In-111-Radioactivity Distribution

| | Equilibrium : % Radioactivity | | | | Final : % Radioactivity | | | |
	Heart	Liver	Spleen	Whole Body	Heart	Liver	Spleen	Whole Body
Mean	7.26	9.1	31.1	100	1.9	28.7	35.6	96.6
% 1SD	1.4	1.2	6.1	0	1.2	8.3	9.7	5.2

Organ radioactivity is expressed as a percentage of whole body activity

The image may also be analyzed by a semiquantitative method.
This is done by calculating the ratio between radioactivity in a
region of interest and the background. The size of these two regions
should be similar. Changes in radioactivity may be expressed as a
time-activity curve or the ratio between radioactivity at equilibrium
and at the end of platelet life-span determined. Radioactivity in the
region of interest may also be expressed as radioactivity counts per
pixel per unit time. These methods are not without inherent
difficulties since it is not easy to standardize the selection of
regions of interest and reference regions. In fact, in our experience
visual inspection of the image gives as much information on platelet
deposition as the calculation of the region of interest: reference
region ratio.

An important concept that has recently come to the fore is that
platelet deposition in a thrombus does not necessarily occur in a
predictable and continuously accumulative manner. Thus, in patients
with aortic aneurysms it is clear that not all aneurysms are
hematologically active. In some, circulating platelets are in dynamic
exchange with those on the surface of the thrombus in the aneurysm.
These platelets are eventually sequestered in the reticuloendothelial
system, and after equilibrium is reached, thrombus In-111-activity
remains constant. This finding may have important implications for
the interpretation of platelet survival curves and assessment of
platelet deposition and turnover in thrombi. This phenomenon may also
occur in other disease states, and we have documented similar findings
in a patient with thrombotic thrombocytopenic purpura.

Recently, William Powers and his group (14) at the Mallinckrodt
Institute of Radiology have applied the technique of dual-isotope
scintigraphy to solve some of the problems of quantification of region
of interest In-111-radioactivity. We also have some limited
experience with this method. In brief, images of a second radiotracer
distributed in the circulating blood (Technetium-99m-labeled blood
cells) are also obtained. This technique permits correction for
variability in the size of different regions based on their relative
blood volumes. One may thus determine actual In-111-platelet
deposition at a given site by subtracting the contribution of radio
labeled platelets still circulating in the blood. This is a sensitive
method for the detection of relatively small thrombus activity and is
probably the method of choice if small organs or regions with a
relatively large blood component are examined.

In summary, the geometrical mean method is currently the best
method available for accurate quantification of the distribution of
In-111-labeled platelets in large organs or regions of interest. The
technique has been validated. Since labeled platelets in the
circulating blood cannot be distinguished from those deposited in a
thrombus, this method may be relatively insensitive in certain
situations. The dual-tracer technique may then prove to be the method

of choice. Expressing regional radioactivity simply as a count rate per unit area, or as a ratio between radioactivities in a region of interest and a reference area, is not particularly sensitive and probably no better than examination of images visually.

Platelet Survival Measurement with the Scintillation Camera

Since it is possible to determine accurately region of interest radioactivity with the gamma camera, it is also feasible to generate time-radioactivity curves which will reflect blood In-111-radioactivity. These may then be used for the calculation of platelet survival. We have done this with some success. Results correspond to those of whole blood measurements. The region of interest selected for this purpose must not contain active bone marrow since sequestration of senescent platelets in the reticuloendothelial component of this organ will affect the slope of the disappearance curve. We have found the mid-thigh and mid-calf suitable.

Comparison of the disappearance of In-111-labeled platelets from the whole blood with that of the lower limb, has opened a new approach to the study of platelet deposition in diseased vasculature. If platelets are utilized in areas of atherosclerosis, or if circulating platelet aggregates are trapped in the microvasculature, the slope of the disappearance curve of In-111-radioactivity from the lower limb changes. The life-span of labeled platelets as determined from blood radioactivity, is thus shorter than that measured with the scintillation camera lower limb image analysis.

We have demonstrated this in patients with aortic aneurysms, in aortofemoral prostheses, and in familial hypercholesterolemia. This approach certainly warrants further investigation.

CONCLUSION

Labeling cells with In-111-chelates provide a new approach to the in vivo study of platelet kinetics. Quantitative and semiquantitative imaging techniques are now within the reach of all investigators and have almost become routine in certain centers. The application of these methodologies has led us to new insights into the role of platelets in disease.

Analysis of platelet survival curves is difficult, and optimum determination of platelet life-span probably requires sophisticated computer facilities. These mathematical expediencies should not, however, be allowed to overshadow the crucial importance of blood sample collection and accurate determination of radioactivity of blood samples in the estimation of platelet life-span.

ACKNOWLEDGMENTS

     The contribution of P.N. Badenhorst, M.G. Lotter, M.C. Minnaar, H.F. Kotze, H. Pieters, P. Wessels, and M. Lock, the members of the Blood Platelet Research Unit is gratefully acknowledged: Mrs. E. Herbst provided expert secretarial assistance.

REFERENCES

1. ISCH panel on diagnostic application of radioisotopes in hematology, Recommended methods for radioisotope platelet survival studies, Blood. 50:1137 (1977).
2. M. L. Thakur, M. J. Welch, J. H. Joist, R. E. Coleman, Indium-111 labeled platelets: studies on preparation and evaluation of in vitro and in vivo functions, Thromb Res 9:345 (1976).
3. A. du P. Heyns, M. G. Lotter, P. N. Badenhorst, O. R. van Reenen, H. Pieters, P. C. Minnaar, F. P. Retief, Kinetics, distribution and sites of destruction of Indium-111-labeled human platelets, Br J Haematol 44:269 (1980).
4. J. H. Joist (Guest Editor), Indium labeled platelets, Sem Thromb Hemostas 9:79 (1983).
5. A. du P. Heyns, M. G. Lotter, P. N. Badenhorst, F. de Kock, H. Pieters, C. Herbst, O. R. van Reenen, H. Kotze, P. C. Minnaar, Kinetics and sites of destruction of Indium-111-oxine labeled platelets in idiopathic thrombocytopenic purpura: a quantitative study, Am J Hematol 12:167 (1982).
6. P. N. Badenhorst, M. G. Lotter, A. du P. Heyns, O.R. van Reenen, C. Herbst, H. Pieters, H. F. Kotze, L. J. Duyvene de Wit, P. C. Minnaar, The influence of the "collection injury" on the survival and distribution of Indium-111-labeled platelets, Br J Haematol 52:233 (1982).
7. E. A. Murphy and M. E. Francis, The estimation of blood platelet survival. II. The multiple hit model, Thromb Diath Haemorrh 25:53 (1971).
8. A. C. Dornhorst, The interpretation of red cell survival curves, Blood 6:1284 (1951).
9. T. L. Simon, T. M. Hyers, J. P. Gaston, L. A. Harker, Heparin pharmacokinetics: increased requirements in pulmonary embolism, Br J Haematol 39:111 (1978).

10. J. M. Paulus, Measuring mean life span, mean age, and variance of
     longevity in platelets, in: "Platelet Kinetics," J. M.
     Paulus, ed. Amsterdam, North Holland (1971).
11. D. G. Meuleman, G. M. T. Vogel, S. M. Stulemeyer, H. C. T.
     Moelker, Analysis of platelet survival curves in an arterial
     thrombus model in rats, Thromb Res 20:31 (1980).
12. A. du P. Heyns, M. G. Lotter, H. F. Kotzke, H. Pieters, P.
     Wessels, Quantification of in vivo distribution of platelets
     labeled with Indium-111-oxine, J Nucl Med 23:943 (1982)
13. A. du P. Heyns, M. G. Lotter, P. N. Badenhorst, Platelet imaging,
     in: "Methods in Haematology," L. A. Harker, ed. Churchill
     Livingstone, London (1983).
14. W. J. Powers, C. J. Mathias, K. T. Hopkins, J. B. Rubin, B. A.
     Siegel, M. J. Welch, Dual radiotracer techniques for improved
     scintigraphic detection of thrombi, in: "Proceedings of the
     Third World Congress of Nuclear Medicine and Biology," C.
     Raynaud, ed. Pergamon Press, Paris (1982).

INDIUM-111 PLATELETS IN THROMBOEMBOLISM:

CAN LABELED PLATELETS BE USED TO EVALUATE

ANTITHROMBOTIC THERAPY?

Kenneth M. Moser

Department of Medicine
UCSD School of Medicine
225 Dickinson Street
San Diego, California   92103

INTRODUCTION

Venous and arterial thromboembolism continue to represent a major cause of morbidity and mortality in populations around the world. For several decades, intense investigative interest has focused upon the major forms of thromboembolism: coronary thrombosis with its potential consequence of myocardial infarction; cerebral thromboembolism with its consequences of transient ischemic attacks and/or cerebral infarction; and venous thrombosis and its potentially lethal complication, pulmonary embolism. Add to these major problems those arterial and venous thromboembolic events which can cause impairment in other body organs—the kidney, the lower extremities, the bowel—and the clinical magnitude of the problem is clear.

But in addition to these old, established problems, medical progress has generated a whole new subset of thrombo-embolic complications involving such procedures and devices as coronary artery and other cardiac grafts; arterio-venous devices implanted for renal dialysis; cardiac valve prostheses; the use of cardiopulmonary bypass; aortic and other arterial grafts; and carotid thromboendarterectomy. New procedures such as coronary artery angioplasty raise special issues. Even social change has escalated the risk, with intravenous drug abuse inducing a sharp increase in septic thromboembolism arising from the cardiac valves and/or infected, thrombosed veins.

Each type of thromboembolism raises unique questions not shared by the others. However, virtually all do share certain common features which serve as barriers to efforts to our understanding of their pathogenesis and, at the clinical level, to developing effective strategies for prevention or treatment. A major barrier has been our poor diagnostic capabilities. In each form of human thromboembolism, some degree of diagnostic uncertainty exists; specifically, uncertainty as to (a) whether thrombosis or embolism has caused the symptoms and signs observed; and (b) the extent, age, and activity of the thromboembolic material. Furthermore, even when relatively reliable invasive diagnostic methods exist, there is often an understandable reluctance to apply these consistently to patients because of the risks involved. Clearly, this diagnostic limitation is pivotal. Unless one knows that thromboembolism is present, pathogenetic insights are difficult to achieve, and evaluation of therapy is problematic.

Furthermore, beyond the diagnosis and treatment of the acute episode, there is need to define the natural history of these thromboembolic events. If the natural history of the disorder is not clear, then the impact of therapy must remain obscure. Therefore, methods suitable for long-term monitoring in patients with documented thromboembolism have been needed.

Another conceptual barrier has been that posed by extrapolating data derived from animal models or in vitro observations to human disease. While there is no question that invaluable insights can be provided by such data, the ultimate resolution of pathogenetic and therapeutic questions requires investigations in humans. Despite the complexity of such investigations, and the admitted untidiness involved in studying sick people, human studies provide the ultimate test. Therefore, any techniques evolved must ultimately be proven applicable to patients.

Radionuclide approaches have been particularly attractive as a means for overcoming these barriers. They are essentially noninvasive, often can be repeated and, in these days of portable imaging devices, can be performed at the bedside if necessary. Over the years, several radionuclide techniques have emerged which have been useful in the diagnosis and management of certain thromboembolic disorders, including the I-125-fibrinogen (I-125-F) leg scan; perfusion and ventilation lung scans; brain scanning and Tl-201 imaging of the heart (1-7). Each of these approaches has enhanced our ability to diagnose patients and evaluate preventive treatment approaches. Unfortunately, each also has certain limitations which have not allowed exploration of important questions. For example, the radiolabeled fibrinogen (RLF) test takes 24 hours to provide useful data and does not allow "imaging" of thrombi. Further, because of the pelvic and pulmonary blood pool, neither thrombi in the upper thigh or pelvis nor pulmonary emboli can be detected.

Lung perfusion scans reliably demonstrate areas of absent or reduced pulmonary blood flow, and ventilation scans can define whether ventilation to such areas is normal. But neither method can document that an embolus is present. Similarly, a Tl-201 myocardial scan can demonstrate areas of absent flow or ischemia; but, again, this procedure cannot demonstrate that thrombosis is the primary, or a contributory, cause.

Thus, while existing procedures have represented a substantial advance, a need has existed for an approach that can provide more definitive information about thromboembolic behavior. Particularly needed are methods which allow assessment of the dynamic behavior of thromboemboli; i.e., whether they are "active," growing, or regressing. In developing such an approach, there are, at the moment, only two viable candidates: (a) direct visualization of the site of interest, using fiberoptic devices; and (b) external imaging of radiolabeled components of thromboemboli.

The first approach is invasive and just now under investigation in man (8). It may prove useful in acute diagnosis but probably will not be a realistic approach, in man, to the long-term observation of thrombus behavior.

However, imaging of the components of thromboemboli does appear quite promising as a method for both diagnosis and monitoring. The two components of special promise are, of course, the two that comprise the bulk of any thromboembolus; namely, fibrinogen and platelets. Of these two, platelets have certain advantages. Most arterial thrombi are so-called "white thrombi," composed chiefly of platelets rather than of fibrin. Thus, in seeking a marker for arterial thrombosis, platelets are a clear first choice. Even in venous thrombosis, platelets appear to be involved not only in pathogenesis, but also in thrombus growth.

The potential of radiolabeled platelets for investigations of thromboembolism, while conceptually obvious, did not become practical until the In-111-oxine method was developed and validated (9). The near ideal characteristics of this label in terms of its gamma imaging potential and half-life were notable from the outset. Over a period of some years, multiple investigators have contributed to the current flowering of research with In-111-labeled platelets by demonstrating that platelets from multiple species could be labeled without major alteration of survival time, of organ distribution, or, most importantly, of their biologic behavior in vitro and in vivo (9-22). The negligible loss of label from dog and human platelets, the acceptable dosimetry, and the development of methods which allowed rapid preparation and reinjection of labeled platelets have added to the investigative potential of this approach.

Not surprisingly, then, the last few years have witnessed a
near-explosion in research with In-111-labeled platelets (In-111-P)
dealing with various aspects of thromboembolism.

The reports available deal with a variety of thromboembolic
disorders. In some instances, the primary goal has been to evaluate
In-111-P as a means for noninvasive diagnosis; and, in others as a
means for evaluating therapy. Perhaps the most orderly review can be
achieved by examining the data regarding specific forms of
thromboembolism.

STUDIES WITH In-111-LABELED PLATELETS

Coronary Thrombosis and Myocardial Infarction:

Indium-111 platelets have been used to address three different
questions regarding coronary artery obstruction: (a) Is a thrombus
present? (b) What events are occurring in the epicardium, myocardium
and/or endocardium? (c) What is the effect of therapy upon the
development of both thrombosis and its consequences?

With respect to imaging thrombi, Riba and associates (23) induced
coronary thrombi in dogs by electrical injury. They consistently
imaged these thrombi (in the left anterior descending coronary artery).
When In-111-P were injected at 1/2 to one hour after the injury
was induced, they were able to image the thrombi at 2 and 22 hours
later. High thrombus-to-blood radioactivity ratios were found (mean:
69/1). However, when In-111-P were injected at 22 hours after
thrombus induction, imaging was not successful.

Recently, Bergman et al (24) have reported results with a canine
model in which coronary thrombosis was induced by inserting a copper
coil. In-111-labeled platelets were injected either 15 minutes before
or 1 hour after thrombus induction. Techneticum-99m-labeled red blood
cells were injected at the same time. Using digital subtraction
technique, the authors were able to detect the coronary thrombi
induced and follow the response of these thrombi to streptokinase
infusion. It is interesting that their calculated thrombus/blood
ratios were almost as high when platelets were injected an hour after
thrombosis as they were with pre-thrombosis injection. This suggests
that turnover of labeled platelets in such thrombi is substantial for
at least 1 hour after thrombus formation. As yet, human
investigations of this type have not been reported.

Other investigators have directed their attention to tissue
events, rather than to the detection of the thrombus per se. Ramson
et al (25) utilized In-111-P and radiolabeled leukocytes to assess the
extent of myocardial injury after 60 minutes of left circumflex
occlusion. They demonstrated that both platelets and leukocytes

increased in the infarcted zone during 24 hours of reperfusion. In animals given ibuprofen (12.5 mg/kg I.V.) starting 30 minutes prior to occlusion and every 4 hours thereafter, myocardial platelet concentration was unchanged, but infarct size and leukocyte infiltration were both significantly reduced.

Laws et al (26) also used In-111-P to monitor the consequences of experimental myocardial infarction in the dog. They demonstrated that platelet deposition occurred on the epicardium and endocardium in the regions of lowest coronary blood flow. Such deposition was maximal 24 hours after reperfusion but had resolved by 48 hours. The authors proposed such imaging with In-111-P as a useful method for quantifying platelet deposition in vivo after infarction.

Such studies have provided several collateral observations which are of substantial relevance to other investigations in which In-111-P images have been used for detecting or following thrombi.

First, any vascular intimal injury causes platelet deposition and can result in a positive image with In-111-P. Thus, a positive image cannot be equated with the presence of thrombosis per se or occlusion of the vessel. Second, experimental induction of coronary (and other) thrombi may not approximate the events which occur in human coronary thrombosis. In most animal models, the coronary arteries are normal just prior to thrombus induction; and thrombosis is induced by an acute electrical or mechanical injury to the arterial wall. Therefore, any extrapolation of animal data utilizing injury models should be cautious indeed. As yet, human investigations in coronary thrombosis are extremely limited. The only major published studies have not involved attempts to image acute thrombi but have dealt with platelet survival in patients with coronary artery disease and the effects of drugs known to alter platelet behavior. These results, to date, have been contradictory (27).

Platelet Deposition in Grafts

Multiple studies have been done which have used In-111-P to address the question of potential thrombosis in various grafts. Repetitively in such studies, the problem arises of distinguishing platelet deposition in grafts from graft occlusion.

For example, Fuster et al (28), demonstrated in dogs, substantial acute platelet deposition in saphenous vein coronary artery bypass grafts. Indium-111-labeled platelets were injected 2 hours after graft replacement. They also demonstrated that a regimen of aspirin and dipyridamole decreased such deposition significantly. However, these data did not examine whether such acute deposition is a predictor of occlusion or of the efficacy of antiplatelet drugs in that regard.

Indeed, there is a reason to question whether platelet deposition in grafts, as determined by imaging, will allow prognostic-therapeutic judgments. For example, Goldman et al (29) have shown that, in man, aortofemoral dacron grafts not only accumulate platelets in substantial quantities soon after placement but that they can be demonstrated to accumulate in grafts up to 9 years old, particularly near anastomotic sites. Since these grafts remained patent, it is evident that deposition alone did not predict graft occlusion.

Subsequently, again using In-111-P, Goldman et al (30) demonstrated that both woven and more porous (velour knitted) aortofemoral grafts in patients accumulated platelets to the same degree soon after surgery and that deposition continued for up to 9 months.

Huang and Harker (31) studied patients with aortofemoral arterial dacron grafts and aortic aneurysms. They were able to image most of these with In-111 platelets. Eleven patients with aneurysms were treated with antiplatelet drugs. In seven, an aspirin-dipyridamole regimen did not reduce platelet deposition, whereas sulfinpyrazone did decrease the intensity of deposition in two of four patients.

Similar studies were carried out by Ritchie et al (32) in 19 patients with several types of vascular access grafts in hemodialysis patients. Again, there was uniformly increased platelet deposition in the grafts when compared with the contralateral extremity. However, there was no relationship between these deposition data and a history of prior graft occlusion. Six patients were given sulfinpyrazone (200 mg tid). Three had a definite decrease in the intensity of deposition; two probably did; one did not. None of the six lost graft function.

Callow et al (33) studied platelet deposition in carotid artery grafts in baboons. They found that dacron grafts showed substantial platelet deposition acutely, whereas polytetrafluoroethylene (PTFE) grafts did not. In a second series of animals, PGI2 infusions (150-200 µg/kg/min) prevented platelet deposition on the grafts and decreased pre-existing deposition. However, no data regarding graft occlusion beyond these acute studies are provided.

Yui et al (34) demonstrated shortened platelet survival times in two patients with aortic grafts and suggested this might play a role in thrombogenesis involving the graft.

Stratton et al (35) also studied In-111-P uptake in the aorto-femoral region in 15 patients with dacron grafts in place and compared the results with those obtained in 13 normal young adults. The grafts had been in place for 9-120 months. They found that all grafts accumulated imageable amounts of In-111-P and, quantitatively, in substantially greater amounts than the controls (who did not have

imageable vessels). In a given patient, the results were
reproducible. They concluded that platelet deposition continues for
long periods after graft placement, implying absent or incomplete
endothelial coverage of the graft. Further, since none of the
patients suffered graft occlusion or distal embolization, they
postulated that a "steady state" of platelet deposition and removal
must be occurring on the graft surface.

Pumphrey and his colleagues (36) studied In-111-P deposition on
freshly placed aortofemoral dacron and politeal femoral PTFE grafts in
patients. They found that dacron grafts consistently became imageable
and that deposition intensity was reduced by preoperative
administration of dipyridamole followed by postoperative
administration of dypyridamole and aspirin. The PTFE grafts showed
minimal deposition that was unaltered by the drug regimen.

Heyns and his associates (37) demonstrated that patients with
aortic aneurysms injected with In-111-P had shortened platelet
survival times, accumulation of radioactivity in the aneurysms,
accumulation of platelets in the lower extremities (presumably due to
microembolism), and spleen-liver-marrow accumulation suggesting
sequestration of damaged platelets. Thus, it appears that both
abdominal aneurysms and the grafts used to replace them can induce
aberrant platelet behavior.

Such aberrant behavior does not appear to involve other plasma or
cellular components. Finkelstein et al (38) induced aortic injury in
rabbits and performed scintigraphic studies and electron microscopic
analyses following injection of In-111-P and In-111 red blood cells
and In-111 plasma protein. They found that In-111-P adherence
generated positive scintiphotographs in animals in which electron
microscopy confirmed extensive regions of denuded endothelium.
Radiolabeled red cells and protein were not deposited in such regions,
and fibrin deposition was not conspicuous.

Agarwal and his colleagues (39) injected In-111-P in nine patients
after placement of grafts connecting a systemic ventricle and the
pulmonary artery. Platelet deposition was imaged in this conduit in
eight patients. Platelet deposition was not altered by administration
of aspirin and dipyridamole.

Thus, it is quite clear that a wide variety of intravascular
grafts are the site of platelet deposition--not only acutely, but
perhaps for years. It is also evident that certain antithrombotic
agents can moderate such deposition, at least for short periods of
time. These same statements appear to apply to naturally occurring or
induced states in which the arterial endothelium is absent or altered.
However, it has not yet been shown that platelet deposition intensity
predicts the thrombogenicity of graft materials in terms of their
long-term patency, nor has it been shown that drugs altering such

deposition can reduce the incidence of occlusion or embolism. These important clinical questions remain to be addressed.

Cerebral Thrombosis

The detection of carotid and cerebral artery thrombosis, of carotid artery stenosis, and of thrombus—embolus generation in stenotic vessels is of great interest in the diagnosis and management of patients with, or of those at risk of, stroke and transient ischemic attacks (TIAs). Not surprisingly, then, multiple investigators have addressed several aspects of this problem.

In a rabbit model, Randall and Wilding (40) induced carotid thrombosis by local electrical stimulation. Platelet—rich thrombi were induced. Animals treated intravenously with a thromboxane synthetase inhibitor (dazoxiben) or aspirin showed substantially less platelet accumulation, as monitored by In-111-P. While aspirin was shown to reduce the blood levels of both thromboxane and prostacyclin, dazoxiben totally inhibited production of the former and increased levels of the latter.

Lusby et al (41) used a canine model in which post-carotid-endarterectomy behavior was followed by In-111-P scintigraphy. All arteries demonstrated In-111-P uptake at 1 hour with little increase over the next 47 hours. Kessler and Trabant reported results of scintigraphic examination after In-111-P injection in 12 patients with cerebral ischemia of various types. Eight of these showed "pathologic platelet accumulations in the appropriate vessel," including two with a history of TIA in whom carotid angiography was normal. However, four symptomatic patients, including two with an occluded carotid artery, had normal scintigrams.

Sinzinger and colleagues (42) injected and scanned 44 patients with signs of carotid artery stenosis in whom stenosis was confirmed by Doppler ultrasound and arteriography. Patients demonstrated reduced platelet half-life when compared with controls. However, in none of the patients was imaging of the carotid arteries successful.

Perhaps the most instructive of these reports was that by Powers et al (43). They studied one hundred patients with suspected cerebrovascular disease. More than half demonstrated increased activity in one or more foci along the course of the carotid arteries. However, there was not significant correlation between the scintigraphic findings and prior or subsequent episodes of TIAs or cerebral infarction. Thus, simple formation of platelet thrombi in the cerebal arteries apparently constitutes only one element in stroke or TIA pathogenesis. Others have reported similar findings (44).

Thrombosis in Other Arteries

Terrier et al (45) demonstrated that In-111-P accumulated, in some patients, at the site of femoral arterial catheterization, though clinical occlusion did not occur. Davis et al (46) detected a thrombus in the right femoral artery prior to the onset of clinical symptoms and found the thrombus/blood radioactivity ratio was 58/1 when it was removed 25 hours after In-111-P injection.

Venous Thrombosis

The participation of platelets in the initiation and growth of venous thrombosis has been recognized for many years. Therefore, there has been substantial interest in utilizing In-111-P as a vehicle not only for enhancing diagnosis but also for following the natural history of the venous thrombus and allowing evaluation of various antithrombotic regimens.

Several animal investigations have indicated that imaging of venous thrombi is readily accomplished when In-111-P are injected prior to, or soon after, thrombus formation. In a dog model in which thrombi were induced by applying electric current to the vein wall, Knight et al (47) injected In-111-P at intervals from one to 72 hours after the injury. They found that positive images were obtained at the site of injury up to 72 hours. However, imaging was much less intense, and thrombus-to-blood ratios fell sharply when injection was delayed beyond 20 hours after injury. All sites of injury were visualized as early as 1-2 hours after injection; but the most positive images were obtained at 12-15 hours. These authors noted that thrombi beyond 24 hours were very small. We have reported that endogenous fibrinolytic dissolution is quite brisk in the dog, so that this animal is not suitable for studies of thrombi aged beyond 18-24 hours. These authors also noted that injury to the venous wall occurred in this model; therefore, deposition of platelets on the vessel wall was present even after thrombus removal.

In this same study, the relative thrombus/blood radioactivity ratios of I-125-labeled fibrinogen and In-111-P were compared. Iodine-125-fibrinogen ratios were lower than those with In-111-P, though they were consistently elevated up to 48 hours. Again, as with In-111-P, the thrombus-blood ratios fell sharply by 20 hours.

Grossman et al (48) used a rabbit ear vein model in which thrombi were induced by placing iron filings in the vein lumen. The filings were kept in place by a magnet. When platelets were injected prior to thrombus formation, imaging was consistently positive at 1 hour after thrombus induction. However, with platelet injection after formation, imaging success fell with the post-thrombosis interval. The longer the interval after formation, the longer it took to image the thrombi, presumably because of less thrombus uptake of In-111-P. Beyond 6 hours, visualization was consistently poor, and no visualization occurred with thrombi aged for 24 hours. These authors also used Tc-99m-labeled platelets with essentially parallel observations.

Moser et al (49) developed a canine model in which leg vein thrombosis could be induced without induction of intimal injury, using transient local venous obstruction plus thrombin injection. They demonstrated that, with In-111-P injection prior to thrombus injection, gamma imaging of the thrombi was consistently possible within 15 minutes after thrombus formation. Further, when such thrombi were released as emboli, all were visualized by lung scanning. When platelets were injected after thrombus formation, a positive image of the thrombus was detectable at 20-90 minutes (average 39 min). Thrombus-blood ratios of activity ranged from 5.5-18.

When thrombi were aged for 2, 4, 8, and 10 hours before In-111-P were injected, the results were the same. Thus, in vivo "aging" up to 10 hours did not alter imaging potential.

In this same study, venous thrombi aged for 2 hours were released as emboli. In four animals in which In-111-P were injected immediately after embolization, imaging of the emboli was successful in only one; when In-111-P were injected 2 hours after embolization, imaging was uniformly successful within 90 minutes. Autopsy data suggested that imaging was least successful when embolism caused total arterial occlusion, thereby preventing access of the embolus surface to circulating platelets.

Finally, the authors demonstrated that heparin administration in high doses (300 i.u./kg bolus; 90 i.u./kg/hr infusion) prevented platelet uptake by venous thrombi, and harvested thrombi had radioactivity below that in blood. However, within 40 minutes after protamine administration, imaging was possible and thrombus/blood ratios of 6-10/1 were observed

Several studies also have indicated that In-111-P allow visualization of lower extremity venous thrombosis in patients. Fenech et al (50) studied 48 patients, 33 of whom had undergone surgical reduction of a fractured femoral neck, while 15 had been admitted with a clinical suspicion of deep venous thrombosis. Imaging of the legs and pelvis was carried out within 24 hours after injection of In-111-P. Results were compared with findings on ascending phlebography. They found no false positive results. Positive In-111-P images were obtained in 24 of 26 limbs with phlebograms and in one patient with an inferior vena caval thrombus. In general, they found that the images reflected the site and extent of thrombosis, although with totally occlusive thrombi, only the proximal ends were labeled. (They suggested that thrombus extent in such circumstances might be estimated by defining a filling defect in the In-111-P venous image). The authors noted that some patients with positive images

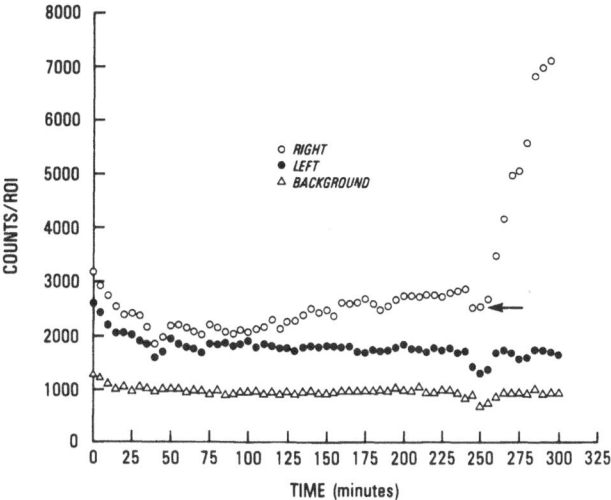

Fig. 1   Study in which venous thrombus was induced in right femoral
         vein. Heparin infusion was begun 30 minutes after thrombus
         formation (at "zero" time). Heparin infusion was discontinued
         at 240 minutes. Note abrupt rise in count rate that
         characteristically occurred over thrombus following heparin
         cessation (arrow). This count rate change correlated with
         time post-heparin at which gamma camera imaging of thrombus
         was observed.

were receiving heparin or oral anticoagulant drugs, although data
regarding the adequacy and duration of such therapy were not given.
One patient with a positive image had a thrombus judged to be 5 weeks
old on clinical grounds.

     Goodwin et al (51) reported successful imaging of venous thrombi
in patients, in some of whom the diagnosis was confirmed by contrast
venography. Some were visualized as early as 4-6 hours, but imaging
was felt to be optimal at 24 hours. Several patients were receiving
anticoagulant drugs at the time of successful imaging.

     Davis et al (52) also have documented imaging of venous thrombi in
six of seven patients with venous thrombosis confirmed by impedance
plethysmography or contrast venography, even though all patients were
receiving intravenous heparin.

     Fedullo et al (53) examined the relationship between heparin
administration and In-111-P accretion in a canine model of venous
thrombosis. They found that high-dose heparin (a 300 i.u./kg bolus
followed by an infusion of 900 i.u./kg/hr) completely blocked platelet
accretion. The activated partial thromboplastin time (APTT) in these
animals averaged eight times control. As lower heparin doses were
examined, it was found that In-111-P uptake was consistently inhibited

so long as the activated partial thromboplastin time was maintained above 2.5 times control values. The only exceptions occurred in animals in which In-111-P were injected within the first 30 minutes after heparin was administered, suggesting that platelet accretion to fresh thrombi may continue for a short period after initiation of heparin. It seems likely that such accretion—and, therefore, thrombus growth—will continue for several days after initiation of prothrombinopenic drug therapy, since full antithrombotic effects of such drugs are not established until the prothrombin time has been in the therapeutic range for several days. These same studies also indicated that platelet accretion occurred rapidly when the APTT fell below 2.5 times control; that accretion occurred within a relatively narrow APTT range (1.3-1.5 x control); and that increasing the heparin dose did not reverse prior accretion.

In those animals in which accretion did occur, it was detectable by imaging within 30 minutes after heparin reversal with protamine sulfate or decline of the APTT spontaneously to the 1.3-1.5 times control range. Thrombus/blood ratios of the thrombi which were imaged ranged from 7 to 22.

In subsequent studies in our laboratories, these observations regarding the antithrombotic effect of certain drugs on venous thrombus behavior have been extended. Fedullo et al (54) have explored, in the same canine model, the effects of various heparin regimens on incorporation of both I-125-F and In-111-P into venous thrombi formed in vivo in both femoral veins. They found that both In-111-P and I-125-F were actively incorporated into the venous thrombi of untreated animals. The ratios of fibrinogen uptake to platelet uptake were not significantly different between the two femoral thrombi in a given animal. When heparin was infused in high doses (APTT 3-4x control), neither In-111-P nor I-125-F were taken up by thrombi. However, at lower levels (APTT = 2-3x control), fibrinogen uptake was completely blocked, whereas, by thrombus/blood ratios, small amounts of platelet accretion occurred. More recently, Czer et al (55) have shown, in this same model, that infusion of prostacyclin (PGI 2), in doses which did not cause adverse hemodynamic effects, completely inhibited platelet accretion. This was true even though in vitro measurements of ADP-induced platelet aggregation did not demonstrate complete inhibition of such aggregation.

Recently, Winter et al (56) have reported a study in patients with antithrombin III deficiency in whom In-111-P were used to monitor the results of treatment with concentrates of antithrombin III. They found this approach promising because of the ability of In-111-P imaging to provide an early indication of thrombus extension.

The puzzle of the high incidence of venous thrombo-embolic incidence in patients with homocystinuria also has recently been addressed using In-111-P. Hill-Zobel et al (57) were unable to demonstrate a difference in platelet kinetics between patients with homocystinuria (pyridoxine responders or nonresponders) and normals. There was no correlation between the severity of homocystinuria and platelet kinetics, nor was there evidence of abnormal adherence to venous walls among the homocystinurics.

Pulmonary Embolism

As noted above, Moser et al (49) demonstrated, in a canine model, that In-111-P could be used to image fresh emboli formed in vivo as venous thrombi. They found that the anatomic characteristics of the embolus (i.e., whether it totally occluded the vessel) appeared to play an important role in imaging success.

More recently, Sostman et al, (58) also in a canine model, have shown that experimental emboli can be imaged consistently for up to 24 hours after release. They also found that injection of In-111-oxine alone allowed some emboli to be imaged. Injection of radiopaque contrast agents did not interfere with In-111-P imaging. Heparin infusion did prevent uptake, but uptake did occur when heparin infusion was discontinued, an observation compatible with findings previously discussed.

Several observations in man regarding the potential imageability of emboli with In-111-P have now appeared. Davis et al (52) were successful in imaging emboli in only one of eleven patients. However, the diagnosis of embolism was made only by lung scan, and all the patients were receiving heparin, an important consideration, as noted previously. Subsequently, Sostman et al (59) and Ezekowitz and colleagues (60) have reported successful imaging of emboli in patients.

However, to date, no systematic study in embolic suspects has been carried out. The ability to image emboli will clearly be influenced by multiple factors, including their age (and "activity"), the extent to which blood (and In-111-P) can gain access to them, the therapy in place, and their location. Such a study is now underway at our institution.

Intracardiac Thrombi

Intracardiac thrombi present a quite special instance of thrombosis in that they may occur in either the right or the left cardiac chambers in several contexts and may result in either pulmonary or systemic arterial embolization. The potential value of In-111-P for detection of intracardiac thrombi in patients has been examined by multiple investigators. Ezekowitz et al (61) reported that left ventricular thrombi frequently demonstrated sufficient uptake of In-111-P to permit in vivo imaging and subsequently compared this approach to (2-D) echocardiography. Among 34 patients with left

ventricular aneurysms in whom the presence or absence of left
ventricular thrombi was determined at surgery or autopsy, In-111-P
detected 71% and there were no false- positives. Echocardiography was
slightly more sensitive and slightly less specific. Incidentally, in
19 patients with mitral valve disease, none had positive In-111-P or
echo demonstration of thrombi, and none was present at surgery.

Stratton et al (62) reported similar results in patients with
cardiomyopathy or a transmural myocardial infarction. There were no
false-positive images, but 8 of 18 patients with thrombi had negative
or equivocal images with In-111-P, which was a substantially lower
sensitivity than with 2-D echo (17/18). The scans were negative at 2
hours, with imaging more successful when done 24, 48, or 72 hours
after In-111-P injection. The authors concluded that thrombi needed
to be "hematologically active" to be successfully imaged.

Taylor (63) studied 11 patients with intracardiac thrombi. All
were imaged with In-111-P. In 5 patients treated with aspirin, in
doses of 300-2400 mg/d, platelets continued to be incorporated into
these thrombi.

DISCUSSION

The data available indicate the substantial promise which the use
of In-111-P offer in the diagnosis and monitoring of thrombotic
disease per se and in evaluating the effects of various antithrombotic
regimens. However, certain potential pitfalls in interpreting such
studies also are evident, as are areas in which further investigation
is necessary. For example, it is clear that successful imaging of
vessels or grafts with In-111-P does not indicate that thrombosis is
occurring or will occur. Therefore, adequate clinical investigations
are required before diagnostic or therapeutic extrapolations to human
disease are made from In-111-P studies in vitro or in animal models.

While the pathology of thrombosis has not been fully defined, our
present understanding indicates that platelets are incorporated into
both venous and arterial thrombi at the time of formation. Beyond
acute formation, it is also clear that circulating platelets are
incorporated into preformed thrombi as they grow and, perhaps, by
exchange between thrombus and blood platelets. However, there are
differences, known and unknown, among various types of thrombi in
terms of platelet participation, differences which may well condition
the information provided by In-111-P injected at a given time in the
natural history of the thrombus.

For example, venous thrombi (and, therefore, pulmonary emboli) are
"red" thrombi in their mature stages. They consist dominantly of red
blood cells enmeshed, along with platelets and white cells, in a
fibrin network. Arterial thrombi, on the other hand, ordinarily are

"white" thrombi, consisting chiefly of platelets, a modest fibrin network, and a significantly less striking component of red blood cells.

Just as there are differences in initial structure, there are differences in natural history among thrombi. Venous thrombi, at least when they remain exposed to flowing blood, grow in length and diameter by the accretion of fibrinogen and platelets, forming the characteristic "lines of Zahn." Some venous thrombi occlude the vessel lumen completely at the time of formation, while others grow rapidly in diameter and cause occlusion. Total occlusion by venous thrombi and other thromboemboli is an important consideration, both clinically and experimentally, because the diagnosis of existing thrombosis using In-111-P (and labeled fibrinogen) requires exposure of the thrombus to circulating blood. If a thrombus totally occludes a vessel, such exposure is nominal and detection unlikely. Thus far, investigations suggest that the vast majority of venous thrombi are not totally occlusive because, when "fresh," most are detectable by systemically-injected In-111-P or I-125-F. Nonetheless, total obstruction is one certain reason that some "false-negative" results will be seen.

As venous thrombi mature, however, they become "inactive." That is, layering and extension of the thrombus ceases. Just when, and why, this occurs in the absence of therapy is not clear. However, when growth stops, the possibility of detection with In-111-P and I-125-F also stops, since incorporation of both agents requires an "active" thrombus. There have been indications that there is some "exchange" of platelets and fibrinogen between the circulating pool and the thrombus that may allow detection in the absence of growth. Whether this occurs in patients, to what degree, and for how long a period are all questions which remain to be defined.

Beyond the acute "active" phase, venous thrombi undergo resolution by fibrinolysis, organization, or a combination of the two. In the dog, such resolution is extremely rapid (64), so that "aging" thrombi in the dog beyond 12-24 hours is not practical--and studies with In-111-P will prove negative because such thrombi are either gone or quite small.

If dissolution is incomplete, venous thrombi and pulmonary emboli undergo organization. This process converts a "thrombus" into a mass of organized, vascular connective tissue which develops a neo-intima. Such "thromboemboli" clearly will not accrete In-111-P.

These considerations condition potential imaging not only of venous thrombi but also of pulmonary emboli and intracardiac thrombi, as the literature reflects. So long as such thrombi remain "active," fibrinogen and platelet accretion will make imaging and detection likely. As they become inactive or organized they will escape detection by such methods.

Similar questions surround the use of In-111-P for detection of arterial thrombi. Clearly, when such thrombi form in the presence of circulating In-111-P, they can be readily imaged. However, in clinical situations, such a sequence would be extremely fortuitous. Most arterial thrombi come to clinical attention only when they have caused sufficient occlusion (often total) to induce symptoms. It should be noted that such patients will rarely be available for such studies until several hours have passed since the thrombotic event. Whether, under such conditions, In-111-P can gain adequate access remains to be demonstrated.

Whatever their ultimate diagnostic value in the face of established venous or arterial thrombosis, In-111-P do have substantial promise for monitoring patients at special thrombotic risk and for evaluating various pharmacologic agents for prophylaxis and treatment.

With regard to the monitoring of patients, In-111-P certainly could be used effectively in this way to follow patients at high risk of lower extremity venous thrombosis--as I-125-F is now used. Indeed, In-111-P have substantial advantages in this regard in that they: (1) should allow imaging of thrombi in the thigh and pelvic veins, including the inferior vena cava; (2) provide an image of the thrombus rather than an inferential diagnosis based on count rates; and (3) allow a diagnosis within hours of injection whereas I-125-F requires a 24 hour period.

The diagnostic potential of In-111-P in other thromboembolic disorders is even more intriguing. In pulmonary embolism the combination of an In-111-P image of the embolus and a perfusion scan demonstrating the consequences of embolic obstruction should allow rapid and reliable noninvasive diagnosis and, with mobile gamma cameras, at the bedside. Particularly promising is the information which In-111-P may offer in detecting patients at risk of coronary, carotid artery, or intracerebral thrombosis or those at risk of thromboembolism due to placement of intravascular valves and grafts. It has already been shown that some atherosclerotic and aneurysmal vessels, as well as a variety of grafts and valves, show accretion of In-111-P in sufficient quantities to allow imaging. What is not known is whether initial or periodic imaging has any value in predicting--directly or indirectly--that a clinical event of importance may occur; e.g., arterial occlusion, graft occlusion or emboli from implanted valves. This is certainly an important and potentially fruitful area for future investigation. Perhaps quantification of the intensity of platelet deposition or the identification or concomitant clinical or laboratory risk factors will be necessary. What certainly will be required is long-term follow-up of such patients to determine the predictive value of In-111-P behavior. For example, we now know that Dacron aortofemoral grafts accrete In-111-P for years after placement;

but does this accretion, or its intensity, correlate with the risk of graft occlusion or distal embolism? In coronary bypass grafts, similar questions need to be answered.

The potential value of the In-111-P approach in evaluating current and future prophylactic treatment approaches also will require long-term, carefully designed studies in patients. Animal models, however reasoned and sophisticated, cannot replicate the complexities encountered in patients. We feel, for example, that In-111-P and fibrinogen accretion to venous thrombi will not occur when a patient is receiving a dose of heparin sufficient to halt thrombus growth; i.e., prevent platelet and fibrinogen accretion. If such accretion is occurring, heparin therapy is, by definition, inadequate. Thus, successful In-111-P imaging of venous (or other) thrombi in patients receiving therapy is, to us, a cause for clinical concern.

Indeed, studies of this type may illuminate our current understanding of what constitutes adequate antithrombotic therapy. One of the major barriers to the development and evaluation of antithrombotic therapy has been the absence of a means for monitoring, in vivo, the behavior of the entity being treated: the thrombus. Therapeutic judgments have been made indirectly. For example, if heparin or coumadin prolong a certain coagulation test to a stipulated degree, therapeutic efficacy is assumed. Or if a given drug inhibits the response, in vitro, of platelets, a therapeutic implication is derived.

It already seems clear that such concepts can be replaced--or at least refined--by testing them against in vivo behavior, as monitored by In-111-P and careful clinical observation.

Our own observations suggest, for example, that monitoring of patients with venous thrombosis using In-111-P will provide valuable insights into appropriate heparin regimens and, ultimately, of the value of prophylactic drugs and devices. Study of the potential utility of prostacyclin, and its likely derivatives, in patients with venous thrombosis and pulmonary embolism also offers substantial promise.

In arterial thrombotic disease, in preservation of graft patency, and in prevention of valve-derived emboli, definition of the value of various drug regimens also awaits appropriate clinical investigation. Will various antiplatelet drugs not only inhibit imaging of arteries, grafts, and valves but also correlate with thromboembolic prevention? Only studies in well-characterized patient groups can answer such questions.

In all these contexts, it is important to emphasize that the availability of the In-111-P technique does not obviate the need for simultaneous application--at least at the outset--of established

diagnostic techniques such as contrast angiography, ultrasound
methods, and others. Indeed, it is imperative that the other methods
be regarded as complementary, not competitive, and as required to
validate the significance of In-111-P images. Unless investigations
with In-111-P incorporate simultaneous observations with established
techniques and rigorous clinical evaluation, the meaning of "positive
In-111-P images" will remain obscure.

Obviously, the In-111-P technique has created an opportunity for
carrying out investigations of fundamental importance to our
understanding of thrombotic disease. It is an opportunity which can
best be exploited by efforts which involve the multiple clinical and
laboratory disciplines pertinent to patients with thrombotic disease.
Such interdisciplinary efforts are essential if the exciting images of
the present are to be translated into the diagnostic and therapeutic
advances of the future.

REFERENCES

1. V. V. Kakkar, C. T. Howe, C. Flanc, M. D. Clarke, Natural History
    of post-operative deep vein thrombosis, Lancet 2:230 (1969).
2. V. V. Kakkar, Fibrinogen uptake test for detection of deep vein
    thrombosis: a review of current practice, Semin Nucl Med
    7:229 (1977).
3. R. Hull, J. Hirsh, P. C. Sackett, P. Powers, A. G. G. Turpie,
    I. Walker, Combined use of leg scanning and impedance
    plethysmography in suspected venous thrombosis, N Engl J Med
    26:1497 (1977).
4. K. M. Moser, M. Guisan, A. Cuomo, W. L. Ashburn, Differentiation
    of pulmonary vascular from parenchymal diseases by
    ventilation/perfusion scintiphotography, Ann Int Med 75:597
    (1971).
5. B. J. McNeil, Ventilation-perfusion studies and the diagnosis of
    pulmonary embolism: concise communication, J Nucl Med 21:319
    (1980).
6. R. D. Hull, J. Hirsh, C. J. Carter, Pulmonary angiography,
    ventilation lung scanning and venography for clinically
    suspected pulmonary embolism with abnormal perfusion lung
    scan, Ann Int Med 98:891 (1983).
7. H. F. Welch, H. W. Strauss, B. Pitt, The extraction of
    Thallium-201 by the myocardium, Circulation 56:188 (1977).
8. D. Shure, K. M. Moser, J. H. Harrell, M. T. Hartman,
    Identification of pulmonary emboli in the dog: comparison of
    angioscopoy and perfusion scanning, Circulation 64:618 (1981).
9. M. L. Thakur, M. J. Welch, H. Joist, R. T. Coleman, Indium-111-
    labeled platelets: studies on preparation and evaluation of
    in vitro and in vivo functions, Thromb Res 9:345 (1976).
10. U. Scheffel, P. A. McIntyre, B. Evatt, J. A. Dvornicky, Jr., T.
    K. Natarajan, D. R. Bolling, E. A. Murphy, Evaluation of

Indium-111 as a new high photo yield gamma-emitting
"physiological" platelet label, Johns Hopkins Med J 140:285
(1977).

11. J. H. Joist, R. K. Baker, M. L. Thakur, M. J. Welch, Indium-
111-labeled human platelets and loss of label and in vitro
function of labeled platelets, J Lab Clin Med 92:829 (1978).

12. B. W. Wistow, Z. D. Grossman, J. G. McAfee, G. Subramanian, R. W.
Henderson, and M. C. Roskopf, Labeling of platelets with oxine
complexes of Tc-99m and In-111. Part 1. In vitro studies and
survival in the rabbit, J Nucl Med 19:483 (1978).

13. W. A. Heaton, H. H. Davis, M. J. Welch, C. J. Mathias, J. H.
Joist, L. A. Sherman, B. A. Siegel, Indium-111: a new
radionuclide label for studying human platelet kinetics, Br J
Haematol 42:613 (1979).

14. K. G. Schmidt and J. W. Rasmussen, Labeling of human and rabbit
platelets with Indium-111-oxine complex, Scand J Hematol 23:97
(1979).

15. K. C. Schmidt and J. W. Rasmussen, Preparation of platelet
suspensions from whole blood in buffer, Scand J Haematol 23:88
(1979).

16. A. Heyns, M. G. Lotter, R. N. Badenhorst, O. R. van Reenen, H.
Pieters, P. C. Minnaar, F. P. Retief, Kinetics, distribution
and sites of destruction of In-111-labeled human platelets, Br
J Haematol 44:269 (1980).

17. I. Klonizakis, A. M. Peters, M. L. Fitzpatrick, M. J. Kensett,
S. M. Lewis, J. P. Lavender, Radionuclide distribution following
injection of In-111-labeled platelets, Br J Haematol 46:595
(1980).

18. J. S. Robertson, M. K. Dewanjee, M. L. Brown, V. Fuster, J. H.
Chesebro, Distribution and dosimetry of In-111-labeled
platelets, Radiology 140:169 (1981).

19. M. L. Thakur, L. Walsh, H. L. Malech, S. Gottschalk, In-111-
labeled human platelets: improved method, efficacy, and
evaluation, J Nucl Med 22:381 (1981).

20. E. M. Hudson, B. B. Ramsey, B. L. Evatt, Subcellular localization
of In-111-labeled platelets, J Lab Clin Med 97:577 (1981).

21. V. Scheffel, M. F. Tsan, T. G. Mitchell, E. E. Camargo, Human
platelets labeled with In-111-8 hydroxyquinoline: kinetics,
distribution and estimates of radiation dose, J Nucl Med
23:149 (1982).

22. J. R. J. Baker, K. D. Butler, M. V. Eakins, G. F. Pay, A. M.
White, Subcellular localization of In-111 in human and rabbit
platelets, Blood 59:351 (1982).

23. A. L. Riba, M. L. Thakur, A. Gottschalk, B. L. Zaret, Imaging
experimental coronary artery thrombosis with In-111 platelets,
Circulation 60:767 (1979).

24. S. R. Bergman, R. A. Lerch, C. J. Mathias, B. E. Sobel, M. J.
Welch, Non-Invasive Detection of Coronary thrombi with In-111
platelets, J Nucl Med 24:130 (1983).

25. J. C. Romson, B. G. Hook, V. H. Rigot, M. A. Schork, D. P.
     Swanson, B. R. Luchesi, The effect of ibuprofen on
     accumulation of In-111-labeled platelets and leucocytes in
     experimental myocardial infarction, Circulation 66:1002
     (1982).
26. K. H. Laws, J. A. Clanton, V. A. Starnes, F. M. Lupinetti,
     J. C. Collins, J. A. Oates, J. W. Hammon, Jr., Kinetics and
     imaging of In-111-labeled autologous platelets in experimental
     myocardial infarction, Circulation 67:110 (1983).
27. A. C. de Boer, P. Han, A. G. G. Turpie, R. Butt, M. Gent, E.
     Genton, Platelet tests and antiplatelet drugs in coronary
     artery disease, Circulation 67:500 (1983).
28. V. Fuster, M. K. Dewanjee, M. P. Kaye, M. Josa, M. P. Metke, J.
     W. Chesebro, Noninvasive radioisotopic technique for detection
     of platelet deposition in coronary artery bypass grafts in
     dogs and its reduction with platelet inhibitors, Circulation
     60:1508 (1979).
29. M. Goldman, H. C. Norcott, R. J. Hawker, Z. Drolc, C. N.
     McCollum, Platelet accumulation on mature dacron grafts in
     man, Br J Surg 69 (Suppl):(1982).
30. M. Goldman, C. N. McCollum, R. J. Hawker, Z. Drolc, G. Slaney,
     Dacron arterial grafts: the influence of porosity, velour and
     maturity on thrombogenecity, Surgery 92:947 (1982).
31. T. W. Huang, L. A. Harker, In-111 platelet imaging for detection
     of platelets deposition in abdominal aneurysms and prothetic
     arterial grafts, Am J Cardiol 47:882 (1981).
32. J. L. Ritchie, A. Lindner, C. W. Hamilton, L. A. Harker, In-111-
     oxine platelet imaging in hemodialysis patients: detection of
     platelet deposition at vascular access sites, Nephron, 31:333
     (1982).
33. A. D. Callow, R. Connolly, T. F. O'Donnell, Jr., R. Gembarowicz,
     Platelet-arterial synthetic graft interaction and its modifi-
     cation, Arch Surg 117:1447 (1982).
34. T. Yui, T. Uchida, S. Matsuda, K. Iwaya, M. Umino, K. Ono, S.
     Muroi, K. Owada, K. Machii, S. Kariyone, Detection of platelet
     consumption in aortic graft with In-111-labeled platelet, Eur
     J Nucl Med 7:77 (1982).
35. J. R. Stratton, B. L. Thiele, J. L. Ritchie, Platelet deposition
     on dacron aortic bifurcation grafts in man: quantitation with
     In-111 platelet imaging, Circulation 66:1287 (1982).
36. C. W. Pumphrey, J. H. Chesebro, M. K. Dewanjee, H. W. Wahner, L.
     H. Hollier, P. C. Pairolero, V. Fuster, In vivo quantitation
     of platelet deposition on human peripheral arterial bypass
     grafts using In-111-labeled platelets, Am J Cardiol 51:796
     (1983).
37. A. D. Heyns, M. G. Lotter, P. N. Badenhorst, H. Pieters, C. J.
     Nel, P. C. Minnaar, Kinetics and fate of In-111-oxine-labeled
     platelets in patients with aortic aneurysms, Arch Surg
     117:1170 (1982).

38. S. Finklestein, A. Miller, R. J. Callahan, J. T. Fallon, F.
    Godley, B. L. Feldman, R. C. Hinton, A. B. Roberts, H. W.
    Strauss, R. S. Lees, Imaging of acute arterial injury with
    In-111-labeled platelets: a comparison with scanning electron
    micrographs, Radiology 145:155 (1982).

39. K. C. Agarwal, H. W. Wahner, M. K. Dewanjee, V. Fuster, F. J.
    Puga, G. K. Danielson, J. H. Chesebro, R. H. Feldt, Imaging of
    platelets in right-sided extracardiac conduits in humans J
    Nucl Med 23:342.

40. M. J. Randall, R. I. Wilding, Acute arterial thrombosis in
    rabbits: reduced platelet accumulation after treatment with
    dazoxiben hydrochloride, Br J Clin Pharmacol 15:49S.

41. R. J. Lusby, L. D. Ferrell, B. L. Englestad, D. C. Price, M. J.
    Lipton, R. J. Stoney, Vessel wall and In-111-labeled platelet
    response to carotid endarterectomy, Surgery 93:424.

42. H. Sinzinger, K. Silberbauer, P. Fitscha, J. Kaliman, Value of
    the detection of arteriosclerotic lesions with labeled
    autologous thrombocytes, Acta Med Austriaca 9:181 (1982).

43. W. J. Powers, D. A. Siegel, H. H. Davis, C. J. Mathias, H. B.
    Clark, M. J. Welch, In-111-platelet scintigraphy in
    cerebrovascular disease, Neurol 32:938 (1982).

44. C. Kessler, R. Trabant, Platelet scintigraphy using In-111, Arch
    Psychiatr Nervenkr 231:449 (1982).

45. E. Terrier, J. Forman, A. Francois, J. Paillard, J. Baillet,
    Detection of platelet thrombi during catheterization for
    intracardiac dynamic studies, Presse Med 12:239 (1983).

46. H. H. Davis, 2nd, B. A. Siegel, M. J. Welch, Scintigraphic
    detection of an arterial thrombus with In-111 autologous
    platelets, J Nucl Med 21:548 (1980).

47. L. C. Knight, J. C. Primeau, B. A. Siegel, M. J. Welch,
    Comparison of In-111-labeled platelets and iodinated
    fibrinogen in the detection of deep vein thrombosis, J Nucl
    Med 19:891 (1978).

48. Z. D. Grossman, B. W. Wistow, J. G. McAfee, G. Subramanian, F. D.
    Thomas, R. W. Henderson, R. F. Rohner, M. L. Roskopf,
    Platelets labeled with oxine complexes of Tc-99m and In-111.
    Part 2. Localization of experimentally-induced vascular
    lesions, J Nucl Med 19:488 (1978).

49. K. M. Moser, R. G. Spragg, F. Bender, R. Konopka, M. T. Hartman,
    P. F. Fedullo, Study of factors that may condition
    scintigraphic detection of venous thrombi and pulmonary emboli
    with In-111-labeled platelets, J Nucl Med 21:1051 (1980).

50. A. Fenech, J. K. Hussey, F. W. Smith, P. P. Dendy, B. Bennett, A.
    S. Douglas, Diagnosis of deep-vein thrombosis using autologous
    In-111-labeled platelets, Br Med J 282:1020 (1981).

51. D. A. Goodwin, J. T. Bushberg, P. W. Doherty, M. J. Lipton, F. K.
    Conley, C. I. Diamanti, and C. I. Meares, In-111-labeled
    autologous platelets for location of vascular thrombi in
    humans, J Nucl Med 19:626 (1978).

52. H. H. Davis, 2nd, D. A. Siegel, L. A. Sherman, W. A. Heaton, M.
     J. Welch, Scintigraphy with In-111-labeled autologous
     platelets in venous thromboembolism, Radiology 136:203 (1980).
53. P. F. Fedullo, K. M. Moser, K. S. Moser, R. Konopka, M. T.
     Hartman, In-111-labeled platelets: effects of heparin on
     uptake by venous thrombi and relationship to the activated
     partial thromboplastin time, Circulation 66:632 (1982).
54. P. F. Fedullo, M. T. Hartman, R. G. Konopka, K. M. Moser, Effect
     of heparin on In-111 platelet and fibrinogen uptake by venous
     thrombi, Circulation (abst., in press).
55. G. Czer, K. M. Moser, R. Konopka, M. T. Hartman, Inhibition of
     In-111 platelet accretion onto venous thrombi in dogs by
     prostacyclin, Circulation Res (in press).
56. J. H. Winter, A. Fenech, M. Mackie, B. Bennett, A. S. Douglas,
     Treatment of venous thrombosis in antithrombin-III-deficient
     patients with concentrates of antithrombin III, Clin Lab
     Haematol 4:101 (1982).
57. R. L. Hill-Zobel, R. E. Pyeritz, U. Scheffel, O. Malpica,
     Kinetics and distribution of In-111 platelets in patients with
     homocystinuria, N Engl J Med 307:781 (1982).
58. H. D. Sostman, R. D. Neumann, S. S. Zoghbi, P. E. Lord, M. L.
     Thakur, P. Carbo, R. H. Greenspan, H. Gottschalk, Experimental
     studies with In-111-labeled platelets in pulmonary
     embolization, Invest Radiol 17:367 (1982).
59. H. D. Sostman, R. D. Neumann, J. Loke, S. S. Zaghbi, M. L.
     Thakur, R. H. Greenspan, A. Gottschalk, Detection of pulmonary
     embolism in man with In-111-labeled autologous platelets, Am J
     Radiol 138:945 (1982).
60. M. D. Ezekowitz, E. R. Eichner, R. Scatterday, R. I. Elkins,
     Diagnosis of a persistent pulmonary embolus by In-111 platelet
     scintigraphy with angiographic and tissue confirmation, Am J
     Med 72:839 (1982).
61. M. D. Ezekowitz, D. A. Wilson, E. O. Smith, R. D. Burow,
     Comparison of In-111 platelet scintigraphy and two-dimensional
     echocardiography in the diagnosis of left ventricular thrombi,
     N Engl J Med 306:1509 (1982).
62. J. R. Stratton, J. C. Ritchie, G. W. Hamilton, K. E.
     Hammerweister, L. A. Harker, Left ventricular thrombi: in
     vivo detection by In-111 platelet imaging and two-dimensional
     echocardiography, Am J Cardiol 47:874 (1981).
63. I. B. Taylor, Failure of aspirin to prevent incorporation of
     In-111-labeled platelets into cardiac thrombi in man, Lancet
     2:440 (1981).
64. K. M. Moser, M. Guisan, E. E. Bartimmo, A. M. Longo, P. G.
     Harsanyi, N. Chiorazzi, In vivo- and post-mortem dissolution
     rates of pulmonary emboli and venous thrombi in the dog,
     Circulation 48:170 (1973).

THE USE OF INDIUM-111 PLATELET SCINTIGRAPHY IN MAN:

COMPARISIONS WITH IN VITRO TESTS AND

IN VIVO PLATELET FUNCTION - A FIVE-YEAR EXPERIENCE

Michael D. Ezekowitz*[†], Edward L. Snyder[+],
Christopher Pope[†], Patricia Ferri*, and Eileen O. Smith*

Yale University School of Medicine
*Dept. of Medicine, [†]Diagnostic Radiology
+Laboratory Medicine
and West Haven VA Medical Center

INTRODUCTION

This paper reports data collected from 540 patients between July
1978 and July 1983. For the first four years, studies were performed
at the Hospitals of the University of Oklahoma Health Science Center
and then continued in the last year at Yale University School of
Medicine and at the West Haven VA Medical Center. A small number of
patients have been studied at St. Raphael's Hospital in New Haven,
Connecticut.

In this review we will be discussing briefly the method we employ
for labeling platelets, emphasizing in vivo and in vitro markers of
platelet function used for quality control of the platelet
preparation. Our first major application of this technique in man was
for the identification of mural left ventricular thrombi. We will
discuss this application of the technique and review the observed
disparity between in vivo and in vitro tests of platelet function in
patients with mural thrombi receiving aspirin. Thereafter we will
deal with the diagnosis of subacute bacterial endocarditis, deep
venous thrombosis, coronary thrombi, left atrial masses, the use of
tomographic imaging, and then make a preliminary statement on the
use of this technique in patients following peripheral angioplasty.
Finally, we shall report changes in platelet function in stored
platelets from normal volunteers.

Table 1. Drugs Administered During Confirmed Positive Platelet Scans

Drugs	Patients																										
	1	2	3	4	5	6	7	8	9	10	11	12	13	14	15	16	17	18	19	20	21	22	23	24	25	26	27
Acetaminophen					+			+		+							+	+	+	+		+	+	+	+	+	
Codine																	+	+		+		+					
Digoxin	+	+	+	+	+				+	+		+	+	+					+								
Dalmane																+		+				+		+			
Furosemide	+	+	+	+	+								+														
Hydrochlorothiazide					+			+				+		+			+								+		
Potassium Chloride	+	+	+	+	+		+	+		+																	+
Multi-vitamins																+	+		+						+		
Nitroglycerine	+	+	+	+	+	+	+	+				+	+	+		+											
Penicillin derivatives					+	+				+						+	+		+							+	
Persantine					+														+		+		+				
Phenobarbitol						+							+							+						+	
Phenytoin												+	+		+												
Prednisone																									+		
Procainamide								+	+																		
Propranolol			+		+	+	+	+		+																	
Quinidine													+														
Valium																	+		+					+			

LABELING METHOD

The proceedings of this meeting are testimony to the fact that multiple methods are currently being used to label platelets. It is important, irrespective of the method finally adopted, to standardize the technique of labeling platelets in order to facilitate the comparison of results from different laboratories. In addition, personnel skilled and experienced in handling platelets and radiopharmaceuticals need to be responsible for the labeling procedure. Good coordination of laboratory and clinical staff is essential. The method that we employ is a modification of that originally described by Thakur et al (1) and later modified by Heaton et al (2). Most of the patients under review were studied for the purpose of imaging thrombosis (83%). Kinetic studies in normal volunteers, which will be detailed later, formed the basis for the remaining studies.

For platelet labeling 43 ml of venous blood is collected in 7 ml of acid citrate dextrose (ACD). The platelets are separated from whole blood by centrifugation at 200 g for 15 minutes and the upper

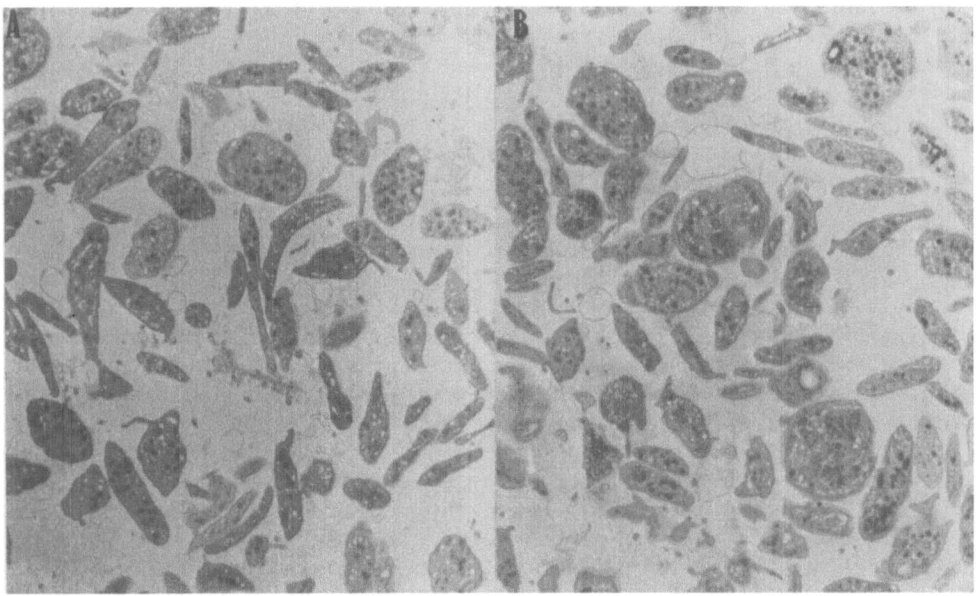

Fig. 1   Transmission electron micrographs of platelets before (panel A) and after labeling (panel B). There is no morphological difference between these two panels. In particular there is no evidence for platelet activation or major pseudopod formation following labeling.

3/4 of the plasma transferred with sterile plastic disposable pipettes
into plastic centrifuge tubes and centrifuged at 2000 g for 5 minutes.
The platelet button so formed is mixed with In-111 complexed to
8-hydroxyquinoline and diluted in ACD: Saline at pH between 6.5 and
6.8 and incubated at room temperature for 20 minutes. The
8-hyroxyquinoline solution is prepared daily in 100% ethanol. After
washing the platelets in plasma to remove the loosely bound surface
In-111, they are finally suspended in ACD- or sodium citrated-plasma
for injection into the patient. Sodium citrated-plasma is used when
in vitro testing of platelet function using adenosine diphosphate
(ADP) and collagen stimulated aggregation is required. A detailed
description of the method has been described elsewhere (3). For
quality control the platelet suspension is characterized by the
following parameters: 1) the response to ADP and collagen
aggregation, 2) the labeling efficiency before and after the plasma
wash, 3) the percentage radioactivity recoverable 15 and 60 minutes
after injection of the platelet suspension, 4) platelet count, 5) the
activity injected, and 6) in selected patients, the ultrastructural
characteristics of the platelet suspension. The latter two are
determined before and after the labeling (Fig. 1).

     In the clinical setting, the potential for drug interference with
the labeling technique and, therefore, the ability to detect active
thrombosis are of significant practical importance. Apart from
specific antiplatelet and anticoagulant drugs which will be discussed
later, a list of drugs co-administered at the time of independently
verified positive studies is provided in Table 1.

IDENTIFYING LEFT VENTRICULAR THROMBI

     Mural left ventricular thrombi form in ventricles with reduced
global and/or regional function. Thrombi are an important source of
systemic embolizaton. Emboli may cause significant morbidity and even
death. Thus treatment should be based on prevention; hence the
importance of identifying mural thrombi (Fig. 2).

     The first data demonstrating the potential value of this
technique for the diagnosis of left ventricular thrombi were presented
at the first conference of radiolabeled cellular blood elements held
in New York in 1979 (4). Later studies were published by us and
others (3,5). In a study designed to determine the diagnostic
accuracy of this technique, all patients had autopsy or surgical
verification of the presence or absence of thrombus. A total of 53
patients were included: 31 were diagnosed to have surgically
resectable left ventricular aneurysms; 3 others died within 10 days of
the platelet study. The remaining 19 were imaged prior to mitral
valve replacement and had normal ventricular function and served as
controls. The mean interval between the injection of labeled
platelets and surgery in the aneurysmautopsy group was 7.5+4.7 days

(mean+1SD). In the mitral valve replacement group platelets were
injected 6.8+4.2 days before surgery. In the entire group of patients
a total of (3.2+2.3)x10⁹ platelets with 407.7+106.4 µCi of In-111
activity and a final labeling efficiency of 70.8+17.4% were injected.
Platelet recovery was 63.4+22 and 43.9+17.2, respectively at 2 and 15
minutes after injection of the platelet suspension. A comparison of
the thrombus positive and negative patients showed no statistical
significance between values by unpaired t-tests. Images were obtained
on at least alternate days in multiple views and interpreted
prospectively by two observers who were blinded to the clinical and
laboratory data. The criteria for a positive reading was an area of
increased activity in the region of the left ventricle which increased
with time against a decreasing blood pool. Equivocal images were
considered negative.

Fourteen patients had left ventricular mural thrombi. In general
the surface of the thrombus was red and friable, and the shape was
either round or oval. Both observers agreed with the interpretation
of the platelet images. The sensitivity of scintigraphy was 71%, and
the specificity was 100% with a diagnostic accuracy of 92%. The
smallest thrombus visualized by this technique had a surface area of

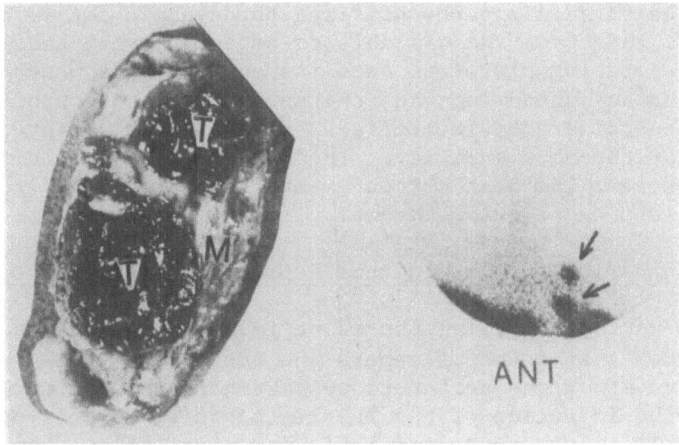

Fig. 2   The scintiphoto in the right panel is in the anterior
           projection and was obtained 72 hours after injection of the
           platelet suspension. The top of the picture is cephalad.
           The arrows mark two thrombi which correspond to thrombus (T)
           in the adjacent surgical specimen. Below the arrowed areas
           are areas of increased activity. On the left is the liver;
           on the right is the spleen (M = myocardium).

approximately one square cm and a depth of 2mm. Thrombi not
visualized on the platelet scan were all greater than two square cm in
surface area and protruded at least 3mm into the left ventricular
cavity. In 12 of 14 patients with positive results, it was possible
to estimate tissue nuclide activity. In patients with positive
scintiphotos, the activity of the surface of the thrombus exceeded the
activity in blood and myocardium (normalized by mass) by at least 9.7
times with a range of 9.7 to 355 with a mean of 108±111. In three of
four patients with a false-negative scintigraphic study, it was
possible to measure tissue In-111 activity from surgical specimens.
The ratio of In-111 activity at the thrombus surface to activity in
surrounding tissue was 4.6, 3, and 16:1, respectively.

Comparison with Two-Dimensional Echocardiography

        The above study was carried out in parallel with an
echocardiographic study in which it was found that 25% of the
echocardiograms were technically poor. These were excluded from the
analysis. In the remaining patients there was observer agreement in
37 of 40 patients (93%). Each reader was asked to consider the
echocardiographic image as positive, negative, or possible for
thrombus. Considering possible thrombi as negative, both observers
agreed that the sensitivity of echocardiography was 77% and the
specificity 96%. The main finding of this investigation was that both
In-111 platelet scintigraphy and two-dimensional echocardiography have
important and complementary roles in the diagnosis of left ventricular
thrombi. Platelet scintigraphy utilized the dynamic nature of the
blood-thrombus interface and was able to detect thrombi that are
actively incorporating platelets onto its surface and thereby
reflected a surface phenomenon and the activity of the thrombus. By
contrast, echocardiography identified a mass and did not provide a
direct index of thrombus activity. These techniques complement each
other and represent the best currently available tests for the
diagnosis of left ventricular thrombi.

Optimum Time and View for Imaging and Reproducibility of the Technique

        The disadvantage of using the above imaging protocol was that it
was time-consuming in terms of camera and technician time. Therefore
a retrospective study was performed to ascertain the optimum time
window following injection of the labeled platelet suspension for
imaging, as well as the most useful imaging views to use (6). In
addition, the study was designed to determine the reproducibility of
this technique. A total of 662 images, obtained from 64 patients was
analyzed retrospectively on two separate occasions by three observers
blinded to the patient identity, views [right anterior oblique (RAO),
anterior, left anterior oblique (LAO), and left lateral (LL)], and
times following injection of the platelet suspension that images were
obtained (1-2, 3-4, and 5-6 days). Images were categorized as either
positive or negative. In every case surgical and/or autopsy

verification of the presence or absence of a left ventricular thrombus
was obtained. The best combination of sensitivity, specificity, and
diagnostic accuracy was found in the 3-4 day LAO view and was 54+5%
(mean+1SD), 98%+1 and 85+2%, respectively. Sensitivity, specificity,
and diagnostic accuracy was not enhanced by adding additional views
(RAO, LL, and anterior) to the LAO view in the 3-4 day time period.
However, using multiple views, localization of the thrombi to the left
ventricle was facilitated. In a second retrospective analysis, a
comparison of day 0 images with images obtained 3-4 days later
enhanced sensitivity and accuracy to 65% (p <.001) and 90% (NS),
respectively. Specificity was unchanged at 99%. The mean inter- and
intra-observer agreement was 91 and 88%, respectively. We conclude
from this retrospective analysis that In-111 platelet scintigraphy is
a reproducible and specific technique for identifying left ventricular
thrombosis and we advise imaging on day 0 and again 3-4 days following
injection of platelet suspension in the RAO, LAO, left lateral, and
anterior views to maximize accuracy and to facilitate localization of
left ventricular thrombosis. Thus a 67% decrease in imaging time with
concomitant lowering of cost was achieved without change in diagnostic
accuracy or specificity. There was, however, a slight decrease in the
sensitivity from 71%-65%.

Left Ventricular Thrombi During Acute Myocardial Infarction

We have performed a study in 41 patients during the acute phase of
myocardial infarction to determine the incidence of active left
ventricular thrombosis. Forty-eight percent of the patients with
anterior transmural myocardial infarction, defined as new Q-waves and
appropriate CK-MB elevations, had left ventricular thrombosis.
Patients with all other kinds of infarcts, i.e., subendocardial or
infero/posterior infarct, were clear of thrombi. Thus it appears that
patients with anterior transmural myocardial infarction are at risk
for thrombus formation and systemic embolization, whereas those with
subendocaridial or transmural infarcts in other locations are at low
risk for systemic embolization.

The Effect of Aspirin on the Incorporation of In-111-Labeled Platelets into Cardiac Thrombi in Man - An In Vitro and In Vivo Correlation

The in vitro and in vivo behavior of platelets was studied in 11
patients with left ventricular aneurysms and mural thrombi (7). Five
patients were on aspirin in doses between 300 and 2,400 mg/day. The
remaining patients were controls. In vitro function was tested by the
aggregation response of platelets to ADP and collagen. In vivo
function was assessed by the incorporation of In-111 labeled platelets
into cardiac thrombi as measured by platelet scintigraphy. Platelets
from patients on aspirin, whether tested before or after labeling,
aggregated less with collagen than controls (18 vs 52%; P 0.1 before
labeling and 13 vs 49%; P .02 after labeling). Second-wave
aggregation induced by ADP was impaired in patients on aspirin. In
all patients scintigraphy showed that the autologous labeled platelets

were incorporated into ventricular thrombi. Thus although platlets
from patients on aspirin aggregated subnormally, in vitro, their in
vivo behavior was not affected. A self-criticism of this paper is
that the scintigraphic images were not quantified and that since
biology is not an all-or-none phenomenon, a second study should be
undertaken using quantification of images to determine if subtle
differences in the aspirin and control groups exist.

BACTERIAL ENDOCARDITIS

Riba et al (8) demonstrated that vegetations may be identified
using labeled platelets. While this animal study was well performed,
the endocarditis produced was particularly aggressive and was not
characteristic of pathology usually seen in humans. We have studied
13 patients with clinically proven bacterial endocarditis and have
found that this technique is insensitive for the diagnosis of
vegetations. However, in the clinical setting, antibiotic therapy
might modify the disease process and result in a less aggressive
lesion, thus reducing the sensitivity of this technique. It is
possible that if background subtraction techniques can be reliably
developed, vegetations may be identified with improved efficacy.

IDENTIFICATION OF DEEP VEIN THROMBOSIS

Indium-111 platelet scintigraphy has exciting potential in the
diagnosis of deep vein thrombosis (Fig. 3) and it is in this situation
that it will probably reach the clinical arena soonest. The half-life
of In-111 and the life-span of platelets allow imaging for at least 5
days following injection of the platelet suspension. Thus if a deep
vein thrombosis is identified, its propagation and/or response to
therapy can be directly monitored for at least 5 days. Also, if a
pulmonary embolus was to occur during this 5-day time period, imaging
of the chest without further injection of radioisotope may allow
identification of the embolic lesion.

We embarked on a study designed to determine the diagnostic
accuracy of platelet scintigraphy using contrast venography as a
reference standard for the identification of deep vein thrombosis.
While this study is ongoing, preliminary results indicate that off
heparin the sensitivity is 100% and the specificity 88% (n=28).

Possible causes of false-positive scan are 1) subcutaneous
hematoma, 2) surgical wounds, 3) inflammatory joint disease, 4)
concomitant arterial disease with arterial thrombosis or platelet
accumulation at the site of prosthetic grafts. Our preliminary
studies indicate that heparin given intravenously, maintaining a
partial thromboplastin time at least 2.5 x control, is effective in
preventing platelet accumulation on thrombi. For the diagnosis of
pulmonary embolus our experience is limited. However, we have been

successful in identifying a patient with pulmonary embolus and have
obtained independent verifications of the accuracy of the diagnosis
(9). The sensitivity and specificity of platelet scintigraphy in man
in heparinized and nonheparinized patients is yet to be determined.

PATIENTS UNDERGOING PERIPHERAL ANGIOPLASTY

Percutaneous transluminal angioplasty has become an increasingly
popular alternative to medical management and major surgery for
symptomatic coronary and peripheral vascular disease. While
angioplasty is a low morbidity nonsurgical procedure and initial
dilatation is usually successful, restenosis following successful
angioplasty constitutes a major problem. We hypothesize that
platelets are involved in the restenosis process and that platelet
deposition which can be visualized using In-111 labeled cells might
predict vessels at risk for occlusion. In a preliminary study
involving six patients, we demonstrated significant platelet
accumulation at the site of balloon dilatation in five (Tables 2 and
3). This was evident within an hour of dilatation and for at least 24

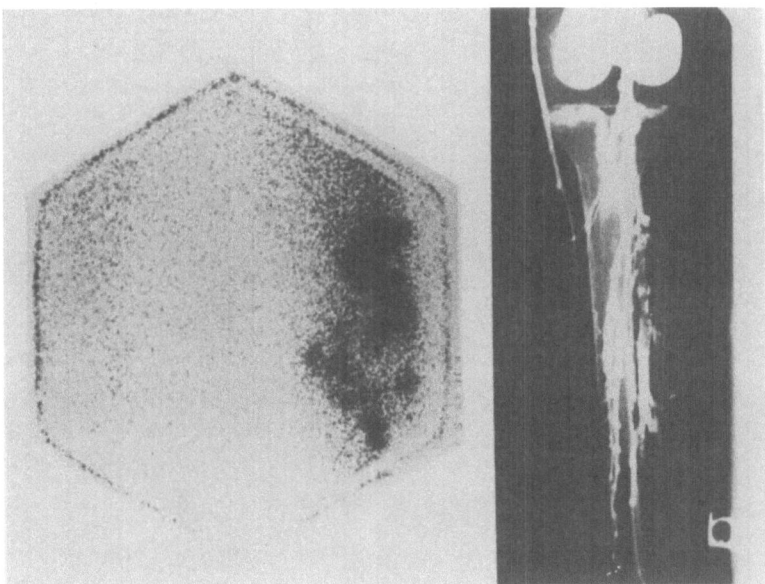

Fig. 3   The panel on the left is a In-111 scintiphoto of both calves
          in a patient with deep vein thrombosis. These images were
          acquired 24 hours after injection of the platelets. Top is
          cephalad. The tortuous area of increased activity
          represents a large venous thrombus. The corresponding
          venogram is shown on the right. Filling defects in the
          corresponding venogram are seen.

Table 2. Patients Undergoing Angioplasty

Patient #	Age	Sex	Artery Dilated	Site of Catheter Insertion
1	53	Male	Left Superficial Femoral	Left Common Femoral
2	54	Female	Left Renal Artery	Right Common Femoral
3	65	Male	Left Superficial Femoral	Left Common Femoral
4	56	Male	Right Common Femoral Right External Iliac	Left Common Femoral
5	48	Male	Right and Left Common Iliac	Left Common Femoral Right Common Femoral
6	49	Female	Left Superficial Femoral Left External Iliac Right External Iliac	Right Common Femoral
Mean ± 1SD	54.2±5.6			

Table 3. Results of Angioplasty

Patient	T	In-111 Scintophotos of Puncture Site 0-24	24	48	72	96	120*	Dilatation Site 0-24	24	38	72	96	120*	H** (I.U.)	S*** (I.U.)	H+	P++ (daily)	A++ (daily)	C++ (daily)
1	0.5	+	+	+	+	+		+	+		+	+		5000	–	–	650 mg	75 mg	5 mg
2	4	+	–	–				+			+		–	5500	20000	1000/hr x 12	–	–	–
3	0.25	+	+	+	+	+		+	+	+	+	+	–	5000	–	1000/hr x 12 500/hr x 8	650 mg	–	–
4	0.5	+	+	+				+	+	+	+	+		5000	–	1000/hr x 12	–	–	–
5	0	+						+						10000	–	–	–	–	–
6	0	+						+	+					14000	48000	5000 stat	–	–	–

+ = focal uptake

– = no uptake

* = hours after dilation

H** = intravenous heparin administered at the time of angioplasty

S*** = intra-arterial streptokinase administered at the time of angioplasty

T = time between dilatation and injection of the platelet suspension (hrs.)

H+ = IV heparin adminstered following angioplasty

++ = administered chronically after angioplasty

P = Persantine

A = Aspirin

C = Coumadin

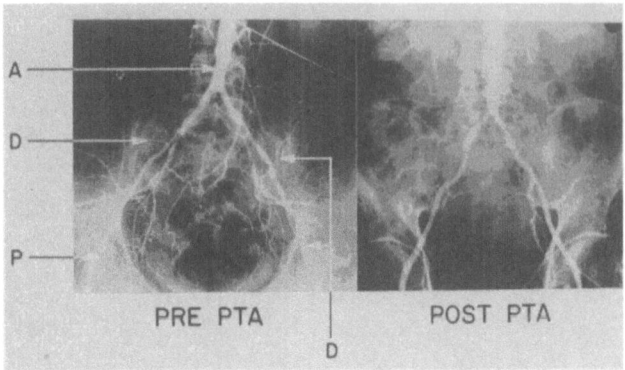

Fig. 4   These are angiograms of the iliac and femoral arteries
          before and after successul percutaneous transluminal
          angioplasty (PTA).

          A = aorta, D = Areas of dilatation, P = Pelvis

          The top is cephalad.

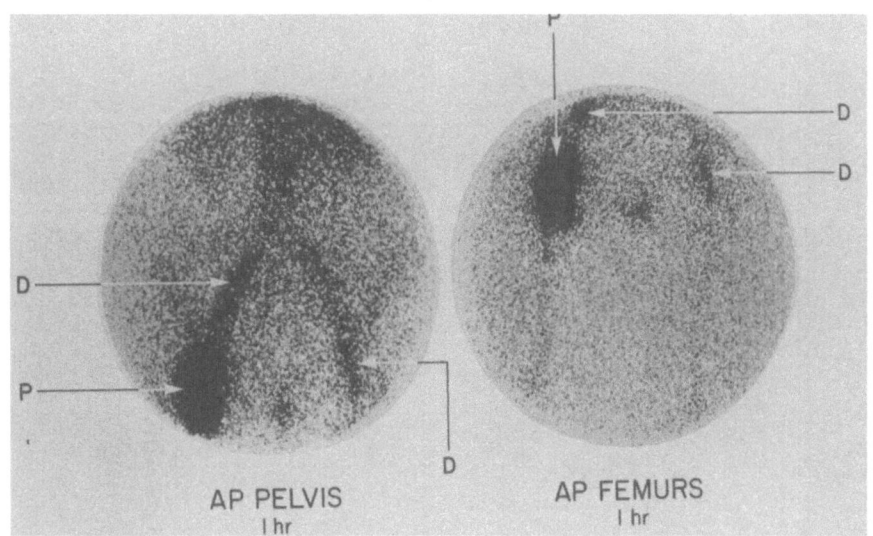

Fig. 5   This is the In-111 platelet scintigraph corresponding to the
          angiogram shown above. The top of both images is cephalad.

          D = Area of dilatation
          P = Puncture site with extravasation of blood
              around that area

hours thereafter (Figs. 4,5). During the period of imaging,
intravenous heparin in doses sufficient to produce a prolongation of
the partial thromboplastin time by at least 2.5 x control as well as

streptokinase were administered. These data indicate that these
agents are not effective in preventing platelet accumulation at the
site of balloon dilatation. The role that platelets play in
restenosis of vessels following dilatation and the possible
therapeutic benefit of antiplatelet and anticoagulation agents at the
time of the procedure needs further evaluation.

Following dilatation there is reconstitution of the lumen of the
vessel at the points marked by the arrows labeled D.

These 40,000 count images were obtained 1 hour following
injection of the platelet suspension using a large field of view gamma
scintillation camera fitted with a medium energy collimator. Both
In-111 peaks were used with 20% windows. Platelet uptake is seen at
sites of dilatation.

DIAGNOSIS OF LEFT ATRIAL MASS

Platelet scintigraphy is thought to be specific for thrombosis.
We reported recently a 46-year-old female whose echocardiographic
examination suggested a left arterial mass (10). Indium-111 platelet
scintigraphy was performed and demonstrated an area of increased
activity in the region of the left atrium. At surgery a left atrial
myxoma was successfully removed. This case illustrates that myxomas
may mimic intracardiac thrombi identified by In-111 platelet
scintigraphy. This, we believe is the first case of a nonthrombus
presenting as a positive In-111 cardiac platelet scintiphoto. The
In-111 platelet images obtained from this patient have certain
distinguishing features from those due to thrombi. The images were
maximally positive 14 to 48 hours following injection of the platelet
suspension and thereafter declined in intensity, indicating that the
kinetics of uptake of platelets on the tumor is somewhat different
from that due to the typical thrombus. Thrombi tend to be maximally
positive 72 hours after injection of the platelet suspension as the
activity of the thrombus increases with respect to the declining
activity in the circulating blood. This difference suggests transient
adherence of platelets to the tumor rather than permanent
incorporation as occurred in a thrombus.

IDENTIFYING CORONARY THROMBI IN MAN

This case report is illustrative of the problems encountered
during imaging of coronary thrombi in man. A 79-year-old white male
with known three-vessel coronary disease presented with an acute onset
of severe retrosternal chest pain which began at 6:00 p.m. on the day
of admission. On admission an acute evolving transmural anterolateral
myocardial infarction was diagnosed on the basis of his admission EKG.
The peak CK was 1985 IU (MB 16%). On the day of admission he arrested
in ventricular fibrillation and was resuscitated successfully with
restoration of normal hemodynamics. Twenty-three hours after the

onset of chest pain In-111 labeled platelets (382 μCi) were injected
intravenously. The scintiphoto with the corresponding coronary
angiogram is shown in Figure 6. The patient died suddenly five days
later of an arrhythmia. At autopsy a recent transmural anterolateral
myocardial infarction was seen. The patient had diffuse coronary
disease with a left dominant system. The right coronary artery was
totally occluded. There was 80% narrowing of the left circumflex 1.5
cms from the origin and two distinct 90% stenoses of the left anterior
descending coronary artery. These corresponded to the coronary
angiogram. No thrombus was seen in any of the vessels. A presternal
subcutaneous hematoma was identified. Careful dissection of the chest
wall and lungs failed to demonstrate a thrombus which might have
produced the increased activity in the region of the left coronary
artery.

We presume that lysis of the thrombus occurred. Thus confirmation of the scintigraphic finding was not possible. The left
ventricular hypertrophy aided separation of the surface coronary
vessel from the cardiac blood pool, allowing imaging of the coronary
artery without background subtraction. The case report illustrates
the following problems concerning imaging of coronary thrombi.

1.   Because of thrombus disaggregation confirmation of scintiphotos
      at autopsy may be difficult.

2.   The background blood pool renders early identification of
      coronary thrombi very difficult. The dual isotope background
      subtraction technique might obviate these difficulties.

SINGLE PHOTON TOMOGRAPHIC IMAGING OF CARDIAC THROMBI

The purpose of this study was to demonstrate the use of the gamma
tomographic scanner (Pho Con 192) to localize cardiac thrombi using
In-111 labeled platelets and to compare the information obtained with
that provided by routine gamma camera imaging.

The Pho Con 192 is a dual-detector system utilizing two
scintillation cameras with multi-hole focusing collimators. The
tracking path is similar to a conventional rectilinear scanner. Each
detector, one anterior and one posterior, moves in unison over the
radioisotopic source. Image reconstruction (Fig. 7) algorithms built
into the camera allow the recognition of 12 images corresponding to a
distinct plane of activity. Six patients with left ventricular
aneurysms were studied by both techniques. Platelets (3.43+1.46 x
10^9), containing 425+19 μCi In-111 activity, with a final labeling
efficiency of 74+10% were injected intravenously. Platelet recovery
at 15 minutes was 65+23%. Platelets aggregated normally with ADP and
collagen (n=4) both before and after labeling. Routine imaging was
performed on the day of injection and daily or on alternate days in
the LAO, RAO, anterior, and LL views. Tomographic imaging was

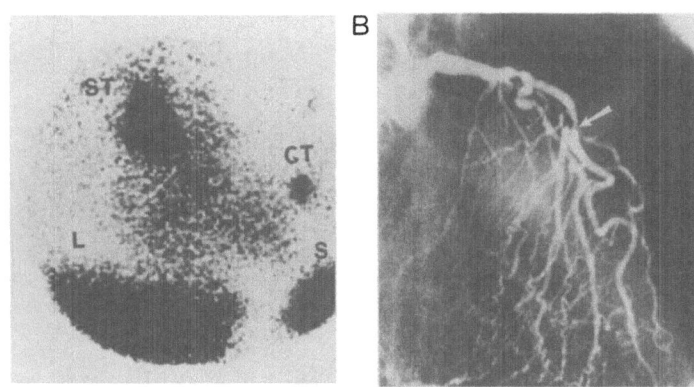

Fig. 6   The figure on the left (A) is a right anterior oblique
          In-111 scintiphoto obtained 24 hours after injection of the
          platelet suspension. The top of the image is cephalad. L =
          liver, S = spleen, CT = presumed coronary thrombus. The
          coronary angiogram (B) is shown adjacent with the arrow
          indicating the narrowing which corresponds to the site of
          increased coronary activity on the scintiphoto. The intense
          area of activity in the scintiphoto in the sternal area was
          a subcutaneous hematoma (ST).

performed with the patient supine and/or in an LAO position. Four
studies were negative by both techniques. Two patients had evidence
of active thrombi, one surgically proven, by both techniques. We
conclude from this preliminary study that 1) single photon tomographic
imaging of cardiac thrombi is possible using a platelet preparation
with the characteristics described; 2) tomographic imaging may
provide more accurate localization of thrombi (from single or multiple
projections) than is possible with conventional gamma camera imaging.

PLATELET FUNCTION OF STORED PLATELETS FROM NORMAL VOLUNTEERS

     Our laboratory investigated platelet function of stored platelet
concentrates in three separate experiments using healthy volunteers as
subjects.

Evaluation of a New Plastic Platelet Storage Container In Vivo
Analysis Only)

     This study was performed to evaluate a new plastic formulation
PL-1240 (Fenwal Laboratories, Deerfield, IL, USA) used for storage of
platelet concentrates. These new storage bags are made of
polyvinylchloride with a trimellitate plasticizer and can be used to

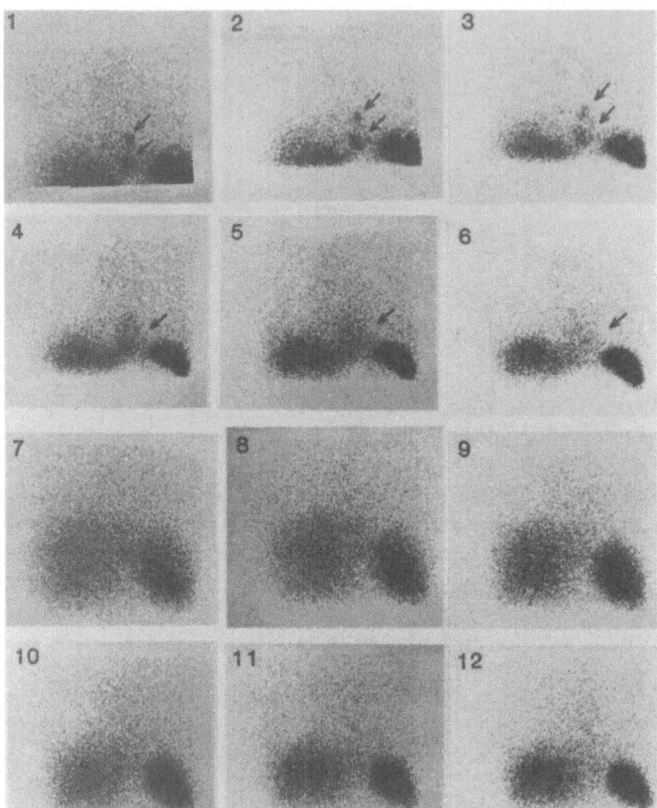

Day 5

Fig. 7   This is a series of 12 tomographic images obtained 5 days
         following injection of the platelet suspension. Each
         scintiphoto is oriented with the top cephalad. The areas
         arrowed represent two separate thrombi within a large
         antero-apical aneurysm. The area of increased activity in
         the left lower zone of each panel is the liver; that in the
         right lower zone is the spleen. Slice 1 is the most
         anterior and slice 12 the most posterior.

store platelets in vitro for up to five days. The standard plastic
bag made of polyvinylchloride with a 2-diethylhexyl phthalate
plasticizer (PL-146) is licensed in the United States for only three
days for platelet storage.

    For this study, platelet concentrates were prepared from
volunteer donors, using either of the two bags, according to a
standard American Red Cross protocol and were stored for 3 days
(PL-146) or 5 days (PL-1240) at 20-24°C on 6 rpm circular rotators.
At the end of storage the platelets were labeled with a maximum of 70

µCi of In-111 oxine and reinfused into the autologous volunteers.
Two changes in the In-111 oxine-platelet labeling technique, described
previously, were necessary to label stored platlet concentrates. First
the starting preparation for labeling was 40-60 ml of platelet concentrate, not whole blood. Thus the concentrate was removed from the storage
bag through a 19 gauge needle. The pH was determined immediately. Labeling was performed if the pH was above 6.0 since below this value platelet
in vivo survival is markedly decreased. Second, the centrifugation
time for the formation of the platelet pellet was increased from 5 to
10 minutes. Platelet recovery was measured at 2 hours postinjection of
the radiolabeled platelets, and platelet survival was determined over
7 days. To estimate mean survival time, the experimental data were
subjected to computer analysis. Three mathematical models were used
for formal curve fitting. A linear model, an exponential model, and a
maximum likelihood estimate of the integer-ordered gamma function
(multiple hit model) (11). Results for PL-1240 were comparable to
those found for In-111 labeled platelets stored for 3 days in standard
(control) PL-146 plastic storage bags (Table 4). These results are
also comparable to those results reported for stored platelets using
Cr-51 as the radionuclide (12,13,14). We conclude that platelets
stored for 5 days in PL-1240 plastic packs show in vivo recovery and
survival values which are comparable with data from standard control
platelet bags stored for only 3 days. Platelets stored for up to 5
days in these new plastic bags should be useful for clinical platelet
transfusions. Furthermore, the ability to store platelets for up to
five days will permit blood banks to maintain a larger inventory of
platelet concentrates, especially over weekends and holiday periods.

Study of Correlation of In Vitro Release of Lactic Dehydrogenase and B-Thromboglobulin with In Vivo Survival of Indium-111 Labeled Platelet Concentrates

Currently, assessment of in vivo platelet function is best
evaluated by platelet survival studies which require use of a
radionuclide, usually Cr-51 or In-111, and the acquisition of blood
samples daily for about 8 days. If a reliable in vitro test of
platelet function which correlates with in vivo survival were found,
cumbersome survival studies would be unnecessary. Thus we attempted
to correlate in vitro release of two platelet proteins lactic
dehydrogenase (LDH) and β-thromboglobulin (β-TG) with in vivo platelet
survival. Platelet concentrates were prepared in either PL-732,
PL-1240 (Fenwal Laboratories, Deerfield, IL, USA), or CLX (Cutter
Laboratories, Berkeley, CA, USA) plastic bags and stored on 1 (E1) or
6 (E6) rpm elliptical (Melco Engineering Corp., Glendale, CA, USA) or
2 (C2) or 6 (C6) rpm circular rotators (Helmer Labs, St. Paul, MN,
USA) for 5 days at 20-24°C. At the end of the fifth day of storage,
the platelet concentrates were assayed for pH, percent release of LDH,
and β-TG, (15), labeled with up to 70 µCi of In-111 oxine and
reinfused into autologous volunteer donors. Recovery and survival
were calculated as described above.

Results (see Table 5) showed that the mean 2 hour recovery for all platelet bag rotator combinations was 39%+11% (1SD; n=52). Platelet concentrates stored in PL-732 bags on the 6 rpm elliptical rotator (E6) showed the poorest two hour recovery (27%+13%). These results concerning PL-732 confirm work previously reported by Murphy et al (12) and Snyder et al (14) on the poor in vitro and in vivo storage characteristics of platelets stored in this bag-rotator combination. All other platelet storage bag rotator combinations showed survival data which were comparable both with each other and with data reported in the literature (12,13). Of interest was the finding that in vivo platelet survival was similar (p > 0.05) among all rotator-plastic-bag pairs even for the PL-732-E6 combination. Apparently those platelets still circulating 2 hours after transfusion are removed from the circulation with a normal survival curve. The γ2 values correlating in vitro percent LDH and β-TG release with in vivo 2 hour platelet recovery were too low or inconsistent for use in predicting in vivo platelet function.

We conclude that platelet concentrates prepared in PL-732, PL-1240, and CLX containers can be stored on any of the platelet rotators studied with the exception of PL-732 which, because of decreased in vivo recovery, should not be stored on 6 rpm elliptical rotators. In vitro release of platelet proteins LDH and β-TG is useful for determining the degree of damage occurring during in vitro storage but does not reliably correlate with in vivo platelet survival.

In Vivo Indium-111 Labeled Platelet Survival After Passage Through an Electromechanical Infusion Pump

Pediatric and neonatal units often employ electromechanical infusion pumps to control the rate and volume of an intravenous infusion. We studied the degree of damage to platelet concentrates caused by an Abbott Model 3 Pump (Abbott Lab, North Chicago, IL, USA) (Fig. 8). Platelet count and percent release of LDH and β-TG were measured in platelet concentrates stored for between 2 and 120 hours both pre- and post-passage through the infusion pump. Control platelets not passed through the pump were passed through a 170 μm standard blood filter. Platelets were passed through the Abbott Pump System and immediately labeled with In-111 oxine. In vivo platelet recovery and survival were determined after injection of 50-70 μCi of In-111 oxine-labeled platelets into autologous volunteer donors. Platelet survival was determined using the computer program as mentioned above.

In vitro results showed no change in platelet count or percent release of either LDH or β-TG after passage through the pump. For in vivo studies, post-pump percent recovery at 2 hours was 38%+12 (1SD) (Table 6). This result was similar to a value of 39%+11% (1SD)

Table 4. In Vivo Indium-111 Oxine Kinetics In Normal Volunteers

| Plastic Bag | N | % Recovery at 2 hours* | Survival (Days) | | |
			Linear*	Exponential*	Multiple Hit*
PL-1240	10	46±16	6.9±1.0	2.6±0.6	5.1±1.8
PL-146	4	35± 8	6.9±1.2	3.1±0.7	5.9±2.5

*For PL-1240 vs PL-146 p>0.05 (NS).
(mean ± SD)

Table 5. In vivo Indium-111 Oxine Platelet Survival after Five Days of In Vitro Storage (Mean Values) Correlation with In Vitro Tests

Bag	Rot	N	pH	LDH[a]	B-TGa	% REC[b]	SURV[c]	r^2 LDH(d)	r^2 B-TG(e)
PL-732	E1	7	7.3	9	32	42	7.6	0.001	0.003
PL-732	C2	7	7.3	7	34	41	6.9	0.05	0.005
PL-732	E6	7	7.4	24*	46*	27*	7.3	0.008	0.04
PL-1240	E1	3	7.4	11	29	46	7.8	0.004	0.49
PL-1240	C2	6	7.2	10	37	43	7.4	0.03	0.09
PL-1240	C6	10	7.2	ND	ND	46	6.9	ND	ND
CLX	E1	6	7.2	7	33	36	7.3	0.80	0.05
CLX	C2	6	7.1	6	31	42	7.1	0.56	0.12

```
*       (P<0.05)
**      E = Elliptical rotator; C=circular rotator; 1,2,
        6=RPM
a -     % release
b -     2 hour recovery
c -     linear survival (days)
r² =    corr. coeff. for 2 hour recovery vs LDH(d)
        and B-TG(e)
```

obtained in our laboratory for platelet concentrates stored for 5 days but not passed through the electromechanical pump before reinfusion.

We conclude that passage of platelet concentrate through the Abbot Model 3 Electromechanical Pump is an acceptable clinical

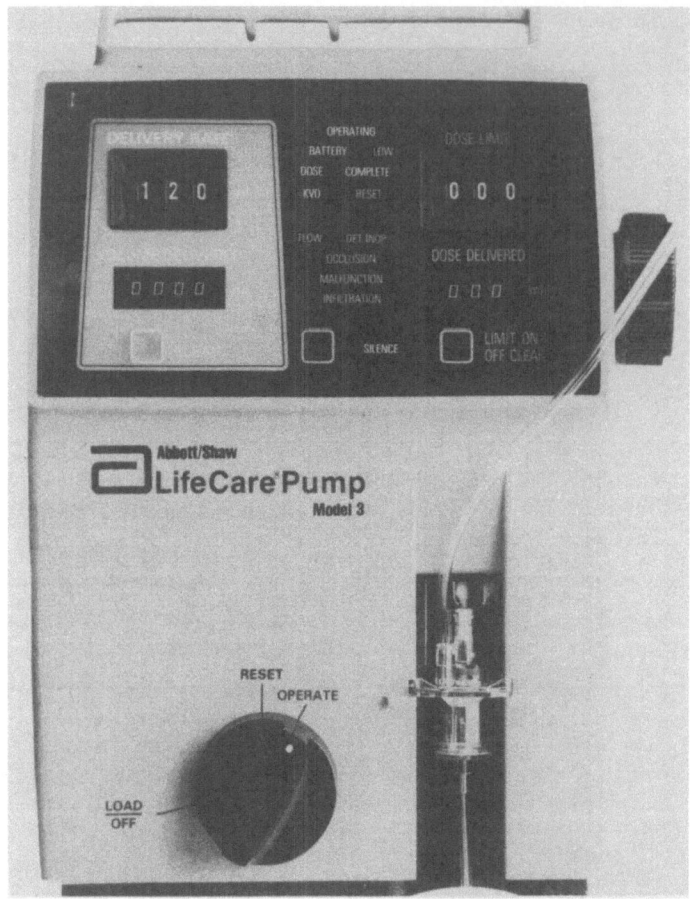

Fig. 8   Abbott/Shaw life care infusion pump model 3.

practice, useful in pediatric or neonatal units when precise volume
and rate control of platelet transfusion is desired.

CONCLUSION

     During the past 7 years platelet imaging has advanced
significantly in the areas of improved platelet labeling methods and
imaging thrombosis in man. It has probably entered the clinical arena
in three areas: the identification of cardiac thrombi, deep vein
thrombosis, and early rejection of the transplanted kidney. Currently
this technique has reached a critical developmental stage and will
undoubtedly enjoy widespread clinical use if shown superior to and
less cumbersome than competing methods. It has already improved our

Table 6.   Effect of an Electromechanical Infusion Pump on Platelet
           Function

	Pre-pump Control (170 um filter)*	Post-Electromechanical Pump*
Platelet count x 10^7/ml	145 + 32	146 + 22
LDH % release	6.7 + 1.1	6.9 + 1.1
B-TG % release	24 + 10	22 + 9
In-111 in vivo % recovery at 2 hours	39 + 11**	38 + 12***
In-111 in vivo survival (days)****	7.3 + 0.3	7.1 + 0.4

*   For all parameters p>0.05 (NS)
**   N=68
***   N=5
****   Linear analysis

understanding of thrombosis and thromboembolism and has defined more
clearly the importance and efficacy of anticoagulation therapy.   With
the introduction of new data acquisition and image-processing
techniques its future should be bright.

ACKNOWLEDGMENTS

     We wish to thank the following collaborating investigators: K.
Barker, G. P. Basmadjian, R. D. Burow, V. Caride, A. C. Cox, R. C.
Elkins, D. C. Galloway, L. H. Harrison, Jr., P. J. Kanaly, D. J.
Kellerman, H. Krous, H. R. Lee, D. E. Parker, M. D. Peyton, S. Rieker,
D. C. Riggs, H. D. Sostman, T. M. Streitz, F. B. Taylor, and D. A.
Wilson. M. J. Welch, C. J. Mathias, and U. Scheffel have at various
times helped Dr. Ezekowitz with the labeling techniques.   We thank
them also.

     We would also like to thank Mary Murray, Debbie Beauvais, and
Linda Latham for preparation of the manuscript.

This work was supported by NIH PPG No 2 PO 1 HC 17812-06 and Merit
Review Grant, Veterans Administration Medical Center.

REFERENCES

  1. M. L., Thakur, M. J. Welch, J. H. Joist, R. E. Coleman,
      Indium-111 labeled platelets: studies on preparation and
      evaluation of in vitro and in vivo functions, Thrombo Res
      9:345, (1976).
  2. W. A. Heaton, H. H. Davis, M. J. Welch, C. J. Mathias, J. H.
      Joist, L. A. Sherman, B. A. Siegel, Indium-111:   a new

radionuclide label for studying human platelet kinetics, <u>Br J Haematol</u> 42:613 (1979).

3. M. D. Ezekowitz, J. C. Leonard, E. O. Smith, E. W. Allen, F. B. Taylor, Identification of left ventricular thrombi in man using Indium-111-labeled autologuous platelets. A preliminary report, <u>Circulation</u> 63(4):803 (1981).

4. M. D. Ezekowitz, E. O. Smith, E. W. Allen, J. C. Leonard, C. W. Smith, G. P. Basmadjian, F. B. Taylor, Identification of left ventricular thrombi in humans using Indium-111 labeled platelets, <u>in</u>: "Proceedings of a State of the Art Symposium on Radiolabeled Blood Cellular Elements: Current Accomplishments, Immediate potential, Future Possibilities", New York, Sept. 14-15 (1979).

5. J. R. Stratton, J. L. Ritchie, G. W. Hamilton, K. E. Hammermeister, L. A. Harker, Left ventricular thrombi: in vivo detection by Indium-111 platelet imaging and two dimensional echocardiography, Am J Cardiol 47:874 (1981).

6. M. D. Ezekowtiz, R. D. Burow, P. W. Heath, T. Streitz, E. O. Smith, D. E. Parker, The diagnostic accuracy of Indium-111 platelet scintigraphy for identifying left ventricular thrombi, Am J Cardiol 51(10):1712, (1983).

7. M. D. Ezekowitz, E. O. Smith, F. B. Taylor,: Failure of aspirin to prevent incorporation of Indium-111 labeled platelets into cardiac thrombi in man, Lancet 29:440 (1981).

8. A. L. Riba, M. L. Thakur, A. Gottschalk, V. T. Andriole, B. L. Zaret, Imaging experimental infective endocarditis with Indium-111 labeled blood cellular components, <u>Circulation</u> 59:336, (1979).

9. M. D. Ezekowitz, E. R. Eichner, R. Scatterday, R. C. Elkins, The diagnosis of a persistent pulmonary embolus by Indium-111 platelet scintigraphy with angiographic and tissue confirmation, <u>Am J Med</u> 72:839, (1982).

10. M. D. Ezekowitz, E. O. Smith, R. Rankin, L. H. Harrison, H. F. Krous, Left atrial mass: Diagnostic value of transesophageal two-dimensional echocardiography and Indium-111 platelet scintigraphy, <u>Am J Cardiol</u> 51:1563, (1983).

11. U. Scheffel, M. F. Tsan, T. G. Mitchell, E. E. Camargo, H. Braine, M. D. Ezekowitz, L. Nickoloff, R. Hill-Zobel, P. A. McIntyre. Human platelets labeled with In-111 8-hydorxyquinoline: kinetics, distribution, and estimates of radiation dose, J Nucl Med 23:149, (1982).

12. S. Murphy, R. A. Kahn, S. Holme, G. L. Phillips, W. Sherwood, W. Davisson, D. H. Buchholz, Improved storage of platelets for transfusion in a new container, <u>Blood</u> 60:194, (1982).

13. T. L. Simon, E. J. Nelson, R. Carmen, S. Murphy, Extension of platelet concentrate storage, <u>Transfusion</u> 23:207, (1983).

14. E. L. Snyder, R. Aster, S. Murphy, W. Davission, P. Ferri, C. Pope, E. Smith, R. Kakaiya, M. Ezekowitz, D. Buchholz, Clinical evaluation of platelets stored for five days in PL-1240 plastic packs, <u>Transfusion</u> (in press).

15. E. L. Snyder, A. Hezzy, A. J. Katz, J. Bock, Occurrence of the release reaction during preparation and storage of platelet concentrates, Vox Sang 41:172, (1981).

16. E. Snyder, T. A. W. Koerner, R. Kakaiya, P. Moore, T. Kiraly, Effect of mode of agitation in storage of platelet concentrates in PL-732 containers for five days, Vox Sang 44:300, (1983).

THE USE OF INDIUM-111-LABELED PLATELETS IN

THE MANAGEMENT OF RENAL TRANSPLANT PATIENTS

Helmut F. Sinzinger and Christian W. Leithner*

Division of Nuclear Medicine
*Second Department of Internal Medicine
University of Vienna, Garnisongasse 13
A-1090 Vienna, Austria

INTRODUCTION

Role of Platelets in Pathogenesis of Renal Transplant Rejection

The endothelium of arteries, capillaries, and veins of transplanted organs represents a frontier between the transplanted organ and the recipient. Its integrity depends on immunological considerations. In a recent publication, Frampton and coworkers (1) stated, "The more acute the rejection, the more profound the platelet involvement."

Hyperacute rejection is characterized by a sudden and severe immunological event occurring during or shortly after transplantation. Numerous studies in several experimental models and in humans have revealed that preformed circulating antibodies activate the complement system and play a dominant role (25, 48, 88). When the endothelium is damaged, platelets accummulate on its surface. In earlier experiments on the inbred rat, we showed, by scanning electron microscopy (52, 54), that one of the first detectable alterations is severe endothelial damage, which may produce desquamation (Fig. 1). Almost immediately platelets deposit at the site of injury (Fig. 2). Platelets also contribute to alterations in the glomerulus in hyperacute rejection by forming aggregates and thrombi (7-11). The activation of the complement system during hyperacute rejection might further enhance platelet aggregation via the release of the peptides C3a and C5a (12). Although Colman and coworkers (8) suggest that hyperacute rejection is a localized reaction, confined to the graft and not characterized by peripheral platelet count decrease, Kohler and Tilney (13) observed microangiopathic hemolytic anemia and thrombocytopenia with hyperacute rejection.

201

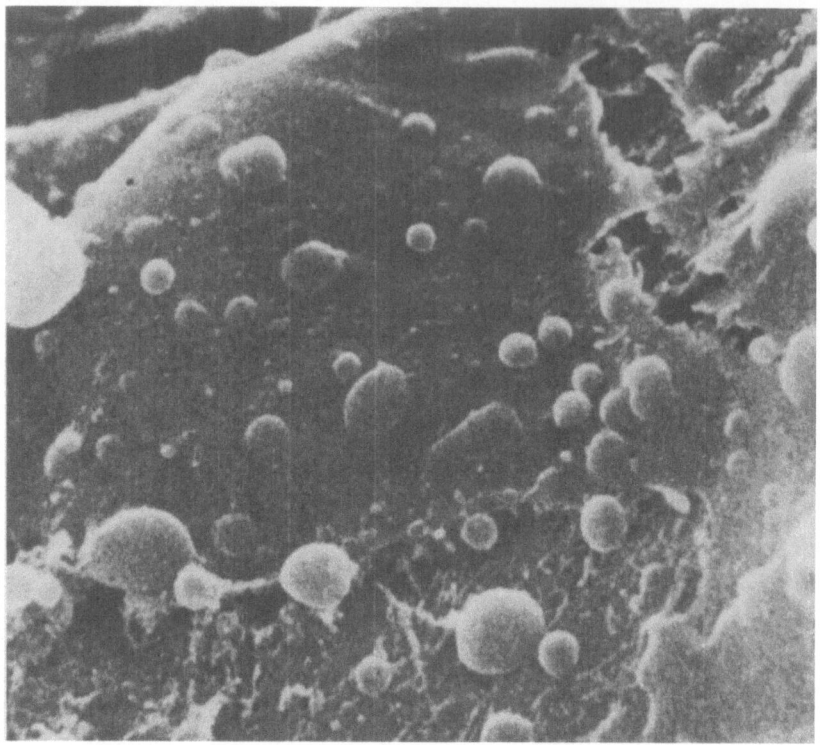

Fig. 1.   Scanning electron micrograph of hyper acute rejection in the
          inbred rat (see Leithner, et al [5]), or (see Ritchie, et al
          [52]).   Two minutes after declamping the vascular
          anastomoses, severe endothelial damage resulting partly in
          desquamation (arrows) can be demonstrated.   Original
          magnification 4650 x.   (Reproduced with permission of Verlag
          Hans Huber, Bern, Switzerland).

    Acute vascular rejection shares several characteristics with the
hyperacute type.   The essential difference involves the rapidity of
the process.   The thrombotic process does not develop within a few
minutes, but usually needs several days.   Platelet aggregates in
glomerular capillaries can be demonstrated in acute rejection by
electron microscopy of well-fixed biopsy specimens but may escape
recognition by routine light microscopy (14-19).   In 1975 Cerilli et
al (20) suggested that antivascular endothelium antibodies may play a
crucial role in pathogenesis of acute rejection.   These findings were
supported by Paul et al (21) and Claas et al (22) who demonstrated
antibodies directed against donor antigen on the endothelial cells.

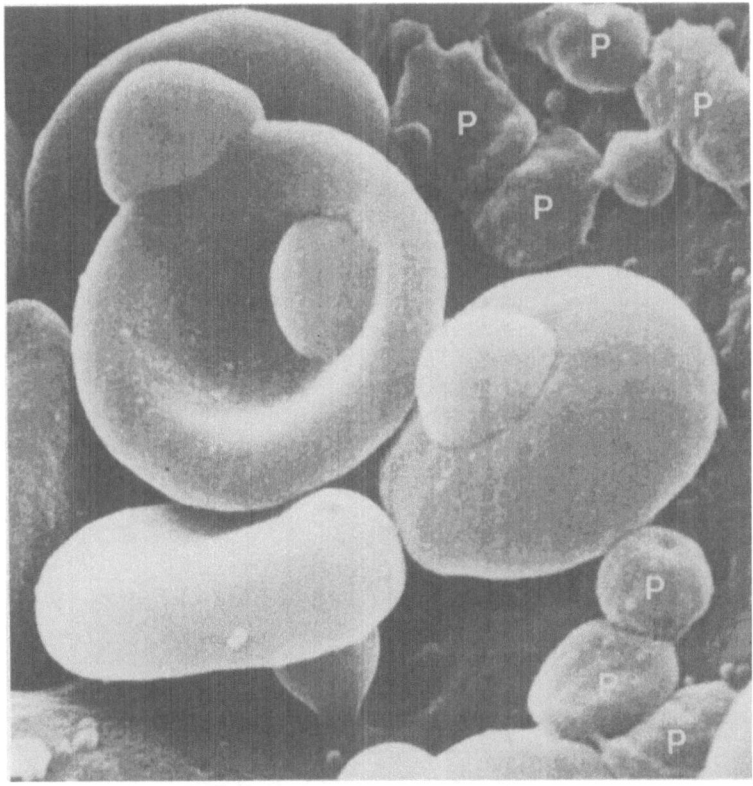

Fig. 2. Deposition of platelets (P) in areas of endothelial damage
(center). Original magnification 3850 X. (Reproduced with
permission of Verlag Hans Huber, Bern, Switzerland).

Monocytes are found in the intima in the majority of patients
suffering from acute rejection. Recently Hendriks et al (23)
concluded from the Eurotransplant data base that HLA-DRw6-positive
recipients of HLA-DRw6-negative renal transplants are particularly
prone to develop acute rejection, probably by producing endothelial
monocyte antibodies (24). Circulating immune complexes also
contribute to endothelial damage in acute graft rejection (25-26).
Besides thrombotic, inflammatory, and, in severe cases, even necrotic
alterations of arteries, arterioles, and glomerular capillaries, acute
rejection is often characterized by obliterative proliferative lesions
of arteries (Fig. 3) and arterioles (14, 15, 27-31). It is not a
proliferation of the endothelial cell layer itself but a thickening of
the subendothelial intimal region by proliferation of modified smooth

muscle cells (Fig. 4) and fibroblasts and an accumulation of ground
substance that cause the problems.   These lesions also can be
encountered in chronic graft rejection.   They develop slowly over
months and years reflected by its rather fibrous nature (15, 27-31).
Since the beginning of renal transplantation, pathologists and
clinicians were puzzled by this obliterative arteriopathy.   Several
mechanisms may play a role in its pathogenesis.   Kincaid-Smith (15,
33) regarded these alterations as a result from repeated organization
of mural thrombi.   Jonasson et al (34) speculated that antigen-
antibody-complement complexes in the vessel walls may serve as an
irritative stimulus to vascular proliferation.   O'Connell and Mowbray
(35) concluded from their experiments that intimal thickening is a
result of the repair process following antibody-mediated endothelial
damage.   Platelets presumably also play an important role (36-38).
They liberate a growth factor that stimulates proliferation of
arterial smooth muscle cells (39) and, therefore, might initiate and
promote obliterative arteriopathy.

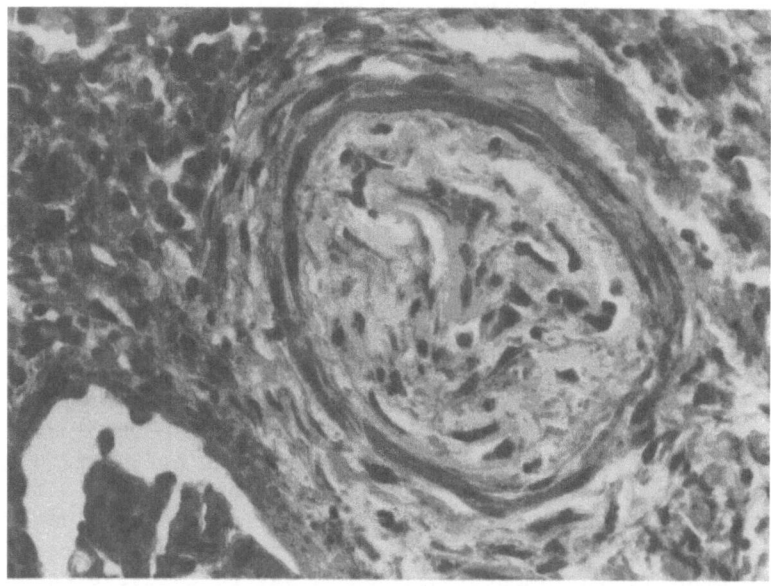

Fig. 3.   Obliterative arteriopathy in acute human renal transplant
          rejection.   The process has resulted in a complete
          obliteration of the vascular lumen; hematoxylin-eosine
          staining, orginal magnification 160 x.

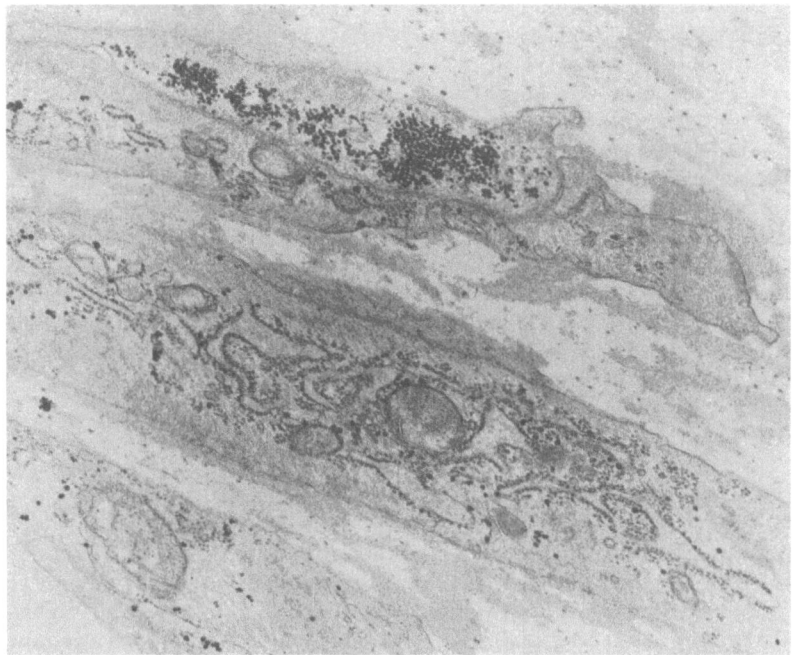

Fig. 4. Transmission electron micrograph of modified smooth muscle
cells proliferating in the subendothelial layer of an
intrarenal artery. Obliterative arteriopathy of a human
renal allograft (original magnification 32.000 x).

PLATELET LABELING IN RENAL TRANSPLANTATION

In 1971 Mowbray and Pariyananda (40) reported that Cr-51-labeled
platelets form thrombi in human renal allografts undergoing acute
rejection. However, the relatively high radiation dose of this tracer
proved an obstacle for repeated clinical application (4). Without
doubt Thakur et al (42, 43) advanced the clinical use of platelet
imaging by creating a new platelet-labeling technique using
In-111-oxine. The tracer has a half-life of 67 hours and gamma
emissions that permit imaging up to a week after injection. The
relatively low radiation dose (44-46) enables repeated administration.
Numerous studies have shown that In-111 platelet labeling is of great
value for scintigraphic detection of arterial and venous thrombi
(47-49) and also for study of platelet kinetics and deposition
(50-52).

In 1979 Smith et al (53) used In-111-labeled autologous platelets
as a diagnostic aid for detection of acute renal graft rejection. A

year later we demonstrated deposition of In-111-labeled platelets in
chronic rejection. It was, however, of lower intensity than in the
acute type (36). The value of platelet scintigraphy for diagnosis of
acute graft rejection was confirmed by several other groups (54-61).
With the calculation of a platelet uptake index (PUI), i.e., the ratio
of graft radioactivity to the radioactivity in the contralateral iliac
fossa, an improved technique was achieved (57, 58, 38, 61). Buckles
et al (54, 55) found the ratio of kidney over aortic arch
radioactivity was more useful. In 1980 Fenech and coworkers (62)
demonstrated accumulation of In-111-labeled platelets in the graft
region due to post-transplant renal hematoma. The differentiation
from acute rejection is easily made by Tc-99m-pertechnetate perfusion
scan which shows that the area of platelet uptake does not correspond
to the transplanted kidney.

ROLE OF PROSTACYCLIN IN KIDNEY TRANSPLANT REJECTION

    The knowledge that platelets and coagulation play a central role
in graft rejection has stimulated many groups to study prevention or
treatment of rejection by the administration of compounds which
inhibit platelet function and coagulation, respectively. Heparin and
sulfinpyrazone were effective in modifying experimental hyperacute
rejection (63, 10). Intra-arterial infusion of citrate or disodium
ethylenediaminetetraacetate (chelating agent) prevented rejection
temporarily (3, 64). This was probably achieved by inhibition of
platelets, complement and coagulation caused by binding free calcium
ions. In 1969 Kincaid-Smith (33) suggested that dipyridamole and
anticoagulants might prevent not only thrombosis in vessels in acute
rejection but also the progressive narrowing of vessels so
characteristic of chronic rejection. In the following years, Mathew
and coworkers (65) tested this hypothesis in a prospective controlled
trial in 92 patients with primary cadaveric renal transplants.
Fifty-four patients completed the trial - 26 in the treatment group
and 28 in the control group. Whereas oral anticoagulation and
dipyridamole had a significantly beneficial effect on the histological
appearance of glomeruli and arteries, no difference of interstitial
alterations could be detected. Although the number of patients with
irreversible deteriorating grafts was higher in the control group (5
versus 2), there was no significant difference in the mean serum
creatinine of remaining patients in the two groups. Kauffman and
coworkers (66) compared aspirin versus dipyridamole in a prospective
randomized trial. It seemed remarkable that dipyridamole therapy was
associated with a significant reduction in the incidence of
post-anastomotic transplant renal artery stenosis. However, aspirin
was not a satisfactory drug in these patients, who were receiving
routine steroids, because of an unacceptable incidence of acute upper
gastrointestinal bleeding.

Synthetic prostaglandins known to inhibit platelet aggregation
were tested in an experimental model. In 1977 Anderson and coworkers
(47) showed that prostaglandin (PG)E1 prolonged viability of murine
skin allografts. A decisive step toward understanding the interaction
between platelets and the vessel wall was made when Moncada and
coworkers (68-69) discovered prostacyclin (PGI2). This compound is
the most potent natural inhibitor of platelet aggregation and is a
very active vasodilator. The chemical synthesis of PGI2 (70) in
sufficient amounts enabled the administration of this fascinating drug
in experimental and clinical studies. It was a surprise even for
researchers familiar with PGI2 when Mundy et al (71, 72) demonstrated
that intra-arterial application of this compound was able to prevent
temporarily hyperacute renal graft rejection in several experimental
models. These studies suggested that the importance of platelets in
hyperacute rejection was more important than previously presumed.

This article will deal with our experience using In-111-platelet
scintigraphy for monitoring renal transplant patients. It will also
investigate the value of PGI2 treatment in chronic and acute graft
rejection.

MATERIAL AND METHODS

Platelet Scan

Ninety-one patients were studied in three groups. The groups were
categorized according to rejection type (groups I and II), and group
III was a control nonrejecting transplant group.

Group I were 33 patients - 20 men and 13 women - aged 32 ± 11 years
($L\bar{x}\pm SD$). They were examined between 4 and 8 weeks after renal
transplantation. The overwhelming majority (n=29) suffered from
chronic glomerulonephritis as underlying renal disease and had
received cadaver grafts (n=30) during their first (n=28)
transplantation. Twenty-seven patients were treated with prednisolone
(beginning with 200 mg/day and then tapering down to 20 mg/day),
azathioprine (0.5-3 mg/kg/day) and horse-anti-thymocyte-globuline (20
mg/kg body weight/day per infusion for 10 days). The rest (n=6)
received cyclosporin A, a generous gift of Dr. T. Beveridge, Sandoz
Ltd., Basle, Switzerland. Cyclosporin A (CyA) was administered
postoperatively in a dose of 15 mg/kg/body weight/day during the first
2 weeks, and then tapered down.

The labeling of autologous platelets using In-111-oxine (100-150
μCi; Research Center Seibersdorf, Austria) was performed according to
Thakur et al (42, 43) usually in weekly intervals, but if necessary
more frequently. The patients were examined two to three times daily
by means of computerized gamma camera equipment (Nuclear Chicago)
during a static study of 5 minutes. Platelet trapping was estimated
by calculation of a PUI, defined as counts in the transplant region

divided by counts in the contralateral region. For measurement of
platelet survival, 3 ml of blood were drawn 2 hours after injection of
radiolabeled autologous platelets, as well as three times daily during
at least the following three days. After separation of platelets,
radioactivity was determined and expressed as percentage of the 2 hour
radioactivity, which was assumed as 100%. Platelet t/2 was estimated
by means of survival curves. In addition, ultrasonic scan (Superscan,
Roche), serological testing for viral diseases, estimation of
biochemical plasma profile (SMAC, Technicon autoanalyzer), blood cell
counts, gas check, and urinanalysis were performed regularly using
standard procedures. Acute graft rejections were diagnosed
clinically, the majority confirmed by percutaneous biopsy.

Group II consisted of 25 patients (16 male, 9 female) aged 19-60
years (37+12). They were studied 12-119 months (61+35) after their
first cadaveric donor transplantation. The underlying kidney disease
which had resulted in terminal renal insufficiency before
transplantation was chronic glomerulonephritis in all cases.
Immunosuppression was carried out using prednisolone (15-25 mg/day)
and azathioprine (12.5-50 mg/day). All suffered from chronic
transplant rejection which was confirmed by percutaneous biopsy. At
the time of study plasma creatinine measured 4.2+1.7 mg/d. Ultrasonic
scan, laboratory tests, and platelet labeling were performed as stated
for group I. The patients were examined three times daily under the
gamma camera for three consecutive days, beginning 6 hours after
labeling.

Group III, which served as control, were 33 patients (18 men, 15
women), aged 9-56 years (38+13) with stable and well-funtioning grafts
(plasma creatinine 1.0-2.0 mg/dl) tested at least 9 months (67+50)
after transplantation. In this group the overwhelming majority had
suffered from chronic glomerulonephritis as primary renal disease and
had undergone their first cadaveric donor transplantation. These
patients were under immunosuppression with prednisolone (5-20 mg/day)
and azathioprine (25-100 mg/day). The platelet labeling, gamma camera
imaging, and estimation of platelet t1/2 procedures were identical as
described for groups I and II. The patients of all groups were free
of medication with aspirin-like drugs.

Prostacyclin Infusion in Chronic and Acute Graft Rejection

Twelve first-transplant patients (8 male, 4 female), aged 24-52
years (37+9) had chronic transplant rejection which was proven by
percutaneous biopsy. Transplantation had been performed 4-106 months
(44+35) before study. For immunosuppression the patients received
10-25mg prednisolone/day and 25-50 mg azathioprine/day. They all had
given their informed written consents before the study according to
the declaration of Helsinki. Autologous In-111-labeled platelets were
injected two weeks before PGI2 treatment as well as two days after the

beginning of PGI2 infusion. The sodium salt of PGI2 (Epoprostenol,
kindly provided by Dr. J. O'Grady, the Wellcome Research Labs.,
Beckenham, Kent, UK) was dissolved in glycine buffer (pH 10.5) and
infused at a rate of 5 ng/kg/min by means of a central vein catheter
for five days.

Six patients (5 male, 1 female), aged 22-56 years, under
immunosuppression with prednisolone and azathioprine, and suffering
from acute rejection were treated with PGI2 for five days. Due to
ethical reservations, this was done in addition to the usual rejection
treatment, namely, the daily infusion of 0.5g methylprednisolone on
three consecutive days. The platelet labeling was performed as
described above.

Statistics: Student's t-test for paired and unpaired data was
performed and values were expressed as $\bar{X}$+SD.

RESULTS

Platelet Scan

In the control group (group III) the grafts usually demonstrated
no or only mild platelet trapping (Fig. 5). Therefore the PUI
measured only 1.09+0.12. The platelet t1/2 of 102.0+18.5 hours was in
the normal range.

In group I, monitored during the first weeks after transplantation, 15 rejection episodes were observed in 14 of 33 cases.
Whereas two rejections occurred in two oliguric-/anuric patients, the
rest of acute rejections were associated with a decline of transplant
function. Several days before rejection the grafts showed no or only
mild platelet deposition. In connection with rejection, however, an
intensive platelet trapping was observed by gamma camera imaging. A
typical example of this change is given in Figures 6 and 7A. The PUI
increased significantly (p < 0.01) from 1.13+0.11 to 1.81+0.3 in 14
episodes, while platelet t1/2 declined from 87.4+17.8 hours to
46.6+23.6 hours (p < 0.01). The PUI increase was observed 0-48 hours
(14.8+12.2) prior plasma creatinine rose (Fig. 8) in 12 rejections of
urine-producing grafts, as well as 24 and 36 hours before ultrasonic
signs of rejection appeared in two oliguric-/anuric cases,
respectively (Fig. 9). The histological examination of biopsy
specimen in 8 rejection episodes showed typical alterations of acute
vascular rejection in 6 cases. One interstitial rejection was
associated with a minimal PUI increase from 1.3 to 1.5. In the
remaining case that underwent biopsy, PUI had increased from 1.2 to a
high value of 2.7, when gamma camera imaging showed a marked platelet
deposition (Fig. 10). However, light microscopy of percutaneous
biopsy showed "only" a severe interstitial rejection. Since PUI
increased further to 4.2, the second biopsy was performed. Now the
histological picture had changed to a severe acute vascular rejection

with extensive thrombotic alterations. A rejection episode detected
by graft biopsy was not diagnosed by the platelet scan, probably due
to insufficient platelet labeling in this thrombocytopenic case. When
rejections responded to steroid pulse treatment, the platelet trapping

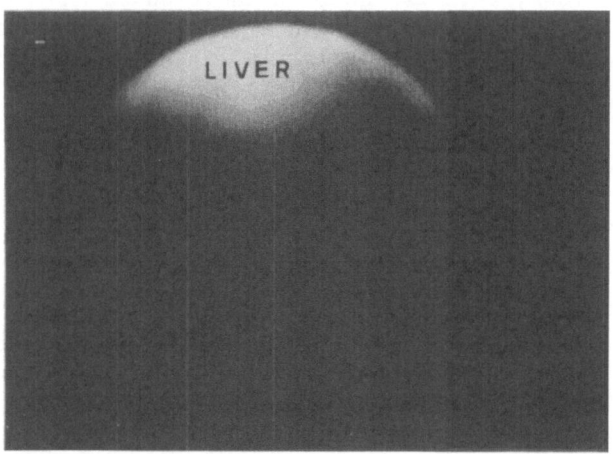

Fig. 5.   Platelet scan of a patient with a stable and well-functioning
           graft tested 2 years after transplantation. In the
           transplant region (left iliac fossa) no platelet deposition
           is observed (PUI 1.0). The tracer has accumulated in the
           liver.

Fig. 6. Platelet scan
in 23-year-old male, 3
weeks after
transplantation, but
one week before acute
rejection. There is
only a minimal
platelet deposition in
the left iliac fossa
(PUI 1.2).

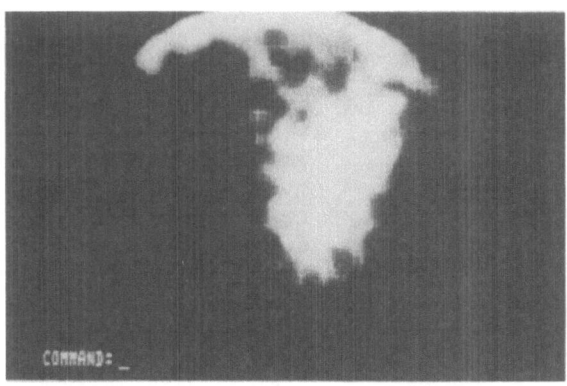

Fig. 7A. The same case as in Figure 6. Twenty-four hours before clinical symptoms of acute rejection the gamma camera image shows an intense platelet deposition in the graft (PUI 1.8).

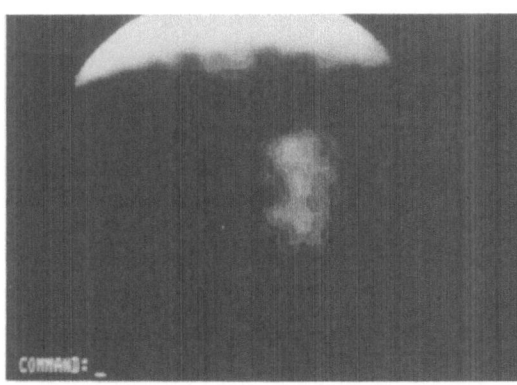

Fig. 7B. The same patient as in Figures 6 and 7A. The acute rejection had responded to steroid treatment. The platelet scan now shows that trapping has declined (PUI 1.4) in comparision with the condition shown in Figure 7A.

decreased to a moderate level (Fig. 7B). This was associated with a redecline of PUI to 1.34 ± 0.10 which had not yet reached the original level. Platelet t1/2 rose to 83.0 ± 26.8 hours again. On the other hand, irreversible rejection resulted in a further increase of PUI to 2.17 ± 1.0 associated by a fall of platelet t1/2 to 30.7 ± 15.2 hours.

A false-positive increase of PUI was observed in the case of a hematoma surrounding mainly the lower graft pole while biopsy excluded rejection (Fig. 11). A second patient had a complicated posttransplantation course. This man had suffered from hemolytic uremic syndrome (HUS) leading to chronic renal insufficiency. The first transplantation under conventional immunosuppression was unsuccessful because of irreversible acute rejection. After the second transplantation, this time under CyA and prednisolone, the patient was monitored by the platelet scan. The first two weeks passed without complications while the graft showed a good function (plasma creatinine 1.2 mg/dl). Then the situation changed. Graft function deteriorated rapidly, and symptoms of HUS, as well as

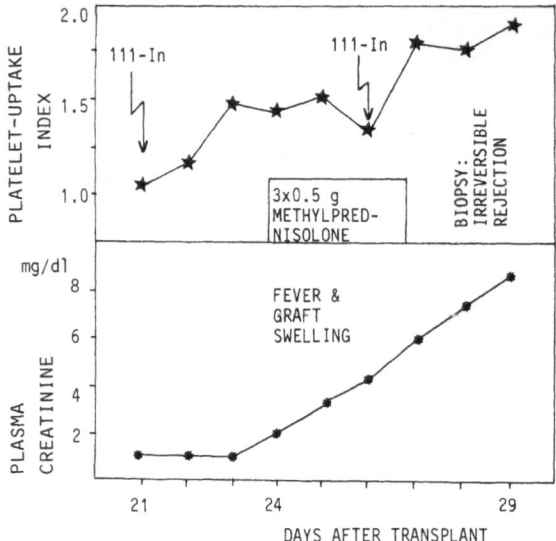

Fig. 8. Increased PUI before deterioration of the transplanted kidney function. The rejection did not respond to cortiocosteroid pulse treatment as indicated by a further increase of PUI. Finally irreversible graft destruction resulted.

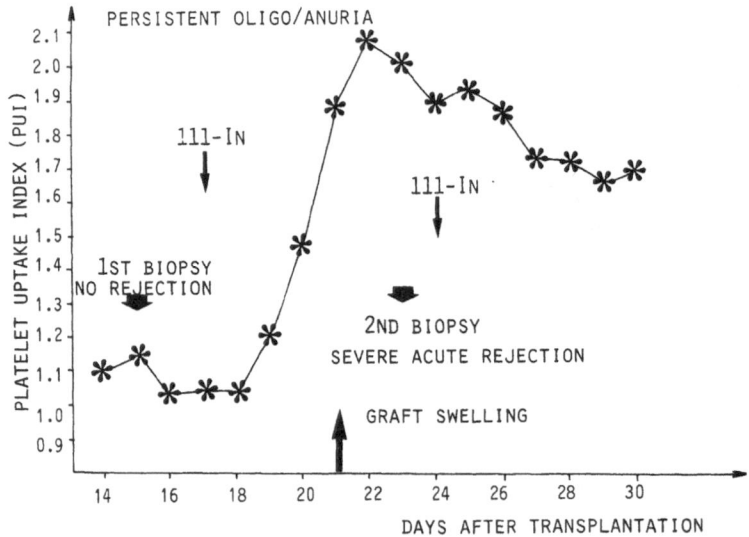

Fig. 9. Monitoring of a patient suffering from acute tubular necrosis after transplantation. An increase of PUI was observed before signs of rejection were detected by ultrasonic scan.

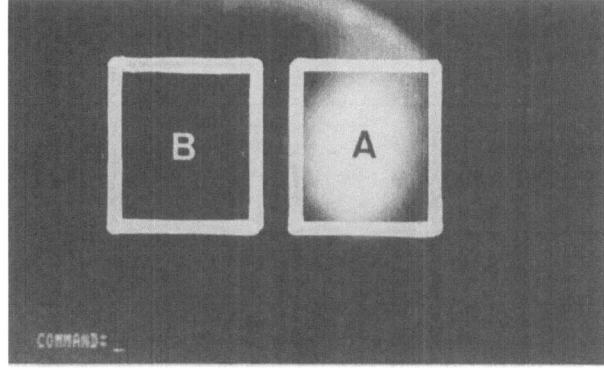

Fig. 10. Marked platelet trapping (PUI 2.7) in a case of severe acute
         interstitial rejection (A).

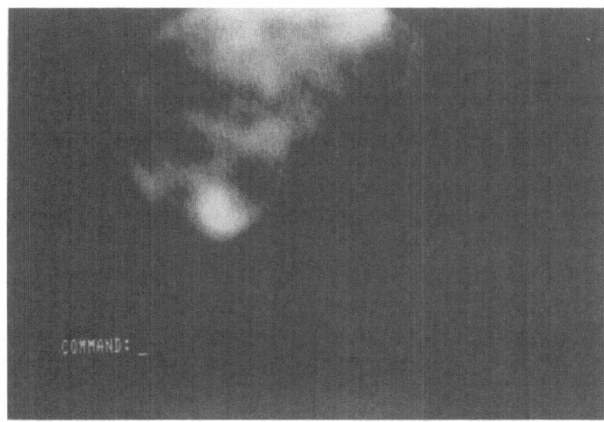

Fig. 11. Deposition of In-111-lableled platelets in the lower part
         of the right In-111-iliac fossa due to a hematoma
         surrounding the lower transplant pole.

hemolytic anemia and thrombocytopenia, reappeared. The gamma camera
imaging showed an intense platelet deposition in the graft (PUI 1.6).
However, percutaneous biopsy revealed interstitial or vascular
infiltration by plasma cells and/or lymphoblasts was absent.
Moreover, the graft was not swollen as demonstrated by ultrasonic
scan. It was concluded, therefore that this phenomenon was not acute
graft rejection but recurrence of HUS after transplantation.

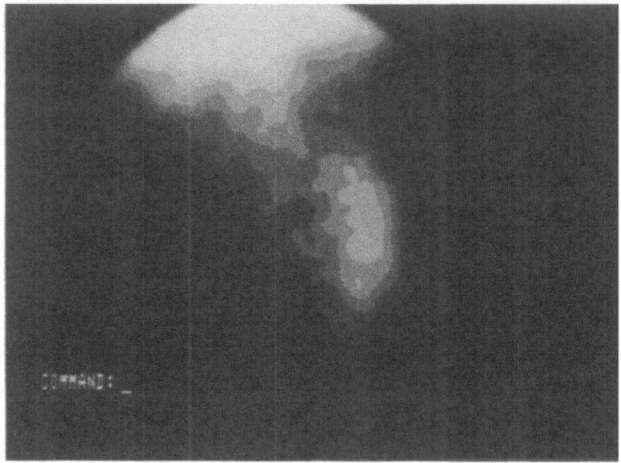

Fig. 12. Moderate platelet deposition (PUI 1.4) in chronic graft
rejection proven by percutaneous biopsy.

Because the liver and spleen take up the tracer these organs might
interfere with the evaluation of the platelet scan. The overlap
between liver or spleen and the graft region was seen in two cases
without rejection episodes. The second case of insufficient platelet
labeling probably due to thrombocytopenia occurred in a patient
without rejection.

In three cases acute rejection was suspected because of worsening
graft function. The PUI, however, was found to be in the normal range
(0/9, 1.1, and 1.0, respectively). Further diagnostic procedures
including biopsy revealed that graft dysfunction was caused by
bacterial infection and septicemia respectively. These cases belonged
to 14 patients without any rejection showing a mild or missing
platelet trapping. PUI and platelet t1/2 measured amounted 1.19 ± 0.14
and 91.1 ± 19.5 hours respectively.

In group II (chronic rejection) the platelet deposition in the
graft was usually mild to moderate (Fig. 12). The PUI of 1.33 ± 0.15
was significantly higher ($p < 0.01$) than in group III. There was a
negative correlation ($r = -0.95$) between PUI and platelet t1/2
(62.2 ± 22.4 hours), a behavior which could not be demonstrated in group
III.

Prostacyclin Infusion in Chronic and Acute Rejection

In chronic transplant rejection the PGI2 infusion treatment caused
a significant decline of PUI, an increase of platelet t1/2, as well as

Table 1. Treatment of Chronic Graft Rejection with PGI2 Resulted in a
          Significant Decrease of PUI, Prolongaton of Platelet tl/2,
          and Decline of Plasma Creatinine.

	PUI	Platelet tl/2	Plasma Creatinine
Before	1.33+0.12	37.1+12.4 hr	4.0+1.2 mg/dl
PGI2 Treatment			
End	1.17+0.11	52.8+10.1 hr	3.4+1.3 mg/dl
paired t-test	p<0.0005	p<0.0025	p<0.0005

a decrease of plasma creatinine (Table 1). However, this beneficial
effect was not seen with oral medication and with 200 mg
sulfinpyrazone and 3x75mg dipyridamole/day, so that graft function
further deteriorated in 8 of 12 cases 6 months later.

     In the six patients suffering from acute graft rejection who were
selected for PGI2 treatment, rejection was associated with an increase
of PUI from the range of 1.0-1.2 to 1.5-2.2. and in a shortening of
platelet tl/2 from 72-102 hours to 28-54 hours. The graft function
worsened in four cases as demonstrated by plasma creatinine increase
from 1.5-5.2 mg/dl to 3.5-8.6 mg/dl. In two oliguric-/anuric cases
the ultrasonic scan had revealed the rejection. These rejections were
proven by biopsy, including two cases without sufficient urine
production. PGI2 treatment for 5 days resulted, in all cases, in a
decline of PUI to a range of 1.1-1.4 (Fig. 13), while platelet tl/2
was prolonged to 46-84 hours. Plasma creatinine decreased in these
four cases to 1.3-5.4 mg/dl and persisted for at least 2 months.
However, the two oliguric-/anuric grafts remained functionless and
were destroyed by later rejections. Side effects of PGI2 treatment
were minimal. We observed a facial and upper body flush as well as a
moderate decline of systolic and diastolic blood pressures (20 mm Hg
each). The latter effect was advantageous during these rejection
episodes because the patients were usually hypertensive.

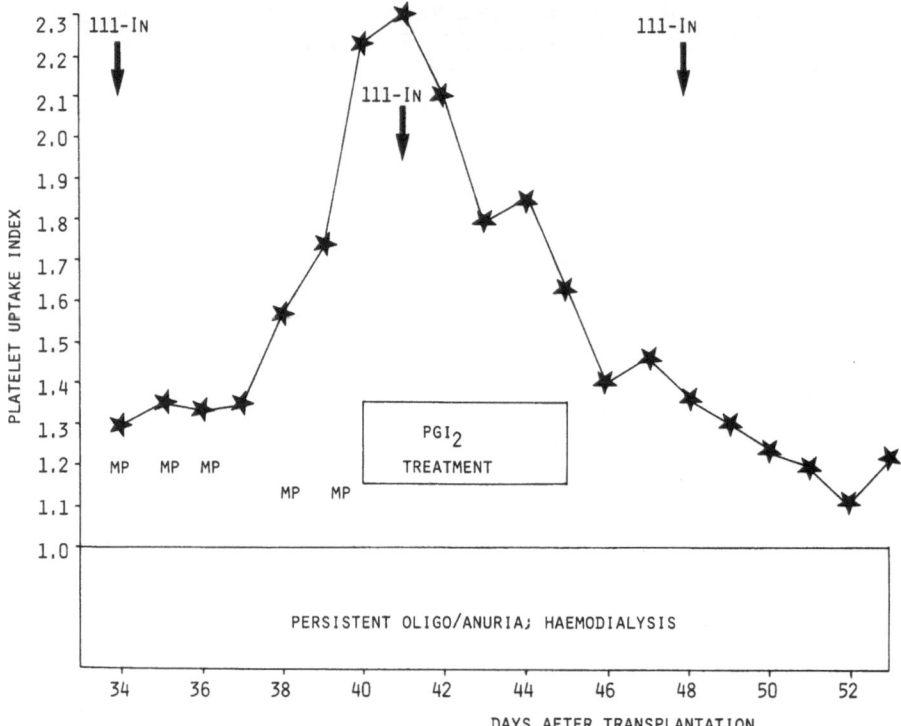

Fig. 13. Treatment of acute graft rejection (proven by biopsy) with
PGI2 In a patient with persistent oliguria/anuria. Although
a declined PUI was observed, the graft never produced urine
and was destroyed later by recurrent rejection; (MP . . . 0.5
g methyl-prednisolone).

DISCUSSION

In our opinion the platelet scanning has contributed markedly in
the understanding of the complex role of platelets in different kinds
of graft rejection. This technique enables the in vivo demonstration
of platelet deposition in the graft. The experience gained with
patients during the first posttransplantation weeks supports the
hypothesis that platelet deposition is an early event in acute graft
rejection, thereby preceding clinical or laboratory signs of acute
rejection. The important question of whether platelets deposit
before, concomitantly with, or after infiltration by lymphocytes is
still to be determined. In a very impressive study on cardia
allotransplantation in the rat, Wang et al (73) demonstrated that

In-111-labeled leucocytes ($\geq$ 92% lymphocytes) and platelets
accumulated simultaneously during acute rejection. However, further
comparative studies on renal transplantation in the human are
necessary to determine which technique is more sensitive and practical
for clinical purposes. Whereas the evaluation of gamma camera images
by means of the naked eye can be associated with subjective
misinterpretations, the insertion of regions of interest and
calculation of these indices has major advantages. Several groups,
ours included, calculate the ratio of graft over contralateral region
radioactivity (37, 38, 57, 58, 61). On the other hand, Buckels et al
(54, 55) prefer the ratio of kidney over aortic arch. In most cases
of renal transplant patients the weekly labeling with In-111 will be
sufficient for monitoring. However, when platelet t1/2 is decreased
as in acute rejection or when thrombosis of dialysators occurred (74),
the frequency of labeling should be increased. All groups with
experience with In-111-platelet labeling in renal transplantation
report that this method provides a valuable tool for detecting acute
graft rejection. However, Griño and coworkers (58) used this
technique only for differential diagnosis in difficult cases. They
suggested that the use of this method was limited by the fact that the
accumulation of radioactive platelets in the spleen might be
oncogenic. Since In-111 has short t1/2, the radiation dose is rather
low (44-46) and the early detection and treatment of acute graft
rejection is often decisive for the future life of the patient, we
affirm the value of the In-111-platelet scintigraphy in the management
of patients with renal transplant.

In our study, 14 of 15 acute rejections were diagnosed by means of
the platelet scan, in the mean 14.8+12.2 hours prior to plasma
creatinine increase, as well as 24 and 36 hours before acute rejection
was diagnosed by ultrasonic scan in oligouria/anuria. It seems to be
remarkable that 1 interstitial rejection was associated with a minimal
PUI increase from 1.3 to 1.5. Recently Jurewitz et al (75) reported
that a case of mild interstitial rejection also escaped diagnosis by
bedside probe counting after platelet labeling. On the other hand, we
observed a sharp increase of PUI from 1.2 to 2.7, when histological
examinations of percutaneous biopsy by light microscopy showed "only"
a severe interstitial rejection leaving the vessel free of
infiltrative or thrombotic involvement. We suggest that the platelet
deposition would have been documented by transmission electron
microscopy. Since platelet deposition aggravated in this case while
the histological picture changed to severe acute vascular rejection,
one might speculate that platelet deposition could be an integral
feature of severe acute interstitial rejection. When acute rejection
responded to steroid pulse treatment, a decline of PUI was observed
which did not reach the baseline level. A progression of acute
rejection to irreversible destruction of the graft was associated with
a further increase of PUI. However, it must be emphasized, that
misinterpretations of the platelet scan are possible at this stage.
In the theoretic case of a severe acute vascular rejection, a complete

graft thrombosis will be associated with a distinct platelet
deposition which can be easily demonstrated by gamma camera platelet
imaging. If the graft is not perfused at this stage, a new platelet
label will not show deposition in the transplant. This can be
misinterpreted as improvement. The theoretical example illustrates
that other diagnostic approaches—particularly ultrasonic scan,
Tc—99m—pertechnetate perfusion scan, and percutaneous biopsy are
valuable supplementary diagnostic approaches. Another limitation of
the platelet scan is that insufficient labeling occurred in two
thrombocytopenic patients, so that one acute rejection was missed. An
overlap between liver or spleen and the graft region inhibited the
evaluation of the platelet scan in two cases. Gamma camera imaging
and PUI were of great help in three cases, when the clinicians
suspected acute rejection, but graft dysfunction was caused by
bacterial infection and septicemia, respectively. In such cases the
platelet scan showing no deposition in the graft argues strongly
against acute rejection. Thereby important information can be gained
which advocates an unnecessary and potentially harmful steroid pulse
treatment.

     The case demonstrating a recurrence of HUS after transplantation
under CyA needs further explanation because it illustrates the complex
interaction between vascular PGI2 synthesis, platelets, and CyA
treatment. The PGI2 synthesis by blood vessels can be regarded as a
defense mechanism against endothelial injury and subsequent
uncontrolled platelet deposition (69, 76). Its synthesis is
stimulated by a plasma factor present in healthy subjects (77).
Patients who are prone to develop HUS or thrombotic thrombocytopenic
purpura lack this plasma factor activity (78), as was shown in our
patient. One can assume that these patients are at risk of
uncontrolled platelet deposition and consumption which are especially
localized to the kidneys as in HUS or more generalized as in
thrombotic thrombocytopenic purpura. It was a surprising finding that
a minority of patients having been treated with CyA for prevention of
graft—versus—host disease after bone—marrow transplantation developed
HUS or related entities (79, 80). Stimulated by these observations
Rocchi et al (81) and Neild and Rocchi (70) studied the effect of CyA
on the PGI2 system in an experimental model. They found that
high—dose CyA inhibited vascular PGI2 synthesis via decrease of plasma
factor activity and suggested, therefore, that CyA might be of danger
in transplant patients who had HUS as their underlying renal disease.
Our case with recurrence of HUS after renal transplantation (83) seems
to support this. However, Hamilton and coworkers (84) reported the
successful renal transplantation using CyA for immunosuppression in a
girl with HUS as primary renal disease. Mihatsch (85) recently
reported on a higher incidence of microangiopathy resembling that of
HUS in renal transplantation under CyA than with conventional
immunosuppression. The problem is that HUS in a renal transplant
cannot always be clearly differentiated from acute rejection. In
spite of these reservations one cannot exclude completely that in

isolated cases the platelet scan may suggest acute rejection when the
pathogenesis of platelet deposition is CyA toxicity.

The important role of platelets in acute graft rejection is
supported by studies on platelet products in biological fluids. Foegh
et al (86, 87) reported on the significant increase in urinary
excretion of immunoreactive thromboxane B2 as an early indicator of
acute rejection. Frampton and coworkers (1) measured intraplatelet
serotonin concentration in 16 patients following transplantation.
Although some patients showed abrupt falls in intraplatelet serotonin
coincident with acute rejection episodes, there was no difference
between well-functioning, stable grafts and those which were destroyed
by repeated rejection. On the other hand, Capitanio et al (88)
observed more evidence for abnormal platelet activation in acute
rejection, which is in accordance with the kinetics of In-111-labeled
platelets as studied by Heyns et al (89) and our group.

In group II (chronic rejection) the platelet deposition was
usually mild to moderate which resulted in a PUI of 1.33 ± 0.15. This
value was significantly higher than the PUI of the controls
(1.09 ± 0.12). There were a few patients of group III having PUI values
of about 1.3. We speculate that these patients can be subjected to
harmful platelet deposition in the grafts which could turn the
clinical course toward chronic rejection in some years.

The knowledge of the harmful role of platelets in development of
atherosclerosis, (39), as well as the evidence of a pathogenetic role
of platelets in chronic rejection presented here, supports the
hypothesis that any disturbance of the balance between platelets and
the endothelium of the graft could put the transplant patient at risk
of development of obliterative arteriopathy. One of the most
important risk factors stimulating proliferative arterial lesions is
without doubt hypertension (15, 30, 33). It is very interesting that
patients are more prone to develop hypertension when treated with CyA
than under conventional immunosuppression (30). This fact leads back
to the speculation on inhibiting of PGI2 synthesis by CyA. The next
important risk factor is hyperlipoproteinemia after transplantation
(90), particularly elevation of serum LDL. This metabolic disturbance
can enhance platelet sensitivity for proaggregatory actions (91) and
reduce the arterial PGI2 synthesis by peroxidation of enzymes (92).
The importance of synergy between rejection and hyperlipoproteinemia
for atherosclerosis in experimental cardiac transplantation was
emphasized by Laden (93) and Alonso and coworkers (94). Since the
local synthesis of PGI2 in the renal vasculature (94.A) is of great
importance for protection of endothelial integrity, any disease
disturbing PGI2 production can change the platelet vessel wall
interaction in a harmful way. Diabetes mellitus is another metabolic
disturbance which is associated with inhibition of vascular PGI2
release (95) as well as with a decrease of platelet sensitivity to the
antiaggregatory PGI2 (50). Cigarette smoking induces a similar

insensitivity of platelets to PGI2 (83). In a summary concerning risk factors, all these alterations can promote generalized atherosclerosis as well as obliterative arteriopathy of the renal transplant. Recently the understanding of these mechanisms was rendered more difficult by the finding that the platelet-derived growth factor is able to stimulate not only vascular proliferation but also local PGI2 synthesis in a kind of rebound phenomenon for inhibition of further platelet aggregation (14). The final outcome of this phenomenon is probably decisively influenced by the situation of the local endothelium whether it can enhance PGI2 synthesis or not.

For the critical evaluation of the effect of PGI2 infusion in acute rejection a controlled or double-blind study would be necessary, because platelet deposition can decrease after reversible rejection treated with steroid pulses, and a lot of other drugs which might modify platelet behavior are often used in these patients. Therefore we began our therapeutic study with patients suffering from chronic rejection. In this disease the graft function is disturbed but relatively stable over several weeks. In addition, it is usually not necessary to change medication, so that the effect of any additional therapy as PGI2 could be more easily detected by parameters as PUI, platelet t1/2, and plasma creatinine. Our results indicate that it is possible to reduce the harmful platelet trapping in the graft, to reduce platelet consumption as indicated by a prolonged platelet t1/2, and to improve graft function temporarily. This PGI2 effect can be evoked by two possible mechanisms. First, PGI2 probably induces clearing of platelet aggregates and deposition from the transplant. This suggestion is favored by measurement of PUI and platelet t1/2. The improvement of graft function might be partly caused by improved perfusion after removal of obstructive platelet deposition. On the other hand, pure hemodynamic effects of PGI2 (98, 99) can contribute to increase of graft function. Similar conditions were experienced by Niwa and coworkers (100), who applied PGE1, a platelet inhibitor and vasodilator too, in patients suffering from chronic renal disease and glomerulonephritis.

Our experience of PGI2 treatment in acute graft rejection suggests that this drug might be of value in overcoming a secondary, but important, phenomenon, namely the platelet trapping. Since these acute rejections are often accompanied by hypertension (101), the antihypertensive effect of PGI2 (102) might be a valuable side effect. For us it was rather disappointing that the beneficial effect of PGI2 in chronic rejection could not be maintained by oral medication of antiplatelet drugs, so that graft function worsened further 6 months later. This fact strongly argues for a controled trial on stable and oral PGI2 analoges.

ACKNOWLEDGMENT

We are indebted to the valuable cooperation of the following
colleagues: Doz. Dr. G. Syre, Department of Pathology; Dr. M. Schwarz
and Dr. E. Pohanka, Second Department of Internal Medicine, University
of Vienna; Prof. Dr. H. Stachelberger and Dr. P. Kruzik, Technical
University, Vienna.

REFERENCES

1. G. Frampton, A. Parbtani, D. Marchesi, P. Duffus, M. Livio, G.
   Remuzzi, and J. S. Cameron, In vivo platelet activation with
   in vitro hyperaggregability to arachidonic acid in renal
   allograft recipients, Kidney Int 23:506.
2. H. Gewurz, D. S. Clark, J. Finstad, W. D. Kelly, R. L. Varco, R.
   A. Good, and A. E. Gabrielsen, Role of the complement system
   in graft rejection in experimental animals and man, Ann NY
   Acad Sci 129:673 (1966).
3. M. Kux, H. J. Boehmig, H. Amemiya, M. Torisu, T. Yokayama, B.
   Launois, M. Popovtzer, C. B. Wilson, F. J. Dixon, T. E.
   Starzl, Modification of hyperacute canine renal homograft
   rejection by the intra-arterial infusion of citrate, Surgery
   70:103 (1971).
4. A. Schilling, W. Land, E. Pratschkle, K. Pielsticker, W.
   Brendel, Dominant role of complement in the hyperacute
   xenograft rejection reaction, Surg Gyn Obstet 142:29 (1976).
5. C. Leithner, H. Piza-Katzer, P. Kruzik, M. Winter, H. Sinzinger,
   W. R. Mayr, H. Stachlberger, Hyperakute
   Nierentransplantat-Abstbung: Rasterelektronenmikroskopie der
   intrarenalen Arterien, VASA 9:137.
6. C. Leithner, H. Piza-Katzer, P. Kruzik, M. Winter, W. R. Mayr,
   H. Stachelberger, H. Sinzinger, A study of hyperacute
   rejection of renal transplants using scanning electron
   microscopy, Dialysis and Transplantation 10:336 (1981).
7. G. J. Busch, A. C. P. Martins, N. K. Hollenberg, R. E. Wilson,
   R. W. Colman, A primate model of hyperacute renal allograft
   rejection, Am J Pathol 79:31 (1975).
8. R. W. Colman, W. E. Braun, G. J. Busch, G. J. Dammin, J. P.
   Merrill, Coagulation studies in the hyperacute and other
   forms of renal-allograft rejection, New Engl J Med 281:675
   (1969).
9. J. C. Rosenberg, R. J. Broersma, G. Bullemer, E. F. Mammen, R.
   Lenaghan, and B. F. Rosenberg, Relationship of platelets,
   blood, coagulation and fibrinolysis to hyperacute rejection
   of renal xenografts, Transplantation 8:152 (1969).
10. H. M. Sharma, S. Moore, H. W. Merrick, M. R. Smith, Platelets in
   early hyperacute allograft rejection in kidneys and their
   modification by sulfinpyrazone (anturan) therapy, Am J Pathol
   66:445 (1972).

11. G. M. Williams, H. J. Lee, R. F. Weymouth, W. R. Harlan, K. R.
     Holden, C. M. Stanley, G. A. Millington, D. M. Hume, Studies
     in hyperacute and chronic renal homograft rejection in man,
     Surgery 62:204 (1967).
12. C. Grossklaus, B. Damerau, E. Lemgo, W. Voght, Induction of
     platelet aggregation by the complement-derived peptides C3a
     and C5a. Naunyn Schmiedebergs Arch Pharmacol 295:71 (1976).
13. T. R. Kohler, and N. L. Tilney, Microangiopathic hemolytic
     anemia associated with hyperacute rejection of a kidney
     allograft, Transplant Proc 14:444 (1982).
14. G. J. Busch, E. S. Reynolds, E. G. Galvanek, W. E. Braun, G. J.
     Dammin, Human renal allografts. The role of vascular injury
     in early graft failure, Medicine 50:29 (1971).
15. P. Kincaid-Smith, Histological diagnosis of rejection of renal
     homografts in man, Lancet ii:849 (1967).
16. M. Lagarde, P. Bericaud, M. Burtin, M. Dechavanne,
     Refractoriness of diabetic platelets to inhibitory
     prostaglandins, Prostaglandins Med 7:341 (1981).
17. R. Lowenhaupt and P. Nathan, Platelet accumulation observed by
     electron microscopy in the early phase of rena allotransplant
     rejection, Nature 220:822 (1968).
18. R. Lowenhaupt and P. Nathan, The participation of platelets in
     the rejection of dog kidney allotransplants: hematologic and
     electron microscopic studies, Transplant Proc 1:305 (1969).
19. K. A. Porter, J. B. Dossetor, T. L. Marchiorio, W. S. Peart, J.
     M. Rendall, T. E. Starzl, P. I. Terasaki, Human renal
     transplants. I. Glomerular changes, Lab Invest 16:153
     (1967).
20. G. J. Cerilli, J. E. Holliday, I. C. Lee, Antivascular
     endothelium antibody in renal transplantation, Surg Forum
     26:341 (1975).
21. L. C. Paul, L. A. Van Es, J. J. Van Rood, A. Van Leeuwen, G.
     Brutel de la Riviere, J. DeGraeff, Antibodies directed
     against antigens of the endothelium of pertubular capillaries
     in patients with rejecting renal allograft, Transplantation
     27:175 (1979).
22. F. H. J. Claas, L. C. Paul, L. A. Van Es, J. J. Van Rood,
     Antibodies against donor antigen on endothelial cells and
     monocytes in eluates of rejected kidney allografts, Tissue
     Antigens 15:19 (1980).
23. G. F. J. Hendriks, G. M. T. Schreuder, F. H. J. Claas, J.
     D'Amoro, G. G. Persijn, B. Cohen, J. J. Van Rood, HLA-DRw6
     and renal allograft ejection, Br Med J 286:85 (1982).
24. G. F. J. Hendriks, F. H. J. Claas, G. G. Persijn, M. D.
     Witvliet, W. Baldwin, and J. J. Van Rood, HLA-DRw6-positive
     recipients are high responder inrenal transplantation,
     Transplant Proc 15:1136 (1983).
25. Y. M. Ooi, B. S. Ooi, E. H. Vallota, M. R. First, V. E. Pollak,
     Circulating immune complexes after renal transplantation.
     Correlation of increased I-125-C1q binding activity with

acute rejection characterized by any fibrin deposition in the kidney, J Clin Invest 60:611 (1977).

26. T. Palosuo, K. Kano, S. Anthone, R. Gerbasi, and F. Milgrom, Circulating immune complexes after kidney transplantation, Transplantation 21:312 (1976).

27. W. J. Dempster, C. V. Harrison, and R. Shackman, Rejection processes in human homotransplantated kidneys, Br Med J 2:969-976 (1964).

28. G. Dunea, J. B. Hazard, W. J. Koff, Vascular changes in renal homografts, JAMA 190:199 (1964).

29. D. M. Hume, J. P. Merril, B. F. Miller, C. W. Thron, Experiences with renal homotransplantation in the human: report of nine cases, J Clin Invest 34:327 (1955).

30. P. Kincaid-Smith, Vascular changes in homotransplants, Br Med J 1:178 (1964).

31. J. J. Trentin, The arterial obliterative lesions of human renal homografts, Ann NY Acad Sci 129:654 (1966).

32. K. A. Porter, K. Owen, J. F. Mowbray, W. B. Thomson, J. R. Kenyon, W. S. Peart, Obliterative vascular changes in four human kidney homotransplants, Br Med J 2:639 (1963).

33. P. Kincaid-Smith, Modification of the vascular lesions of rejection in cadaveric renal allografts by dipyridamole and anti-coagulants, Lancet ii:920 (1969).

34. O. Jonasson, H. J. Winn, M. H. Flax, P. S. Russell, Renal biopsies in long-term survivors of renal transplantation: immunofluorescent studies, Transplantation 5:859 (1967).

35. T. X. O'Connel and J. F. Mowbray Arterial intimal thickening produced by alloantibody and xenoantibody, Transplantation 15:262 (1973).

36. C. Leithner, H. Sinzinger, P. Angelberger, G. Syre, Indium-111-labeled platelets in chronic kidney transplant rejection, Lancet ii:213 (1980).

37. C. Leithner, H. Sinzinger, M. Schwarz, Treatment of chronic kidney transplant rejection with prostacyclin - reduction of platelet deposition in the transplant; prolongation of platelet survival and improvement of transplant function, Prostaglandins 22:783 (1981).

38. C. Leithner, H. Sinzinger, M. Schwarz, E. Pohanka, G. Syre, Increased deposition of Indium-111-labeled platelets in chronically rejected kidney transplants, Clin Nephrol 18:311 (1982).

39. R. Ross, J. Glomset, B. Kariya, L. Harker, A platelet-dependent serum factor that stimulates the proliferation of arterial smooth muscle cells in vitro, Proc Natl Acad Sci USA 71:1207 (1974).

40. J. F. Mowbray, and A. Pariyananda, Platelet thrombi in rejection of renal allografts, in: "Immunological Mechanisms in Blood Coagulation, Thrombosis and Haemostasis," R. Duckert, K. M. Brinktous, S. Hinnom, eds. Schattauer Verlag, Stuttgart (1971).

41. International Committee for Standardization in Hematology,
     Recommended methods for radioisotope platelet survival
     studies, Blood 50:1137 (1977).
42. M. L. Thakur, M. J. Welch. J. H. Joist, R. E. Colemna,
     Indium-111-labeled platelets: studies on preparation and
     evaluation of in vitro and in vivo functions,
     Thromb Res 9:345 (1976).
43. M. L. Thakur, L. Walsh, H. L. Malech, A. Gottschalk,
     Indium-111-labled human platelets - improved method,
     efficacy, and evaluation, J Nucl Med 22:381 (1981).
44. U. Scheffel, M. F. Tsan, T. G. Mitchell, E. E. Camargo, H.
     Braine, M. D. Ezekowitz, E. L. Nickoloff, R. Hillzobel, E.
     Murphy, P. A. McIntryer, Human platelets labeled with In-111
     8-hydroxy-quinoline-kinetics, distribution, and estimates of
     radiation dose, J Nucl Med 23:149 (1982).
45. O. R. Van Reenen, Radiation dose from human platelets labeled
     with Indium-111, Br J Radiol 54:1011 (1981).
46. O. R. Van Reenen, M. G. Lotter, A. Du, P. Heyns, F. De Kock, C.
     Herbst, H. Kotze, H. Pieters, P.C. Minnaar, P. N. Badenhorst,
     Quantification of the distribution of In-111-labeled
     platelets in organs, Eur J Nucl Med 7:80 (1982).
47. H. H. Davis, W. A. Heaton, B. A. Siegel, C. J. Mathias, J. H.
     Joist, L. A. Sherman, M. J. Welch, Scintigraphic detection of
     atherosclerotic lesions and venous thrombi in man by
     Indium-111-labeled autologous platelets, Lancet i:1185
     (1978).
48. A. Fenech, J. K. Hussey, F. W. Smith, P. P. Dendy, B. Bennett,
     A. S. Douglas, Diagnosis of deep vein thrombosis using
     autologous Indium-111-labeled platelets, Br J Med 282:1020
     (1981).
49. R. L. Lantieri, D. A. Goodwin, D. Guthaner, F. Conley, M. L.
     Goris, P. W. Doherry, Static and dynamic studies of deep
     veinous thrombosis and atherosclerosis in humans with
     Indium-111-labeled platelets, Br J Radiol 53:9222 (1980).
50. J. H. Joist, and R. K. Baker, Loss of In-111 as indicator of
     platelet injury, Blood 58:350 (1981).
51. K. A. Peterson, M. K. Dewanjee, M. P. Kaye, Fate of Indium-
     111-labeled platelets during cardiopulmonary bypass performed
     with membrane and bubble oxygenaors, J Thorac Cardiovasc Surg
     84:39 (1982).
52. J. L. Ritchie, J. R. Stratton, B. Thiele, G. W. Hamilton, L. N.
     Warrick, T. W. Huang, L. A. Harker, Indium-111 platelet
     imaging for detection of platelet deposition in abdominal
     aneurysms and prostetic arterial grafts, Am J Cardiol 47:882
     (1981).
53. N. Smith, S. Chandler, R. J. Hawker, L. M. Hawker, A. D. Barnes,
     Indium-labelled autologous platelets as diagnostic aid after
     renal transplantation, Lancet ii:1241 (1979).
54. J. A. C. Buckels, S. Chandler, R. Hawker, A. D. Barnes,
     L. N. McCollum, The early diagnosis of acute renal transplant

rejection using In-133-labeled platelets, Br J Surg 69:285
(1982).

55. J. A. Buckles, S. Chandler, R. Hawker, C. N. McCollum, A. D.
Barnes, The early diagnosis of acute renal transplant
rejection using Indium-111-labeled platelets, Transplant
Proc 15:1192 (1983).

56. J. Chandler, J. Buckles, R. Hawker, N. Smith, A. D. Barnes, C.
N. McCollum, Does platelet deposition predict renal
transplant failure due to rejection? Br J Surg 68:801
(1981).

57. A. Fenech, A. Nicholls, F. W. Smith, Indium-111-labeled
platelets in the diagnosis of renal transplant rejection:
preliminary findings, Br J Radiol 54:325 (1981).

58. J. M. Griño, J. Alsina, J. Martin, M. Roca, A. Castelao, R.
Romero, A. Caralps, Indium-111-labeled autologous platelets
as a diagnostic method in kidney allograft rejection,
Transplant Proc 14:198 (1982).

59. M. R. Hardeman, J. B. Vanderschoot, J. Vreeken, The early
diagnosis of kidney graft rejection with Indium-111-
oxine-labeled thrombocytes, Br J Radiol 53:923 (1980).

60. A. J. W. Hilson, C. Lazarus, A. Parbtani, J. S. Cameron, M. N.
Maisey Indium-111 platelets in early detection of kidney
graft rejection in: "Radionuclides in Nephrology," A. M.
Joekes, A. R. Constable, N. J. G. Brown, W. N. Taoxe, eds.
Academic Press, London (1982).

61. J. H. Ten Veen, E. A. Van Royen, S. Surachno, J. B. Van der
Schoot, M. Hardeman, J. Vreeken, J. M. Silmink, Diagnostic
use of Indium-111-labeled platelets in kidney transplant
rejection, 8th Int.Congr. of Nephrol., Athens (abstr.)
(1981).

62. A. Fenech, B. Bennet, G. R. D. Catto, N. Edwards, A. S. Douglas,
F. W. Smith, P. F. Sharp, D. Parry-Jones, Diagnosis of
post-transplant renal haematoma with autologous
Indium-111-labeled plateletes, Lancet i:1250 (1980).

63. G. J. Busch, A. C. P. Martins, N. K. Hollenberg, R. C. Moretz,
R. E. Wilson, R. W. Colman Successful short-term modification
of hyperacute renal allograft rejection in the primate.
Intrarenal effects of phenoxybenzamine and methylprednisolone
combined with heparin, Am J Pathol 82:43 (1976).

64. P. Belitsky, M. Popovtzer, J. Corman, B. Launois, K. A. Porter,
Modification of hyperacute xenograft rejection by
intra-arterial infusion of disodium ethylenediaminetetraacetate, Transplantation 15:248 (1973).

65. T. H. Mathew, P. Kincaid-Smith, D. H. Clyne, B. M. Saker, R. S.
Nanra, P. J. Morris, V. C. Marshell, A controlled trial or
oral anticoagulants and dipyridamole in cadaveric renal
allografts, Lancet i:1307 (1974).

66. H. M. Kauffman, M. B. Adams, L. A. Herbert, P. M. Walczak,
Platelet inhibitors in human renal homotransplantation;
randomized comparision of aspirin versus dipyridamole,
Transplant Proc 12:311 (1980).

67. C. B. Anderson, B. M. Jaffe, R. J. Graff, Prolongation of murine skin allografts by prostaglandin El, Transplantation 23:44-447 (1977).

68. S. Moncada, R. Gryglewski, S. Bunting, J. R. Vane, An enzyme isolated from arteries transforms prostaglandin endoperoxides to an unstable substance that inhibits platelet aggregation, Nature 263:663 (1976).

69. S. Moncada, E. A. Higgs, J. R. Vane Human arterial and venous tissues generate prostacyclin (prostaglandin X), a potent inhibitor of platelet aggregation, Lancet i:18 (1977).

70. R. A. Johnson, F. H. Lincoln, J. L. Thompson, E. G. Nidy, S. A. Mizsak, U. Axen, Synthesis and stereochemistry of prostacyclin and synthesis of 6-ketoprostaglandidn Fl2, J Am Chem Soc 99:4182 (1977).

71. A. R. Mundy, Prolongation of cat to dog renal xenograft survival with prostacyclin, Transplantation 30:226 (1980).

72. A. R. Mundy, M. Bewick, S. Moncada, J. R. Vane, Short term suppression of hyperacute renal allograft rejection in presensitised dogs with prostacyclin, Prostaglandins 19:595 (1980).

73. T. S. T. Wang, S. Oluwole, R. A. Fawwaz, M. Woff, N. Kuromoto, K. Satake, M. A. Hardy, P. O. Alderson, Cellular basis for accumulation of In-111-labeled leukocytes and platelets in rejecting cardiac allografts: concise communication, J Nucl Med 23:998 (1982).

74. J. L. Ritchie, A. Lindner, G. W. Hamilton, L. A. Harker, Indium-111 platelet imaging for detection of platelet deposition in abdominal aneurysms and prosthetic arterial grafts, Am J Cardiol 47:882 (1982).

75. W. A. Jurewitz, J. G. A. Dykes, R. J. Hawker, S. T. Chandler, B. K. Gunson, A. D. Barnes, Tubular damage or rejection: bedside method for differential diagnosis in patients on cyclosporin, Lancet i:998 (1983).

76. H. Sinzinger, P. Clopath, K. Siberbauer, M. Winter, Is the variation in the susceptibilty of various species to atherosclerosis due to inborn differences in prostacyclin (PGI2) formation, Experientia 36:321 (1980).

77. D. E. MacIntytre, J. D. Pearson, J. L. Gordon, Localisation and stimulation of prostacyclin production in vascular cells, Nature 271:549 (1978).

78. G. Remuzzi, D. Marchesi, G. Mecca, R. Misiani, M. Livio, G. DeGaetano, M. B. Donati, Uraemic syndrome: deficiency of plasma factor(s) regulating prostacyclin activity? Lancet i:871 (1978).

79. J. M. Hows, P. M. Chipping, S. Fairhead, J. Smith, A. Baughan, E. C. Gordon-Smith, Nephrotoxicity in bone marrow transplant recipients treated with cyclosporin A, Br J Haematol 54:69 (1983).

80. H. Shulman, G. Striker, H. J. Deeg, M. Kennedy, R. Storb, E. D. Thomas, Nephrotoxicity of cyclosporin A after allogeneic

marrow transplantation: glomerular thrombosis and tubular
injury, New Engl J Med 305:1392 (1981).

81. G. Rocchi, L. Imberti, F. Fumagalli, G. Neild, G. Remutti
     Cyclosporin-A induces glomerula thrombosis possibly affecting
     vascular PGI2-production, V.Int.Conf. on Prostaglandins,
     Florence (abstr.), 759 (1982).

82. G. H. Neild and G. Rocchi, Effect of cyclosporine on
     prostacyclin synthesis by vascular tissue in rabbits, First
     International Congress on Cyclosporin (abstr.), Houston,
     Texas, (1983).

83. C. Leithner, H. Sinzinger, E. Pohanka, M. Schwarz, G.
     Kretschmer, G. Syre, Recurrence of haemolytic uraemic
     syndrome triggered by cyclosporin A after renal
     transplantation, Lancet i:1470 (1982).

84. D. V. Hamilton, R. Y. Calne, D. B. Evans, Haemolytic-uraemic
     syndrome and cyclosporin A, Lancet i:151 (1982).

85. D. V. Hamilton, D. J. S. Carmichael, D. B. Evans, R. Y. Calne,
     Hypertension in renal transplant recipients on cyclosporin A
     and corticosteroids and azathioprine, Transpl Proc 14:497
     (1982).

86. M. L. Foegh, M. Zmudka, C. Cooley, J. F. Winchester, G. B.
     Helfrich, P. W. Ramwell, G. E. Schreiner, Urine i-TXB$_2$ in
     renal allograft rejection, Lancet ii:431 (1981).

87. M. L. Foegh, J. F. Winchester, M. Zmudka, G. B. Helfrich, P. W.
     Ramwell, G. E. Schreiner, Aspirin inhibition of thromboxane
     release in thrombosis and renal transplant rejection, Lancet
     i:48 (1982).

88. A. Capitanio, P. M. Mannucci, C. Ponticelli, F. Pareti,
     Detection of circulating released platelets after renal
     transplantion, Transplantation 33:298 (1982).

89. A. D. Heyns, M. G. Lotter, H. Pieters, F. H. Pauw, P. N.
     Badenhorst, P. Wessels, P. C. Minnaar, A quantitative study
     of Indium-111-oxine platelet kinetics in acute and chronic
     renal transplant, Clin Nephrol 18:174 (1982).

90. P. Ghosh, D. B. Evans, S. A. Tomlinson, R. Y. Calne, Plasma
     lipids following renal transplantation, Transplantation
     15:521 (1973).

91. S. J. Shattil, R. Anaya-Galindo, J. Bennett, R. W. Colman, R. A.
     Cooper, Platelet hypersensitivity induced by cholesterol
     incorporation, J Clin Invest 55:636 (1975).

92. A. Szczeklik and R. J. Gryglewski, Low density lipoprotiens
     (LDL) are carriers for lipid peroxides and inhibit
     prostacyclin (PGI2) biosynthesis in arteries, Artery 7:488
     (1980).

93. A. J. K. Laden, Experimental atherosclerosis in rat and rabbit
     cardiac allografts, Arch Path 93:240 (1972).

94. D. R. Alonso, P. K. Starek, C. R. Minick, Studies on the
     pathogenesis of atheroarteriosclerosis induced in rabbit
     cardiac allografts by synergy of graft rejection and
     hypercholesterolemia, Am J Pathol 87:415 (1977).

95. K. Silberbauer, G. Schernthaner, H. Sinzinger, H. Piza-Katzer,
     M. Winter, Decreased vascular prostacyclin in juvenile-onset
     diabetes, New Engl J Med 300:366 (1979).
96. H. Sinzinger and A. Kefalides, Passive smoking severely
     decreases platelet sensitivity to anti-aggregatory
     prostaglandins, Lancet ii:392 (1982).
97. S. R. Coughlin, M. A. Moskowitz, B. R. Zetter, H. N. Antoniades,
     L. Levine, Platelet-dependent stimulation of prostacyclin
     synthesis by platelet-derived growth factor, Nature 288:600
     (1980).
98. J. G. Gerber, A. S. Nies, G. C. Friesinger, J. F. Gerkens, R. A.
     Branch, J. A. Oates, The effect of PGI2 on canine renal
     function and hemodynamics, Prostaglandins 16:519 (1978).
99. J. A. Oates, A. R. Whorton, J. Gerber, J. Lazar, R. A. Branch,
     J. W. Hollifield, Prostacyclin and the kidney, in:
     "Prostacyclin," J.R.Vane and S. Bergström, eds., Raven
     Press, New York (1979).
100. T. Niwa, K. Maeda, H. Asada, M. Yamamoto, K. Yamada, Beneficial
     effect of prostaglandine El in rapidly progressive
     glomerulonephritis., New Engl J Med 308:969 (1983).
101. M. M. Popovtzer, W. Pinggera, F. H. Katz, J. L. Corman, J.
     Robinette, B. Lanois, C. G. Halgrimson, T. Starzl, in
     arterial blood pressure after kidney transplantation.
     Relation to renal function, plasma renin activity and the
     dose of prednisone, Circulation 4:1297 (1973).
102. J. R. Weeks and D. M. Sutter, An antihypertensive effect of
     prostacyclin, in: "Prostacyclin," J.R. Vane and S. Bergström,
     eds., Raven Press, New York (1979).
103. R. R. Lindquist, R. D. Guttmann, J. P. Merrill, G. J. Dammin,
     Human renal allografts. Interpretation of morphologic and
     immuno-histo-chemical observations, Am J Path 53:851
     (1968).

INDIUM-111-PLATELETS IN BYPASS GRAFTS:

EXPERIMENTAL AND CLINICAL APPLICATIONS

Mrinal K. Dewanjee

Sections of Diagnostic Nuclear Medicine and
Cardiovascular Surgical Research
Mayo Clinic and Mayo Foundation
Rochester, MN   55904

INTRODUCTION

Carrel and Guthrie made original and significant contributions in
arterial and venous grafting in 1906 in dogs (1). Carrel was awarded
the Nobel Prize in 1908 for these vessel transplantations with
autologous, homologous, and heterologous blood vessels. Advance in
surgical techniques (2-14) and understanding the cell biology of
endothelial, smooth muscle cells, and fibroblasts; mechanism of
activation of platelet and of coagulation cascades; risk factors for
loss of protective endothelial cells and subsequent thrombosis and
atherosclerosis of native vessels and vein grafts (15-54);
development of platelet inhibitors; anticoagulant drugs and new
diagnostic techniques for evaluation of graft patency (54-96) led to
the present status of arterial and venous reconstructions.

The platelet labeling with In-111-oxine by Thakur et al (56) made
possible quantitation of platelet thrombosis in acute and chronic
phases in atherosclerosis and thrombosis (62,78). About 550,000
vascular grafts are implanted annually in the United States. Of
these, 200,000 are autogenous vein grafts, 50,000-80,000 peripheral
vein bypass grafts, 120,000-150,000 aortocoronary bypass grafts; the
rest are synthetic vascular grafts (greater than 5mm or larger in
diameter). Most of the grafts in the high blood flow and low
resistance regions of the aorta, iliac, and proximal femoral artery
perform quite satisfactorily. About 15 percent of the graft failure
in this category is due to angulation, dilatation, formation of mural
thrombus of vascular grafts, fragmentation and embolization, formation
of pannus or neointima or pseudoneointima (basically made of fibro-

229

blast, collagen, and fibrin, also called fibrous hyperplasia),
stenosis at anastomotic lines, infection, fabric failure with
fragmentation and disruption of polyester strands. Teflon is less
vulnerable to biodegradation than Dacron (knitted or woven) or
polyurethane. None of the grafts stay patent below 4 mm in internal
diameter (4 mm barrier), although a combined aggressive approach of
immobilization of drug on graft surface for inhibition of
calcification, infections, coagulation, and platelets, along
with oral medication and endothelial seeding, might solve some of
these problems in the future. Types of grafts used in vascular
surgery are shown in Figure 1. Polyurethane and collagen in synthetic
vascular grafts and tissue conduits contain multiple functional groups
for anchoring drugs, although Teflon is very inert for this purpose.
Most of the small vessel prostheses (4 mm I. D.) are under
experimental and limited clinical investigation in arterial systems of
high blood flow (subclavian artery). Considering tissue reaction,
biodegradability, suturability, bending and tensile strength, both
Dacron and Teflon grafts appear suitable above the 4 mm diameter. The
high porosity of Dacron fibers necessitates graft preclotting by
soaking in blood and oven baking before graft interposition. The low
porosity of Teflon fiber appears less thrombogenic than Dacron at 5-7
mm diameter. Although smooth surfaces of polydimethylsiloxane and
polyethylene retain fewer platelet aggregates, most of the thrombi
that are formed might be dislodged from the smooth graft surface, and
the emboli are trapped in distal organs. So far other biopolymers do
not appear suitable for the development of synthetic vascular grafts.
A list of graft biomaterials (synthetic polymers) used in clinical and
experimental investigations are shown in Figure 2. The textures of
several types of fabrication of Dacron fibers (woven, knitted,
knitted-woven, and nonwoven) and Teflon (microfibers with nodes) are
visible in photographs (Figures 3A and 3B). The mechanism of platelet
activation leading to thrombus formation on foreign surface of
vascular graft and distal embolization is shown in Figure 4.

Autogenous saphenous vein is widely used as a vascular conduit in
the reconstruction of small and medium-sized arteries due to its
compliance and kink resistance, expendability, accessibility,
sufficient length, and, most importantly, patency rate. Varicosities,
phlebosclerosis, fibrosis of vein valve, inadequate size or length, or
previous surgery of saphenous vein necessitate the use of cephalic,
external jugular, or homologous vein or synthetic vascular graft.
Weisel and associates (8) and Darling and Linton (13) analyzed the
patency rates of saphenous vein, human umbilical vein, and extended
polytetra-fluoroethylene grafts by location of implantation, by
clinical indication, and by the number of patent runoff vessels
demonstrated by angiography. In this comparative evaluation the
grafting was performed between femoral artery, to the popliteal artery
above the knee, to the popliteal artery below the knee, and to the
tibial-peroneal arteries. Graft patency measured by angiography at 36
months was identical above the knee, about 60 percent for EPTFE and

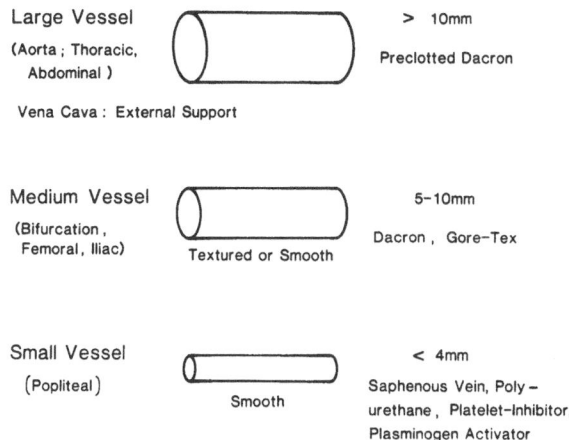

Large Vessel

(Aorta ; Thoracic, Abdominal)

Vena Cava : External Support

> 10mm

Preclotted Dacron

Medium Vessel

(Bifurcation , Femoral, Iliac)

Textured or Smooth

5-10mm

Dacron , Gore-Tex

Small Vessel

(Popliteal)

Smooth

< 4mm

Saphenous Vein, Poly – urethane , Platelet–Inhibitor, Plasminogen Activator

Fig. 1. Types of vascular prostheses and anatomical sites. Four mm microporous EPTFE grafts are undergoing clinical trials in the high-flow area of subclavian artery (120-150 ml/min).

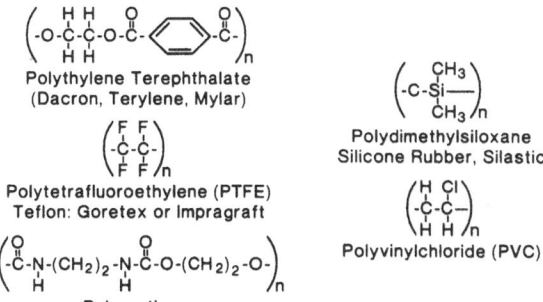

Polythylene Terephthalate (Dacron, Terylene, Mylar)

Polytetrafluoroethylene (PTFE) Teflon: Goretex or Impragraft

Polyurethane Esthane, Ostamer

Polydimethylsiloxane Silicone Rubber, Silastic

Polyvinylchloride (PVC)

Fig. 2. Monomers used in synthetic vascular graft. Dacron grafts are made by a variety of weaving techniques from Dacron fibers. Teflon grafts are made by extrusion into tube form. Extended PTFE surface of different porosity is made by expanding heated Teflon tube longitudinally.

less below the knee, and 20 percent for Dacron graft; higher patency rate of 72 percent indicates that saphenous vein is the best graft. Ninety-eight percent patency rate was obtained with Dacron aortic bifurcation graft at 10 years. The higher thrombotic rate of large diameter Dacron grafts may be due to thick gelatinous and loosely adherent lining of the pseudoneointima (6-14). Causes of early graft failures were technical errors of harvesting, suturing, limited outflow, limited distal arterial bed, diameter of graft, and reduced inflow. Causes of late failures were progressive accumulation of platelet-fibrin-collagen mass and intimal hyperplasia at anastomoses and midsection of graft. Sauvage et al (9) coined the words "thrombotic threshold velocity" to identify the velocity at which thrombosis begins in graft surface (about 8 ml/cm^2. sec). When flow drops below this level, due to the above-mentioned complications, graft closure is threatened by platelet-fibrin thrombus.

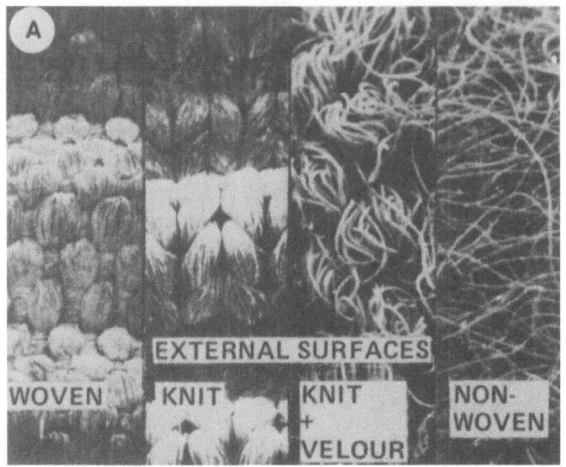

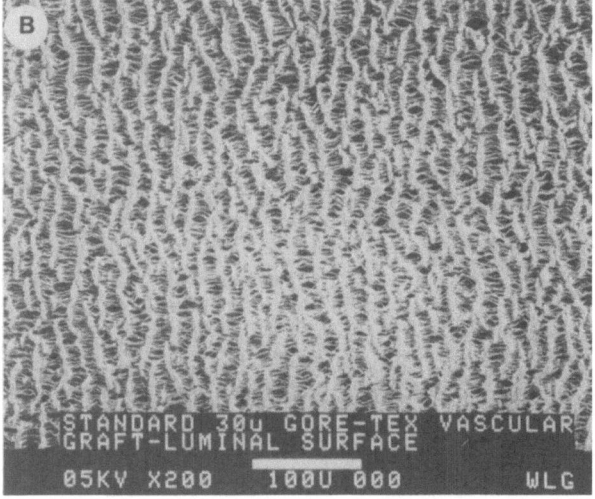

Fig. 3. Photograph of different type of (A) Dacron and (B) Gore-Tex graft (Teflon).

In general both Dacron and Teflon cause tissue reactions. The cell types that were present around the graft sites were predominantly lymphocytes, plasma cells, granulocytes, macrophages and occasional giant cells. Although glutaraldehyde fixation of umbilical vein graft may not totally eliminate graft antigenicity in human, the Dacron support mesh itself may generate a host response.

Despite the preventive measures, graft infection will occur in 1-2 percent of implanted grafts. The most common type is aortofemoral graft infection and aortoiliac graft. Clinical complications are abscess, septic embolization, graft occlusion, false aneurysm, retroperitoneal or femoral hemorrhage, sinus tract formation and gastrointestinal bleeding (graft-enteric fistula). Diagnosis by culture of biopsied material, imaging with In-111-white cells or

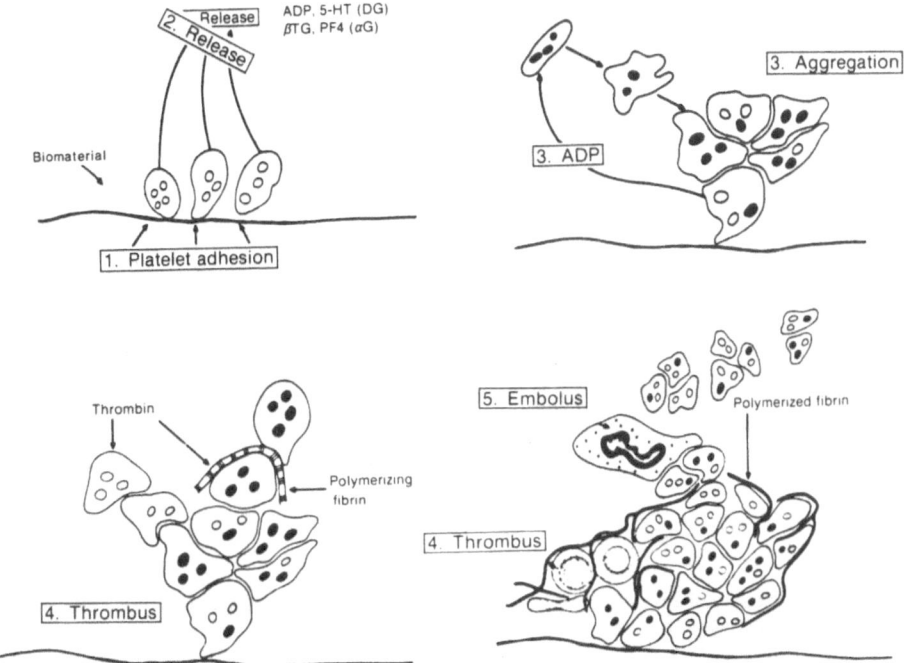

Fig. 4.   Adsorption of fibrinogen Von Willebrand (VWd) factor,
fibronectin and other proteins on the surface of synthetic
vascular graft occurs within a few seconds of perfusion.
Activation of platelet on this adsorped protein surface leads
to release of proaggregating substances (ADP, Serotonin) from
dense granules (DG) and stabilization of platelet aggregates
by fibrin mesh. At high flow and shear rate smaller
fragments of platelet-thrombus are dislodged and are trapped
in the distal perfusion bed of different organs αG (alpha
granule).

Ga-67-citrate have been used (40-42). Early diagnosis, removal of
infected graft, antibiotic treatment, and replacement are essential
for proper patient management.

We have recently used Ficoll-Hypaque separation technique for
quantification of In-111-distribution in In-111-labeled white cells
and platelets. In In-111-labeled white cell preparation only (22 $\pm$
14%) of total In-111 is bound to white cells (PMN and monocyte); (25 $\pm$
16%) to red cells, (46 $\pm$ 20%) to platelets and lymphocytes, and (7 $\pm$
5%) bound to plasma. On the other hand, in In-111-labeled platelet
preparation, (92 $\pm$ 7%) was bound to platelets, (2 $\pm$1%) to RBC, and (5
$\pm$ 3%) to plasma. This study clearly indicates that In-111-labeled
mixed white cell preparation used for graft infection studies are

questionable since the total In-111 uptake in graft is due to
migration of In-111-white cells, to the site of graft infection, and
In-111-labeled platelets due to graft thrombosis, In-111-labeled RBC
and platelets in blood pool and In-111-labeled lymphocytes, due to
graft tissue reaction, although In-111-labeled platelet contamination
is the major issue of concern.

Dardik et al (4,5) used glutaraldehyde-stabilized human umbilical cord vein as a vascular prosthesis as a femoral popliteal
bypass and tibial, peroneal reconstructions, and arteriovenous (A-V)
shunt for hemodialysis patients. Human umbilical veins were
harvested, fixed in glutaraldehyde; to avoid aneurysm of the graft,
an external Dacron mesh support was used as a composite graft.
Patency rate was similar to that of EPTFE graft, although biodegradability was higher. A similar graft of glutaraldehyde-fixed bovine
aortic artery supported with Dacron mesh was used as an A-V shunt in
renal failure patients where multiple access sites for hemodialysis
are necessary before a suitable donor is available for kidney
transplantation.

To avoid rest-pain, claudication (limping), nonhealing ulcers,
gangrene and amputation, femoral-popliteal graft of autologous vein
graft, heterologous glutaraldehyde-fixed umbilical vein graft and
EPTFE grafts have been used with some success. Aneurysm was one of
the complications of the fibrocollagenous tube (ureter, pericardium,
durameter). None of the physical properties of surface, e.g., surface
tension, surface free energy, roughness of surface or texture,
porosity, hard and soft segments of polymer or surface density of
atoms exposed to blood have any direct correlation with
thrombogenicity of graft materials.

A significant advance has been made in noninvasive measurement of
graft thrombogenicity with In-111-labeled platelets. In the past,
Harker et al and Hanson et al (64,75) used Cr-51-platelet and
correlated reduction in platelet survival time in baboon and human
with increased platelet consumption. With In-111-labeled platelets,
amount of thrombus formation, embolization, and platelet survival time
could simultaneously be measured. Due to increase in plasma-bound
In-111 in the late phase of In-111-platelet administration, it is
essential to correct for plasma-bound In-111 for accurate value of
platelet survival time.

ESTIMATION OF PLATELET THROMBOSIS ON VASCULAR GRAFTS

Two approaches have been taken for quantifying platelet thrombosis on
different types of vascular grafts and effect of drugs. Qualitative
estimation of platelet thrombosis was made with a gamma camera at
24-120 hours after administration of In-111-platelets and 5-10 days
post graft implantation and subsequent follow up of platelet

consumption with measurements of platelet survival time and platelet deposition on graft. Unlike all animal models, endothelial coverage occurs only at proximal and distal anastomoses in humans, the rest of the area is covered with fibrin-collagen matrix with a few fibroblasts. This surface is thrombogenic, and platelet deposition could be observed at 10-15 years post graft implantation. Higher platelet deposition in the acute phase suggests that graft may be occluded in the near future. Within 1-2 hours of perfusion a dumbbell shape of platelet deposition is observed, more deposition at anastomosis and less in the middle. The early decline in platelet radioactivity on graft may be due to embolization of platelet aggregates. In the latter phase platelet deposition is observed in the midsection of unendothelialized graft surface; endothelial cells extend from normal vessel around the proximal and distal anastomoses only to a small extent in humans. In previous studies blood pool activity was subtracted with Tc-99m-labeled red cells (80) and In-111-excess in graft was calculated; nonuniform distribution of Tc-99m at anastomoses and midgraft section due to RBC incorporation in thrombus (thrombus on most surfaces looks red), amount of plasma-bound In-111 which increases to 5-22% in the 1-5 day interval post administration, proximity of vein, bone marrow to graft and motion artifacts make proper background subtraction difficult. In some studies platelet deposition in graft was compared with that in the contralateral femoropopliteal vessel. Atherosclerosis, phlebosclerosis, variocosity, blood pool and marrow activity also lead to error in estimating radioactivity ratios. Stratton et al (76,77) evaluated graft thrombogenicity by measurement of graft In-111 activity with a gamma camera and blood activity with a gamma counter. Graft to blood ratios were higher in thrombogenic surface than in normal. In general, due to continuous platelet sequestration, this graft/blood ratio increases in normal vessel, more so in graft and other thrombogenic surface. In our studies we have determined the number of In-111-counts per pixel per microcurie at a definite anatomical site (82). Smaller number of counts per pixel necessitates longer imaging time. Pericardial or Teflon patches are used in repair of septal defects in congenital heart disease. Dacron grafts with valve (valved conduit) were used in pulmonary atresia. We have imaged platelet deposition in conduits in children (86); like other synthetic vascular grafts, higher platelet deposition occurs at 2-3 days post administration of labeled platelets.

Thrombosis and its quantification on interposition, bypass grafts or valved conduits with In-111-labeled platelets for arterial and venous reconstructions at different sites, experimental evaluation of small vessel prosthesis will be described as follows:

1. Carotid artery bypass graft and endarterectomy (patch graft)
2. Aortic arch, thoracic and abdominal aorta interposition graft
3. Aortocoronary saphenous vein bypass graft
4. Aortic bifurcation and femoral artery graft

    5. Venous reconstruction: vena caval graft
    6. Experimental study with small vessel prostheses (femoral and
        carotid artery)

    In the experimental evaluation of biomaterials, platelet
inhibitor drugs, and the development of small vessel prostheses, we
rely more upon estimation of adherent platelets at anastomoses and
midgraft sections per unit area (62,74). In spite of meticulous
surgical techniques and measurements of plasma level of drugs, we
observed a great deal of variation in adherent platelet density. This
complication might be due to embolization, syntheses of thromboxane
(TxA_2) and prostacyclin, and other rheological and blood flow factors.
Due to these variations, measurement of retained thrombus on graft in
carotid artery and trapped embolus in distal organ bed (brain) might
be more appropriate for the proper evaluation of platelet-inhibitor
drugs; an effective drug-graft combination should reduce both
thrombosis and embolization. These studies suggest that Dacron graft
is more thrombogenic than Gore-Tex graft, and some platelet-inhibitor
drugs were found to reduce platelet deposition. The minor changes in
platelet deposition due to various medications could not be accurately
measured with the above-mentioned imaging techniques.

Carotid Artery Bypass Graft

    Ulceration of atherosclerotic carotid arterial lesion results in
platelet adhesion, aggregation and thrombotic occlusion or leads to
hemispheric ischemia or infarction of brain tissue. Embolization of
platelet aggregates leads to transient ischemic attack (TIA);
continued worsening leading to stroke could be avoided by removal of
adherent plaque and thrombus by endarterectomy or carotid artery
bypass graft. Due to dislodgement of fragments of plaque and adherent
thrombus, angioplasty is not used much in this critical artery.
Experimental carotid artery bypass grafts have been used for
evaluation of retained thrombus and quantification of total cerebral
embolus and effect of platelet-inhibitor therapy. Ulcerating carotid
plaque could be occasionally imaged with In-111-labeled platelet
(69,70). We have imaged the site of endarterectomy quite well,
although we have had difficulty in imaging ulcerating carotid plaque.

Aortic Arch, Thoracic and Abdominal Aorta, Interposition Graft

    The major complications of aneurysms of thoracic and abdominal
aorta are fatal ruptures and should be repaired soon after diagnosis.
Surgical replacement of aneurysmic segments with preclotted Dacron
graft and reattachment of the visceral arteries lead to high patency
rate due to high flow and low resistance. Although different types of
weaving techniques are used for making Dacron grafts of different
porosities and textures, the roles of velour, weaves, and knits on
rheology of platelet activation have not been studied well (6-12). Ten
to fifteen years after surgery the midsection of the graft is found

thrombogenic, and platelet deposition could be followed by sequential imaging (71,76,77). Thromboembolic events have been reduced with platelet-inhibitory therapy.

Endothelial seeding of Dacron and expanded Teflon graft was successfully performed by Graham et al, Herring et al, and Burkel et al (94,95,96) and development of nonthrombogenic surface in thoracoabdominal bypass graft was followed by sequential imaging with In-111-labeled platelets in the dog model. In this process, endothelial cells (5-30 cell chunks) are harvested from a jugular vein, with trypsin-EDTA-collagenase treatment and transferred to a dish containing nutrient media. These cells were mixed with blood and double-velour Dacron graft (6mm I.D. Meadox), clotted with this endothelial cell preparation (0.2 - 0.5 x 10^6 cells) and implanted in the thoracoabdominal aorta. The endothelial cell coverage of 80 percent of graft surface was reported at 30 days. Lack of platelet deposition in endothelial cell seeded graft was evaluated with In-111-labeled platelets. The presence of Weibel-Palade body, factor VIII antigens and prostacyclin production in the intima of explanted Dacron graft suggest endothelial cell growth and coverage. Endothelial seeding is not useful for 4 mm I.D. synthetic graft. Platelet thrombus occluded the lumen (1-3 hours) before effective endothelial coverage (14-20 days).

Aortocoronary Saphenous Vein Bypass Graft

A major limitation of aortocoronary bypass (ACB) graft with autologous saphenous vein is the high frequency of graft occlusion by platelet thrombosis in the early postoperative phase and by proliferation of intimal smooth muscle cells in the late phase. All vein grafts develop some morphologic changes with time (15-24), but the degree and functional significance of the changes may differ because of several factors--endothelial cell loss during harvesting and implantation of the graft, deposition of platelets or fibrin, graft-wall ischemia, intraluminal pressure, vortexing and shear stress at distal anastomoses and vein valves, and functional alteration and repair of damaged endothelium (22-29). During harvesting and implantation of femoral-vein grafts in dogs, the grafts become partially de-endothelialized as shown by scanning electron-microscope studies; and consequently graft permeability and cholesterol uptake increase (25-27,30-31). In addition, increased uptake of labeled lipoproteins by venous graft suggests high metabolic activity at three months. The evolution of venosclerosis of ACB graft is shown in Figure 5.

With use of cardiopulmonary bypass, and autologous segment of undistended, reversed femoral vein was implanted as a bypass vessel from the aorta to the left anterior descending coronary artery in 50 mongrel dogs weighing 28 to 26 kg (27,66,67). The design of the connecting graft is shown in Figure 6 and at 24 hours before sacrifice

in the dogs to be sacrificed on later days. Five untreated dogs and
five treated dogs were sacrificed on each of five sacrifice days (1,
3, 7, 30, and 90 days) after surgery. They were heparinized (4mg/kg)
and killed with an overdose of barbiturate. Protocol for the study
and medication level are shown in Figure 7.

Preparation of specimens. After sacrifice of each animal, a 3 ml
sample of its blood was collected into an EDTA tube. The graft,
including the two anastomoses, was excised. Each anastomosis, with 2
to 3 mm of vessel on each side of the suture line, was removed, and
the remaining central graft was divided into three segments—proximal,
middle, and distal. All five segments were opened and rinsed with
isotonic saline. A 4 cm segment was taken from the femoral vein
contralateral to the graft source; and this control-vein specimen was
carefully cleaned of adventitia and tied proximally and distally. To
avoid adhesion of platlets to the abluminal side, blood in the
control-vein specimen was withdrawn by a 27-gauge needle before
residual blood was flushed out with heparinized saline.

Quantification of platelets in specimens. The platelet count in the
whole blood samples was determined with a Coulter counter. The samples
were weighed in a microbalance, and radioactivity per unit weight of
blood was calculated. Blood samples in duplicate were centrifuged to
remove plasma, and the radioactivity (In-111-cpm) in plasma was used

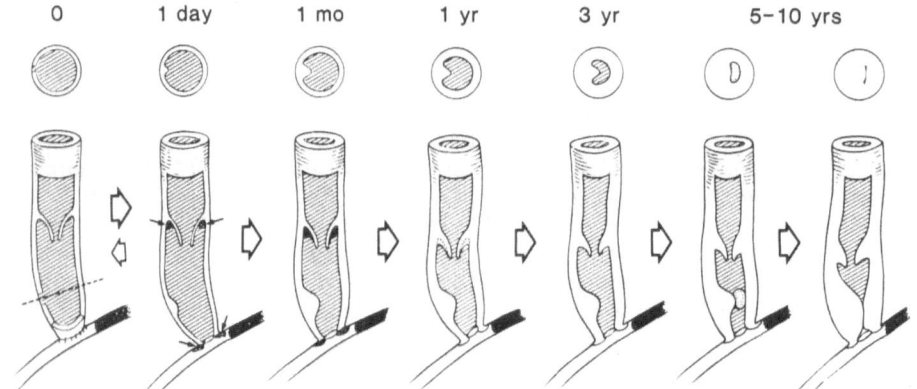

Fig. 5. Schematics of evolution of venosclerosis of aortocoronary
         femoral vein bypass graft. Early denudation is repaired
         within a week. Extensive denudation is covered by a carpet
         of platelet fibrin layer, and organization of thrombus leads
         to narrowing of graft lumen. Vortexing of blood at the
         venous valve and at distal anastomoses leads to platelet
         activation, aggregation, and stasis. Migration of smooth
         muscle cells in intima, effect of growth factor from platelet
         on proliferation of smooth muscle cells, and thrombus
         organization narrow the graft lumen at distal anastomoses,
         and the graft might ultimately be occluded by a platelet
         thrombus.

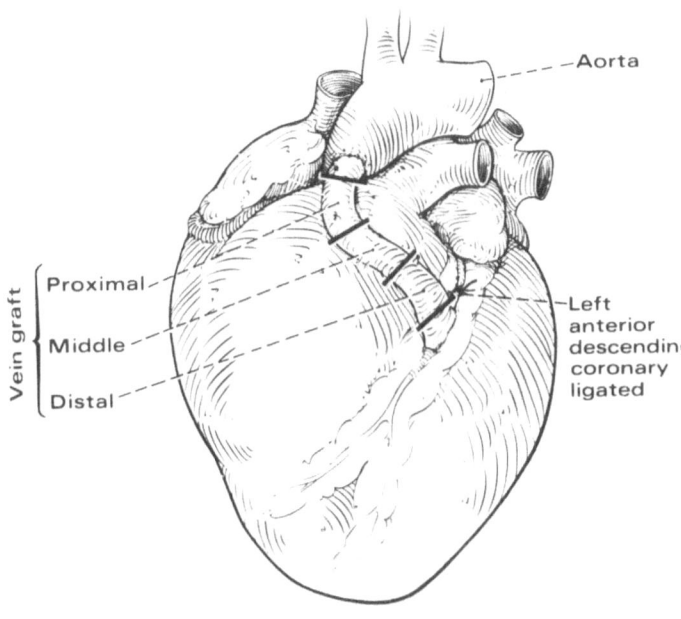

Fig. 6. Design of femoral vein bypass graft connecting aorta to left anterior descending coronary artery (LAD). The proximal LAD above distal anastomosis was ligated.

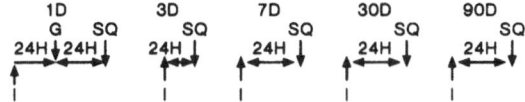

Dipyridamole (2.5 mg/kg/day) orally 2 days prior to surgery
  Aspirin (15 mg/kg/day) orally immediately after surgery and
  medication continued until sacrifice

I.V. Injection (I)   G = Vein grafting
^{111}In-Labeled Platelet: S.Q. = Sacrifice & quantitation
    300-400 μCi (autologous)   D = Day Postgrafting

Fig. 7. Protocol for aortocoronary bypass graft surgery, medication, and evaluation of platelet thrombosis and histochemical evaluation of vein graft.

to determine the percentage of free In-111. With that, the percentage of radioactivity bound to platelets was determined, and next the platelet-bound In-111-cpm/ml blood:

$$\frac{\text{Platelet-bound In-111-cpm}}{\text{blood (ml)}} = \frac{\text{Whole blood In-111-cpm}}{\text{blood (ml)}} \times (100 - \% \text{ free In-111})$$

Then, from In-111-cpm/ml of blood and the platelet count in EDTA blood, the number of platelets/cpm was calculated:

$$\frac{\text{No. platelets}}{\text{In-111-cpm}} = \frac{\text{No. platelets/blood (ml)}}{\text{In-111-cpm/blood (ml)}}$$

Graft and saphenous vein. Each graft and anastomotic segment and each control-vein specimen was spread on linear graph paper and outlined carefully, and its area was determined by counting the number of squares. With the data then available--the area (cm^2) of each segment, radioactivity (cpm) in each, and number of platelets/cpm--the number of platelets adhering to each square centimeter of surface of blood vessel and anastomosis was calculated:

$$\frac{\text{No. platelets}}{\text{area } (cm^2)} = \frac{\text{In-111-cpm in specimen}}{\dfrac{\text{In-111-cpm}}{\text{no. platelets}} \times \text{area } (cm^2)}$$

For determination of thrombogenicity of grafted vein with respect to that of control femoral vein, with treatment and without, the mean In-111-cpm/mg from each of the three central graft segments was expressed as a ratio to the mean from the control-vein specimens.

Scanning electron-microscope examination. To determine the status of the endothelium, graft and control-vein specimens were prepared by pressure-perfusion (120 mm Hg) with HEPES-buffered glucose (4.6% at pH 7.4), HEPES-buffered silver nitrate (0.2%), and 0.1 Mole phosphate-buffered glutaraldehyde (3%). Sections of the fixed veins were coated with gold-palladium alloy. The intimal surface was examined with a scanning electron microscope (ETEC Autoscan), and representative sections were photographed.

Course of platelet deposition. At 24 hours after the grafting (Fig. 8), the quantity of adherent platelets per unit area (platelets/cm^2) of the five graft and anastomotic segments did not differ significantly between the untreated and treated groups (paired t test). At day three, however, significant differences between the untreated and treated groups appeared in the platelets/cm^2 of each of the five segments ($P < 0.01$). Except for the distal graft segment, these differences persisted at day seven ($P < 0.01$) and likewise at day 30 ($P < 0.05$). But by day 90 the platelets/cm^2 of the two groups were virtually equal in every segment--and lower than at day one by almost two orders of magnitude. At day one, in treated and untreated groups alike, the proximal and distal anastomoses had more platelets/cm^2 than any of the three central graft segments; but the difference was not significant and it tended to decrease with time. At day 90, again in each group, the platelet density on the five graft and anastomotic segments was nearly identical. Scanning electron-microscope examination of graft specimens revealed de-endothelialized surface at one and seven days after grafting and almost total re-endothelialization and absence of platelet adhesion at 30 and 90 days (Fig. 9A, 9B, 9C, 9D, and 9E).

Similar endothelial cell loss was observed in angioplasty. Angioplasty of occluded (partial or complete) artery by balloon-catheter
(polyethylene) dilatation caused endothelial cell loss, cracking of
internal elastic lamina, medial necrosis, and exposure of blood to
subendothelial thrombogenic exposed contents of plaque. In medium
size vessel, subsequent platelet thrombosis could occasionally be
imaged with In-111-labeled platelets (79), although this may not be
possible in coronary artery or in other small size vessels.

Vein graft preservation. Haudenschild et al (20), used
papaverine before harvesting and preserved vein grafts by papaverine
incubation. In the past, vein grafts were kept in isotonic saline in
crushed ice with the idea that at low temperature the metabolic
demands of endothelial, smooth muscle cells and fibroblast would be
reduced and cells in vein grafts would maintain viability. Recent
scanning electron micrographs of saline-ice preserved veins showed
major endothelial cell loss due to contraction and expansion of medial
smooth muscle cells. We have recently preserved canine vein grafts in
ACD-plasma containing papaverine (100 µg/ml) and verapamil (22 µg/ml)
for 60 minutes at 25°C (78). These grafts were implanted in
aortocoronary position. After three hours of perfusion in dogs, vein
grafts were harvested; and radioactivity in blood, graft sections, and
anastomoses was determined. The values (mean ± SD) of platelets per
unit area of graft surface are shown in Figure 10. Papaverine and

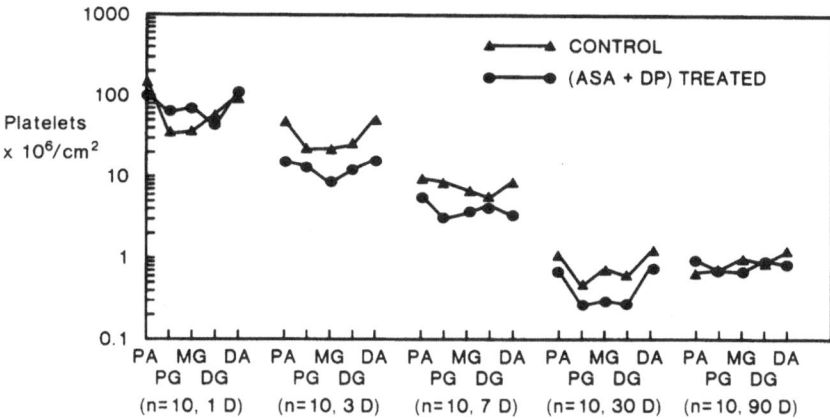

Fig. 8. Course of platelet deposition. Platelets/cm² on surface of
         graft segments and anastomoses: means of data from untreated
         ("control") dogs and dogs treated with dipyridamole and
         aspirin--five of each group sacrificed at 1, 3, 7, 30, and 90
         days after grafting. In each group of five joined points,
         the points--left to right--represent means from proximal
         anastomosis, proximal graft segment, mid-graft segment,
         distal graft segment, and distal anastomosis.

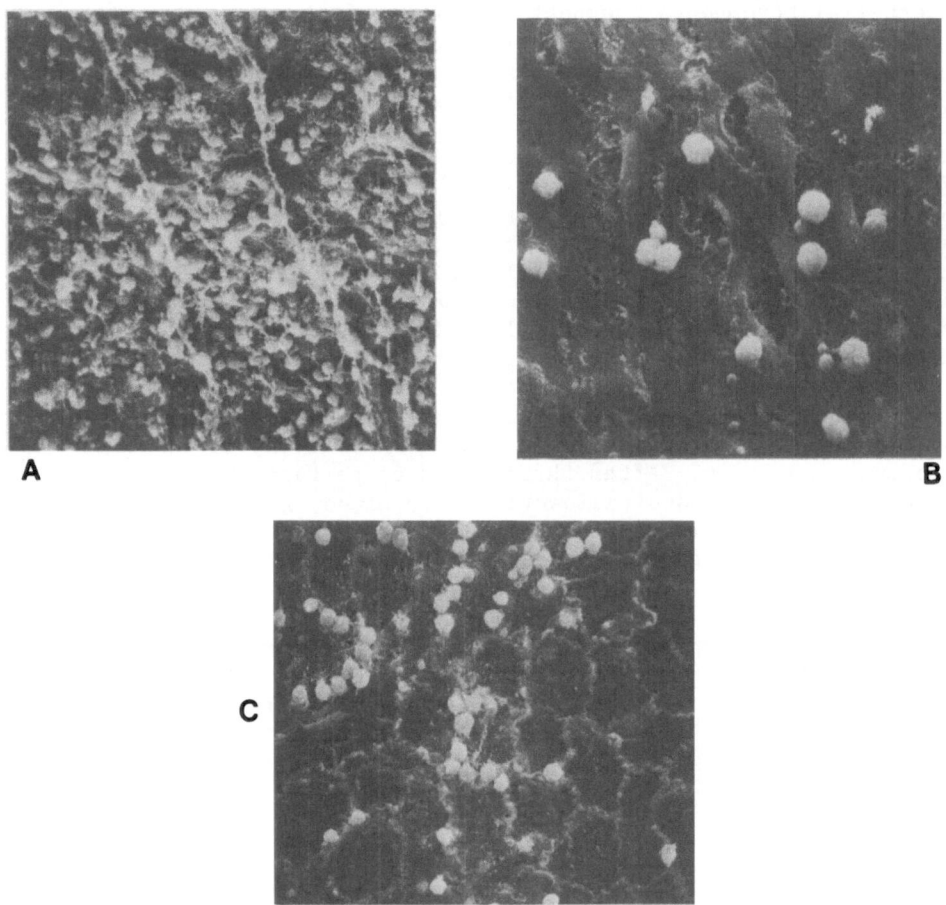

Fig. 9.   A. B. C. D. E.   Midgraft surface in early post-operative phase:
          de-endothelialized, with many adherent platelets (scanning
          electron-microscope views). A, Day 1 (x600); B, Day 2 (x1000);
          C, Day 7 (x600); D, Day 30 (x300), re-endothelialized, with few
          adherent platelets; E, transmission electron micrograph of re-
          endothelialized vein graft at 90 days post aortocoronary bypass
          graft surgery in dog. Note total endothelial coverage in
          intima. Fewer smooth muscle cells and collagen fibers are
          noted in the graft of the treated group.

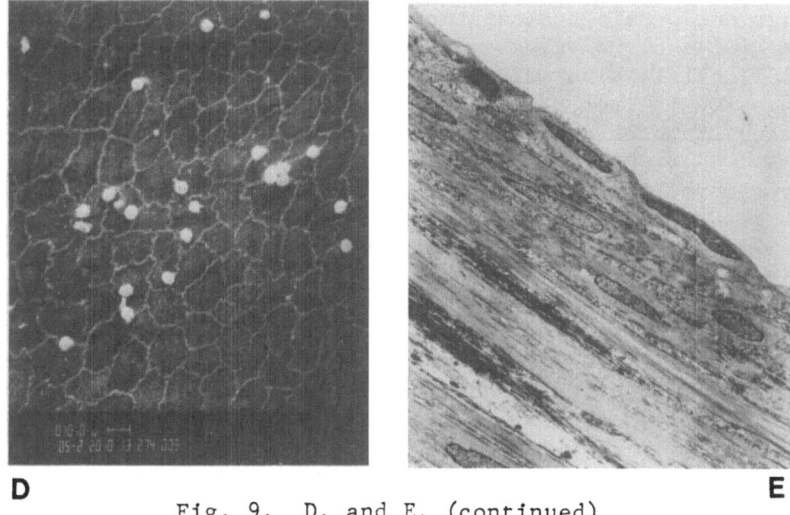

D   **E**

Fig. 9.   D. and E. (continued)

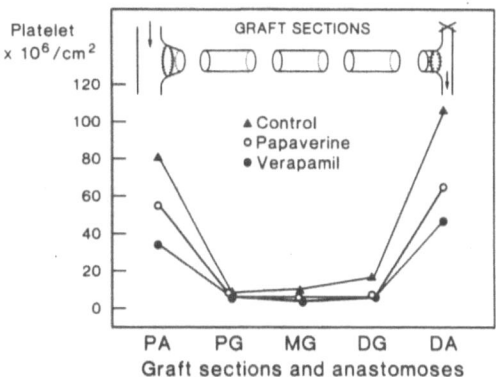

Fig. 10.   Quantitation of retained platelet thrombus in proximal
and distal anastomoses and three midsections of ACB graft of
control, papaverine, and verapamil preserved vein at three hours
of perfusion.

verapamil treatments tend to reduce de-endothelialization and platelet
deposition in vein graft; this may lead to higher graft patency.

   Pathologic and therapeutic mechanisms.   Aspirin inactivates
platelet cyclo-oxygenase by acetylation and thus inhibits the
formation of potent aggregating agents PGG_2, PGH_2, and thromboxane A_2
from arachidonic acid (30-38).   Its inhibition of platelet

cyclo-oxygenase persists about three times longer than the inhibition of endothelial cyclo-oxygenase. Prostacyclin synthesis by the endothelial cell is also affected by aspirin, although vessel-wall cyclo-oxygenase is less sensitive to it. These effects are better demonstrated at three days than at one day after bypass grafting.

Furthermore, prostacyclin production in the vascular endothelium is compromised by atherosclerosis and by the partial de-endothelialization of the edematous vessel wall. Dipyridamole is an inhibitor of phosphodiesterase and thus increases cyclic AMP in platelets. Drugs that increase platelet cyclic AMP also suppress platelet activation. Physiologically, its potentiation of prostacyclin inhibitory effects may be the most important action of dipyridamole. In addition, vasodilation by dipyridamole may reduce endothelial sloughing during the harvesting of vein grafts.

Although increased platelet deposition at the anastomotic sites had been demonstrated previously by light-and electron-microscope techniques, investigation had been limited by lack of a suitable technique for a calculation of the number of platelets per unit of surface area. We have recently quantified platelet lysis and platelet oxygenator consumption during cardiopulmonary bypass in the dog (61-63). Approximately 20% of the platelets are consumed in the bubble oxgenator and another 10-18% in repair of blood vessels. Therefore, for one-day study reported here, the In-111-labeled platelets were administered 24 hours before surgery, so that they would undergo the same trauma as unlabeled platelets during open-heart surgery.

Our recently developed process has been applied for evaluation of drugs intended to inhibit platelet deposition in bilateral femoral-vein and Gore-Tex graft implants in dogs (62). Electron-microscope techniques are useful in demonstrating de-endothelialization and the adherence of single and multiple platelets to de-endothelialized surface; but pressure-perfusion, fixation, and other manipulations involved in preparing specimens for such examination lead to loss of surface-adherent platelets and underestimation of platelet thrombosis. The radioactivity-measurement technique is useful for studying platelet thrombosis in the arterial system as well as for evaluation of platelet-inhibitor drugs in bilateral femoral graft implant models.

In regard to the amount of platelet deposition we found on graft sections at one day after grafting, the insignificant difference between the control and treated groups of dogs might be due to platelet trauma during cardiopulmonary bypass and contamination of the local adventitial surface by blood oozing from the anastomosis.

Clinical implications of the findings. In management of patients who have undergone aortocoronary bypass grafting, it is essential to prevent thrombotic occlusion of the graft without inducing excessive postoperative bleeding. Drug side effects and variability of intolerance and compliance complicate decisions as to how long the treatment should be continued--and drug cost also is a factor. Although the data obtained from our studies suggest that a relatively short period of therapy is necessary, the studies are limited by the use of almost healthy dogs. Their healthy coronary arteries and healthy veins might benefit less from prolonged inhibition of platelet deposition than do the diseased veins and coronary arteries of man. Nevertheless, this study of the course of platelet deposition during the early and late post-grafting periods in untreated and treated dogs indicates that significant platelet deposition occurs within a few days after aortocoronary bypass surgery. It is essential that inhibitors of platelet adhesion be present in this critical period. Such treatment may not be necessary for longer than 30 days after aortocoronary bypass grafting, since, at least in dogs, the graft re-endothelializes, as demonstrated by our scanning electron-microscope studies. The advantage of the tracer technique is that it estimates total platelet depositon over a wider surface area, whereas the electron-microscope reveals only the superficial layer of platelets on a smaller selected area and underestimates total number of platelets adhering. As the graft re-endothelializes, both platelet adhesion and graft permeability to blood-borne molecules decreases, and hence the likelihood of thrombosis and atherosclerosis decrease. Consequently, we suggest that inhibition of platelet deposition is indeed beneficial during the intraoperative and the immediate postoperative period. As re-endothelialization takes place, the need for antiplatelet therapy wanes. At the time of complete re-endothelialization, it should be possible to discontinue administration of antiplatelet drugs without consequent graft occlusion.

Aortic Bifurcation and Femoral Artery Graft

Interposition graft at femoral artery provides a suitable model (50-80 ml/min flow) for measurement of graft thrombogenicity in the acute and chronic phase and evaluation of platelet-inhibitor drugs (76,77,82,88). We and others have evaluated the effect of aspirin and persantine on preclotted double-velour Dacron (Microvel, 12-16 mm I.D.) in the aortofemoral position and Gore-Tex graft (6 mm I.D.) in the femoropopliteal position. Imaging with a computerized gamma camera at 48, 72, and 96 hours delineates the site of platelet deposition. The aspirin-persantine combination tends to reduce platelet thrombosis on Dacron graft.

Venous Reconstruction and Vena Caval Graft

Reconstructive venous surgery has lagged behind the arterial repair; good patency was obtained when this procedure was properly conducted in suitable patients. Palma and Esperon's success (2) in crossover bypasses of saphenous to contralateral femoral vein have

regenerated interest and initiated several experimental and clinical
investigations in patients with phlebitis and external pressure on
vein by a growing tumor. Thrombogenicity of synthetic graft materials, low velocity of blood flow, and venous wall collapsibility have
been suggested as the main factors for high occlusion rate of vascular
prostheses placed in the venous system. We have use In-111-labeled
autologous platelets and I-125-labeled canine fibrinogen to evaluate
these factors on EPTFE graft in infrarenal vena cava in the canine
model (76,85), e.g., effect of increased flow on venous thrombosis
using arteriovenous fistula and effect of air-removal from EPTFE graft
with heparin and external ring support on graft patency. Purified
canine fibrinogen was labeled by the iodogen technique (55). These
studies indicate that a flow-increasing arteriovenous fistula is
necessary to maintain patency of EPTFE grafts; external ring support
of EPTFE grafts does not improve early patency. Removing air by
wetting-out the grafts with heparin reduces graft thrombogenicity. In
addition, we also evaluated the effect of platelet inhibitor-drug
ibuprofen on platelet and fibrin deposition in vena caval graft.
Microscopic and tracer techniques demonstrated a significant reduction
in platelet and fibrinogen with oral ibuprofen treatment. Fibrinogen
depositon results were expressed as microgram of fibrinogen per unit
area of graft surface (Fig. 11A, 11B, 11C). It is interesting to note
the difference in platelet deposition in arterial (small vessel) and
superior vena cava grafts. There are fewer numbers of adherent
platelets at both proximal and distal anastomoses than at the
midsection of the graft. Similar values were obtained in abdominal
aorta with gelatin-coated Dacron graft showing the effects of site of
grafting, flow, pressure, velocity, and graft diameter and
platelet-inhibitor on platelet deposition.

Experimental Study with Small Vessel Prostheses

Long incisions made for harvesting of vein, painful, slow healing
process, and higher infection rate in the groin are generating tremendous interest in the development of small vessel prostheses. The
early failure of small diameter ($\leq$ 4mm) synthetic arterial grafts
after surgical implantation results from the formation of an occlusive
thrombus. The thrombus growth begins with platelet adhesion to the
surfaces of biomaterial followed by platelet aggregation and subsequent interaction with the coagulating proteins to form the arterial
thrombus (32-39). Three independent biochemical pathways are responsible for the process of platelet activation, namely, cellular
release of adenosine diphosphate (36), the prostaglandins: cyclic
endoperoxides and thromboxane A_2 (37) and the platelet aggregating
factor pathway (38). The relative importance of these different
pathways in arterial thrombosis remains to be established (39). They
ultimately result in a decrease in cyclic adenosine 3'-5'
monophosphate, an increase in cytoplasmic calcium, and a subsequent
shape change and release (43). Reduction of this flux of calcium into
the cytoplasm might limit or abolish the platelet response,

irrespective of the nature of the primary stimulus for activation.
Thus, the calcium antagonist drugs, nifedipine and verapamil, may have
a potential role as clinically useful platelet-inhibitor drugs.
Dipyridamole is often prescribed as a platelet-inhibitor, particularly
in combination with aspirin. It appears to act by blocking the
platelet phosphodiesterase enzyme, elevating cyclic adenosine 3'-5'
monophosphate (35,36), and so preventing intracelluar mobilization of
calcium and subsequent platelet activation. In this study, the

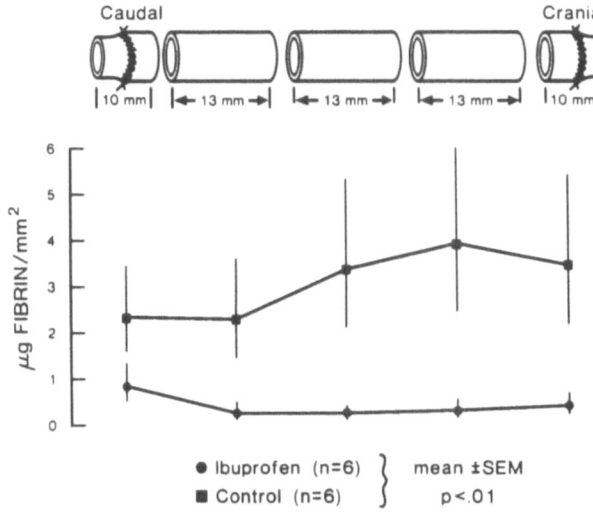

Fig. 11 A. Fibrin deposition on reinforced PTFE vena cava grafts in control and ibuprofen-treated dogs after three-hour perfusion. High concentrations of fibrin are seen in the control group with increasing deposition from caudal to cephalad segments.

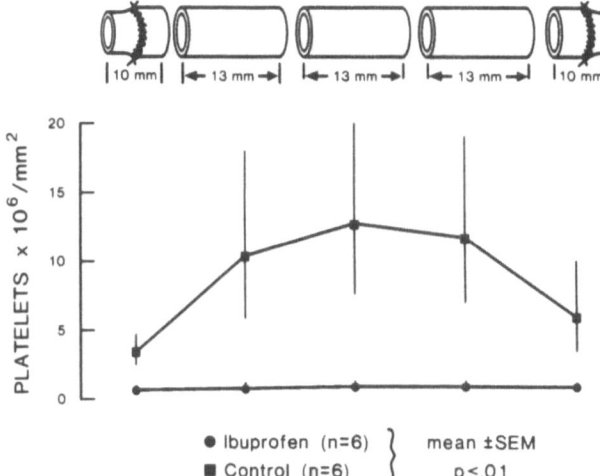

Fig. 11 B. Platelet deposition on reinforced PTFE vena cava grafts in control and ibuprofen-treated dogs after three-hour perfusion.

CONTROL   IBUPROFEN

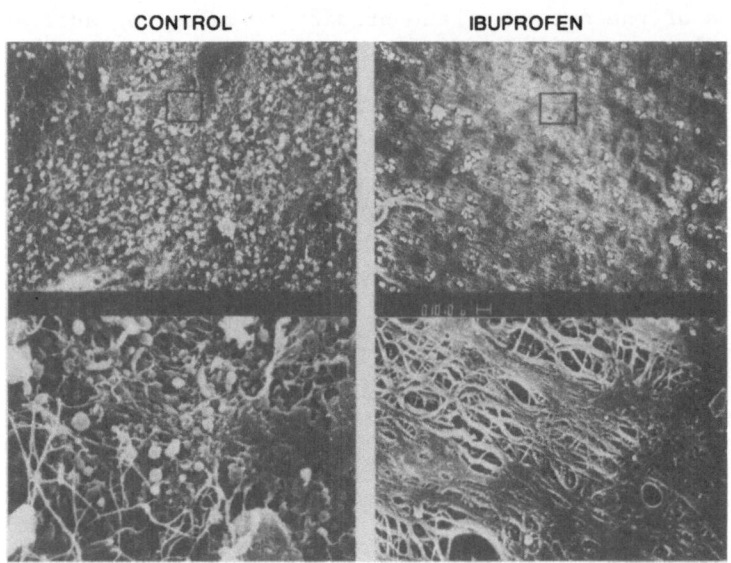

Fig. 11. C. High and low power cross sections of control and
             ibuprofen-treated reinforced PTFE grafts after three-hour
             perfusion. Control group has uniform deposition of
             thrombus along the graft surface with only a few red
             blood cells penetrating into the graft substance.
             Treated group has a fine layer of thrombus on the graft
             surface with many red blood cells penetrating the graft
             substance superficial structure (hematoxylin and eosin
             stain; original magnification x3.5 and x10).

antithrombotic effect of a calcium blocker was found to be similar and
equivalent to that of dipyridamole but less effective than aspirin,
aspirin-dipyridamole combined, and ibuprofen. Aspirin inhibits
cycloxygenase, and ibuprofen inhibits cycloxygenase and lipoxygenase
in platelet and white cell (39).

     In vitro studies of the interaction of platelet and coagulation
factors with biomaterial surfaces provide incomplete information, in
that only individual parts of these complex, synergistic processes can
be studied at one time and studies in the past have not elucidated
their possible effectiveness in vivo. In our laboratory, the in vivo
deposition of platelets and the formation of thrombus have been
detected and quantitated with the use of In-111-labeled autologous
platelets (28,62) in a canine model. In this article we describe an
in vivo model of arterial thrombosis on expanded polytetrafluoroethylene (EPTFE) and vein graft in the acute phase.

A variety of drugs demonstrated inhibitory effects on platelet activation; these inhibitory effects had been studied in the past by both in vitro and in vivo systems (48-52). The chemical structures of several platelet-inhibitor drugs under clinical and experimental investigations are shown in Figure 12. We have chosen aspirin, dipyridamole (Persantine), ibuprofen (Motrin), ticlopidine, and two calcium-channel blockers: verapmail (Isoptin) and nifedipine and have studied their effects on platelet thrombosis at one and three hours of perfusion in dogs, determined the rate of deposition at the proximal and distal anastomoses and mid-graft sections, and established the potency ranking of these drugs in the acute phase. The platelet deposition was also compared with the least thrombogenic material, i.e., autologous vein graft.

The schematics of the bilateral femoral implant model are shown in Figure 13A. The protocol for platelet labeling with In-111-tropolone drug therapy, surgical implantation of grafts, harvesting of grafts, and quantitation of platelets deposited on grafts is shown in Figure 13B. One hundred mongrel dogs weighing 20-25 kg were used in this study. The platelet density at 60 and 180 minutes of perfusion is shown in Figures 14 and 15, respectively.

Labeling of autologous platelets with Indium-111-tropolone (56-58). Indium-111-tropolone was prepared by mixing 250-450 µCi of In-111-Cl$_3$ (Medi-Physics Inc.) with 25 µg of tropolone in a 25 microliter sterile saline solution. Eighteen to 24 hours before graft implantation, platelets were harvested from 43 ml of blood in 7 ml of acid-citrate-dextrose (ACD) anticoagulant and labeled with In-111-tropolone in ACD-saline (1:7) medium, washed of free In-111, and resuspended in ACD-plasma. Indium-111-labeled autologous platelets (250-400 µCi) were injected within three to four hours of blood collection. Platelet labeling efficiency was 85-95%.

Surgical implantation of grafts in dogs. Under pentobarbital anesthesia (30mg/kg) both femoral arteries on each side were exposed (Fig. 12A). After anticoagulation with heparin (100 U/kg IV, beef lung heparin, Upjohn Company), 4 cm of femoral artery segment replaced in an end-to-end fashion with either 5 cm of reversed autologous femoral vein or 5 cm of (4 mm I.D.) EPTFE graft (Gore-Tex, W. L. Gore and Associates). After all the anastomoses at both legs were completed, the four clamps were simultaneously released. Heparin was not reversed. One and three hours after resumption of blood flow, the dogs were killed with an overdose of pentobarbital and the grafts were removed.

Processing of EPTFE and vein grafts. After the grafts were removed, they were gently perfused with isotonic saline and any blood was cleaned off the exterior surface with saline-soaked, cotton-tipped swab. The grafts were then sectioned into five pieces for EPTFE and four for vein graft in order to assess thrombus formation in different parts of the graft. Each anastomotic section consisted of 2 mm of native artery and 2 mm of EPTFE or vein graft. The middle sections consisted of three equal lengths of EPTFE graft (1.5 cm) and two for vein graft (1 cm).

Treatment groups. One hundred dogs in 11 groups were used in this study. Platelet deposition for a perfusion period of 60 and 180 minutes was studied in three and eight groups, respectively, with and without medication. The dosage of medication was selected on the basis of published reports. Both verapmail and nifedipine were given in isotonic saline immediately after induction of anesthesia before surgery was begun, and the infusion was continued until the grafts were removed. The median verapamil concentration was 186 ng/ml (range 128 to 220 ng/ml) at three hours; this regimen caused a median 7% (range, 4 to 13%) increase in femoral artery blood flow measured by a two-channel, electromagnetic flowmeter (Carolina Medical Electronics Inc.), 15% (range, 10 to 18%) increase in heart rate, and no alteration in mean aortic pressure. The delivery syringe and tubing were covered with aluminum foil to avoid photodecomposition.

Dipyridamole was given orally in two doses during the 24 hours before the study. The second dose, given one to two hours before operation, resulted in a median plasma dipyridamole concentration of 370 ng/ml (range, 235 to 550 ng/ml) at three hours. No hemodynamic measurements were made, and as with the control dogs, isotonic saline (100ml/hr. was given intravenously to these dogs during the study). This volume of saline was similar to that given to the dogs treated with verapamil and nifedipine. Ticlopidine was administered orally in two doses during the 24 hours before the study, the first dose 16 hours before and second dose one hour before surgery. Aspirin was given orally perioperatively alone or in combination with dipyri-damole. Ibuprofen was administered orally or intravenously (bolus).

The effect of the drugs was assessed by weight of thrombus formed and by the number of platelets deposited/cm^2. Formation of thrombus was estimated from the weights of the graft section; the dry weight of each 5 cm length of EPTFE graft was found to be (257-264 mg); the increased weight of the graft sections represented thrombus formation. The radioactivity on the segments of graft anastomoses and three blood samples (3 ml each) taken from each animal at the time of graft removal was measured in a gamma well counter (Beckman Gamma 8000). The gamma-ray spectrometer was adjusted to include the 174, 247, and 421 (sum peak) KeV peaks of In-111 radionuclide. From the platelet count in EDTA blood (measured on a sample taken before hepariniza-tion), the number of platelets deposited per unit area of graft surface was calculated as before.

Platelet distribution in thrombus and EPTFE surface. After weighing the graft and thrombus together, the body of each thrombus

was removed and weighed separately with a microbalance. From the
radioactivity of each graft and that of the thrombus removed from it,
the number of platelets located at the graft-thrombus interface and
within the body of the thrombus was measured.

Analysis of data on thrombus weight and platelet deposition. For
each dog studied a close correlation existed between both the weight
of thrombus (r = 0.88, P < 0.001) and the number of platelets (r =
0.89, P < 0.001) deposited in each of the two grafts. This finding
suggested that variations in surgical technique did not contribute
importantly to either of these variables but that the dog's response
to treatment principally determined the thrombogenicity of the grafts.
Consequently, the results of thrombus weight and platelet deposition
from the two grafts in each dog were averaged and the average values
were analyzed. For statistical analysis the Mann-Whitney U test for
differences between independent samples was used, and P values for the
difference of the median number of platelets deposited/cm^2 between
control and treated dog groups were determined.

The degree of occlusion observed in the grafts in this short-term
exposure varied; there was usually either a white lining layer of
platelets streaked with red cells or partial occlusion of the graft by
protruding red thrombus extending along its length. In six control
dogs and in two dogs who had received dipyridamole, verapamil, and
nifedipine, however, the grafts were occluded.

Thrombus weight. The median weight of the thrombus from the
treated dogs was significantly lower than the median weight of those
from the control dogs. No difference in median weight was observed
among the various treatment groups of dipyridamole, verapamil, and
nifedipine. There is a significant reduction in thrombus weight with
other drugs. The reduction in weight occurred along all three middle
sections of the grafts, but not at the anastomoses.

Platelet deposition on sections of graft surface and anastomoses
and platelet distribution in thrombi. Overall, significantly fewer
platelets were deposited in the grafts of the treated dogs than in
those of the control group (see Figures 14 and 15). In the latter
group, platelet accumulation was greatest at the distal anastomoses;
the platelet accumulation at the proximal anastomoses was only half
that at the distal anastomoses. The effect of the drugs was most
significant on the three middle sections, and although all of the
drugs had an effect at the distal anastomosis, only verapamil reduced
platelet accumulation at the proximal anastomosis. The median radioactivity of thrombus removed from grafts was 27% (range, 5 to 49%) of
that of the total graft.

This in vivo bilateral femoral implant model of thrombosis in a
high-pressure, high-flow system has advantages over in vitro tests;
that is, there is no unphysiologic separation of the sequence of

events which take place during thrombosis. Due to production of prostacyclin and fibrinolytic systems by the endothelial cells, autologous vein is the best choice for arterial graft. For effective evaluation of orally administered drugs, it is essential to measure plasma
drug level directly with high-pressure liquid chromatography, spectro-
photometry, or indirectly by clotting times. All the drugs tested had
some effect on platelet deposition – more at the distal anastomoses
than at the proximal anastomoses. The best result was obtained with
intravenous ibuprofen at 180 minutes and platelet deposition was equal

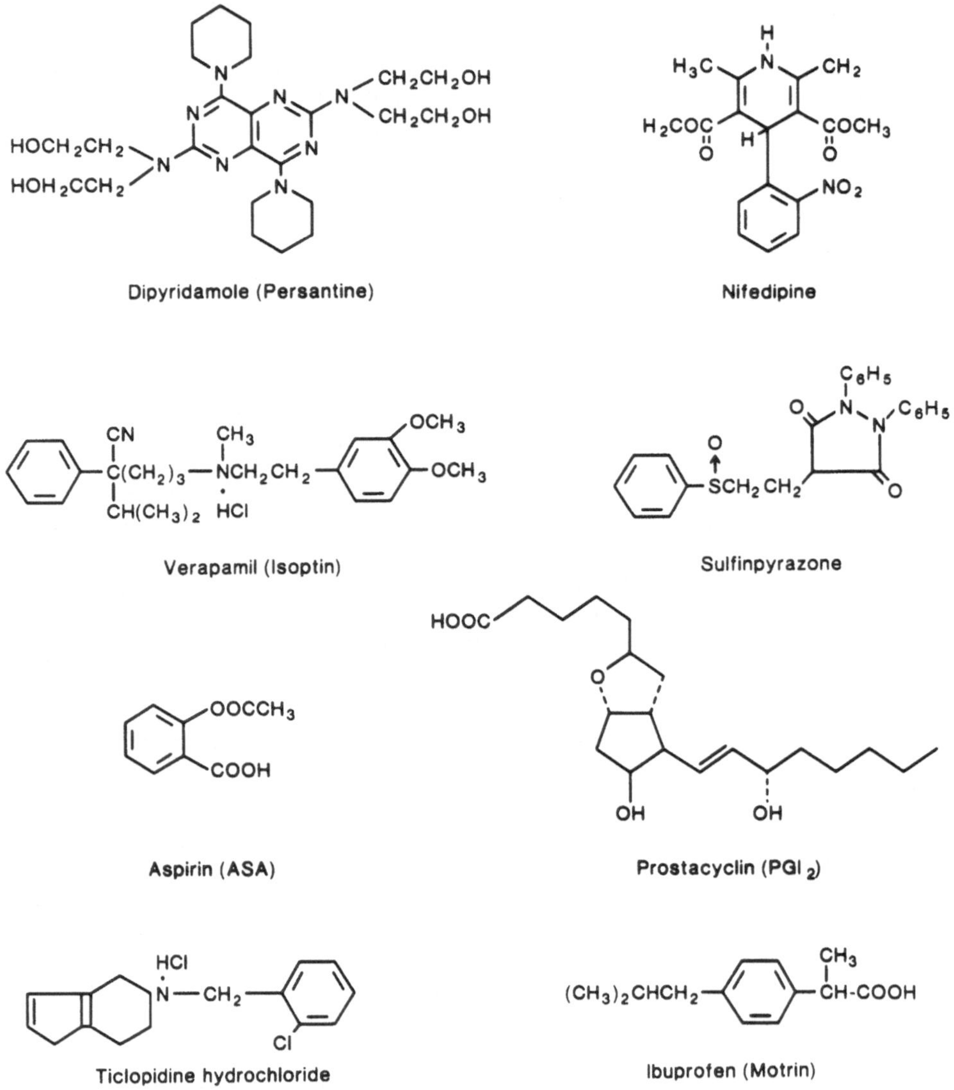

Dipyridamole (Persantine)

Nifedipine

Verapamil (Isoptin)

Sulfinpyrazone

Aspirin (ASA)

Prostacyclin (PGI$_2$)

Ticlopidine hydrochloride

Ibuprofen (Motrin)

Fig. 12. The structures of platelet-inhibitor drugs under clinical
and experimental evaluation.

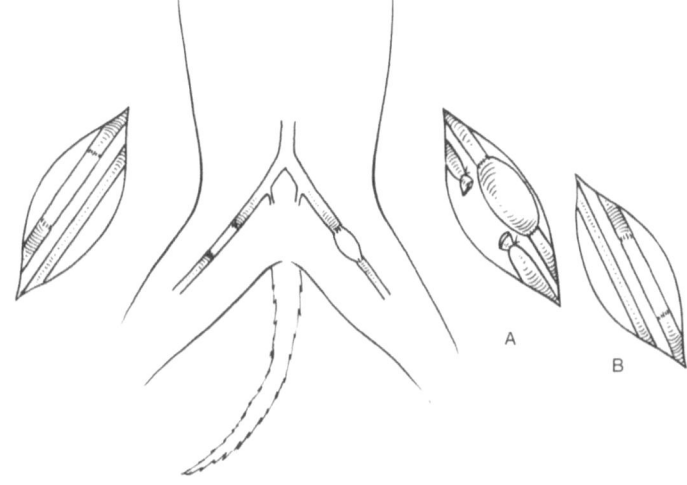

Fig. 13 A. Scheme of bilateral graft implantation in dogs. (A) EPTFE graft and vein graft, (B) EPTFE - EPTFE graft at one and three hours of perfusion.

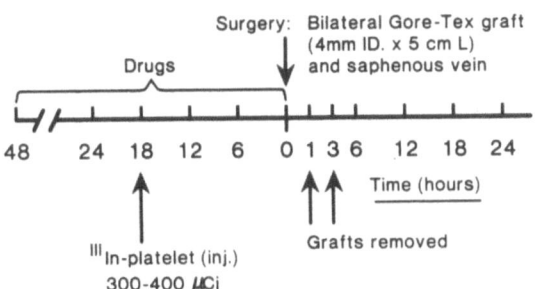

Fig. 13 B. Protocol of platelet labeling and administration, graft implantation and quantitation of platelet deposition on vein and EPTFE graft at one and three hours of perfusion in dogs.

to that of autologous vein at 60 minutes; the worst result was that from nifedipine and verapamil perfusion. These two drugs appear as weak platelet inhibitors. Oral ibuprofen at 25 mg/kg day was four times less effective, and plasma level of 2 µg/ml was obtained at 4 hours after oral administration. On the other hand, 15 minnutes after I.V. administration (12.5mg/kg/day) a blood level of 50 µg/ml of ibuprofen was obtained. Due to the reduction of GI motility in anesthetized dogs, the absorption of ibuprofen and hence platelet-inhibition after oral administration was reduced. Unlike aspirin, ibuprofen might inhibit platelet deposition in a dose-dependent fashion. Decrease of platelet deposition with aspirin alone (7 mg/kg/day) was similar to that of the aspirin and dipyridamole combination.

In the two control groups the rate of platelet deposition was highest at the distal and proximal anastomoses, but the situation changed within the next two hours, when the higher rates of deposition were observed at distal graft, middle graft, and proximal graft sections. Without treatment, the platelet deposition at distal anastomoses was twice as much as that of proximal anastomoses. The treatment with drugs maintains the shallow pan-shaped distribution (see Figures 14 and 15) of platelets. Without treatment, platelet deposition demonstrates a linear increase from proximal to distal anastomoses. Platelets become activated more at the anastomoses, and treatment lets a large fraction of activated platelet or microaggregates escape the graft region near the distal anastomoses without being trapped or preventing their formation.

Frequently no protruding mural thrombus was present on the grafts of the treated animals, and only a lining of white layer of platelets streaked with red cells was observed. It is possible that a threshold number of platelets has to accumulate on the 4 mm I.D. graft surface to trigger the formation of the red-cell-fibrin network, of which the bulk of luminal thrombi are made. The drugs may reduce this platelet accumulation and thereby prevent thrombus formation. The anastomoses of the grafts are important foci for thrombogenesis. The presence of sutures, tissue thromboplastin, vessel injury, and the difference in compliance of the artery and the graft result in turbulence in blood flow, which lead to platelet activation, thrombus formation, and propagation. The effect of the drugs on these areas was less clear.

The number of platelets on the anastomoses included those plugging the suture line and those trapped in extravascular thrombus adherent to the sutures, which is difficult to remove completely during the preparation of the graft for analysis.

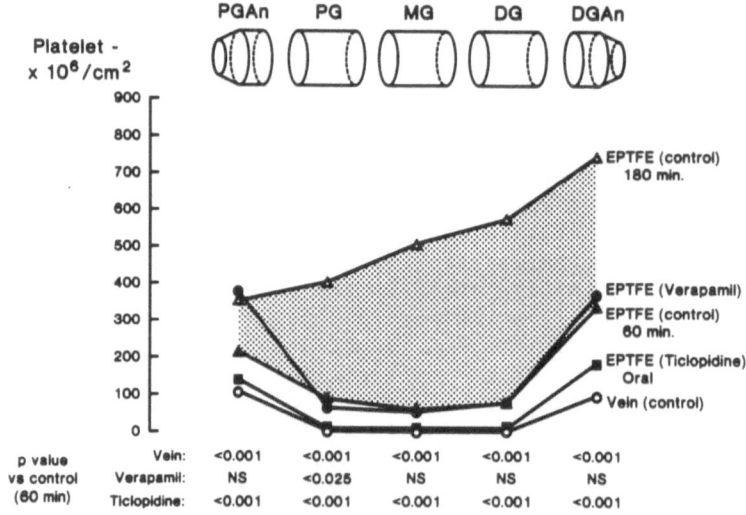

Fig. 14.   Platelet deposition on vein graft and EPTFE graft at 60 minutes of perfusion in control and treated dogs. The shaded area represents deposition of platelets during the two-hour interval from 60 to 180 minutes of perfusion; platelet deposition rate is higher at distal anastomosis.

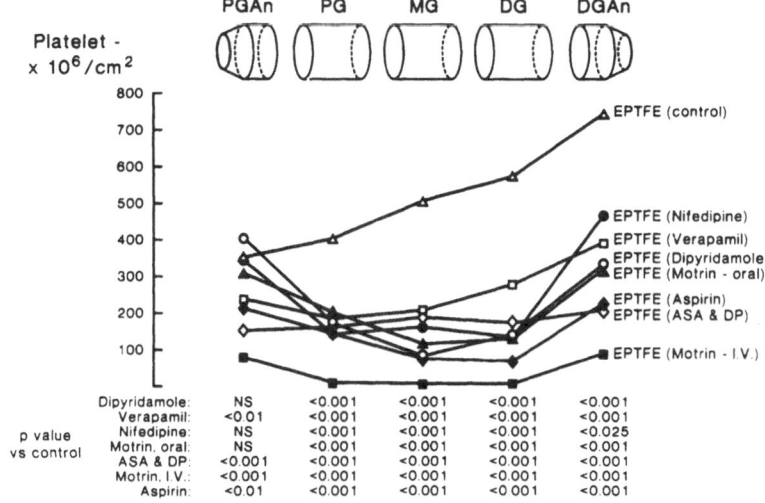

Fig. 15. Platelet deposition on vein graft and EPTFE graft at 180 minutes of perfusion in control and treated dogs. I. V. motrin reduces thrombogenicity of EPTFE graft to that of femoral vein graft.

Although dog platelet and fibrinolysis systems are more reactive than other species (72), we could observe the thrombotic graft occlusion and the effect of platelet-inhibitor drugs at an earlier time frame than other animal models. Considering the availability, low cost, and potency of platelet-inhibition on EPTFE surface, intravenous ibuprofen appears to be a good choice. In the chronic phase this could be followed up by oral administration. Our chronic phase studies with EPTFE graft implanted in dogs also demonstrated increased graft-patency with oral ibuprofen.

In conclusion, In-111-labeled autologous platelets had been used successfully for quantitation of platelet deposition on vascular graft sections and anastomoses of native vein and EPTFE graft. The change in the rate of platelet deposition is different in these segments at 60 and 180 minutes of blood flow. Change is dramatic at both distal segment and distal anastomoses. Platelet inhibition with several drugs produced changes in arterial grafts of equal significance. These quantitative techniques could be used with grafts of Dacron, polyurethane, and other biomaterials in other animal models with and without drug intervention.

Future development in synthetic vascular graft. Several laboratories are working on the development of small caliber synthetic

vessels to cross the 4 millimeter barrier. The synthetic vascular
graft without endothelial cell in intima needs immobilization of
thrombin inhibitor, plasminogen activator, platelet inhibitor with
future possibilities of inclusion of calcification inhibitors and
antibotics. A nonbiodegradable, nonthrombogenic elastic surface with
microenvironment of collagen, proteoglycan, and fibronectin anchored
to the polymer matrix for the attachment of the endothelial cells
might be the ideal, although difficult, choice, mimicking an approach
which Mother Nature worked on for millions of years to keep us nour-
ishing, growing, and thinking.

ACKNOWLEDGMENTS

I greatly appreciate the technical assistance of Messrs. S.
Chowdhury and O. Arlan Hildestad and the surgical expertise of Drs. M.
P. Kaye, P. Gloviczki, G. Platte, M. Tago, M.J. Hosa, and L. Hollier of
Cardiovascular Surgical Research. I also thank Ms. Judy A.
Ashenmacher for typing the manuscript. The dipyridamole (Persantine)
was generously supplied by Boehringer-Ingelheim, Ltd., the ibuprofen
(Motrin) by Upjohn, Kalamazoo, Michigan, and Gore-Tex graft by W. L.
Gore and Associates Inc., Flagstaff, Arizona.

This investigation was supported in part by Research Grants HL-24602
and HL-28974 from the National Institutes of Health, Public Health
Service.

REFERENCES
1. A. Carrel and C. C. Guthrie, Uniterminal and biterminal venous
    transplantations, Surg Gynecol and Obstet 2:266 (1906).
2. E. C. Palma and R. Esperon, Vein transplants and grafts in the
    Surgical treatment of the postphlebitic syndrome, J Cardiovasc
    Surg 1:94 (1960).
3. J. A. DeWeese, C. G. Rob, R. Satran, D. O. Marsh, R. J. Joynt, E.
    O. Lipchik, D. N. Zehl, Endarterectomy for atherosclerotic
    lesions of the carotid artery, J Cardiovasc Surg 12:299
    (1971).
4. H. Dardik, I. M. Ibrahim, I. I. Dardik, Glutaraldehyde-stabilized
    human umbilical cord vein as a vascular prosthesis, in:
    "Graft Materials in Vascular Surgery," H. Dardik, ed.,
    Yearbook Medical Publishers, Chicago (1978).
5. H. Dardik and I. Dardik, Successful arterial substitution with
    modified human umbilical vein, Ann Surg 183:252 (1976).
6. M. E. DeBakey, E. S. Crawford, H. E. Garrett, et al., Surgical
    considerations in the treatment of aneurysms of the
    thoraco-abdominal aorta, Ann Surg 162:650 (1965).
7. E. S. Crawford, D. M. Snyder, G. C. Cho, J. O. F. Roehm, Jr.,
    Progress in treatment of thoracoabdominal and abdominal aortic
    aneurysms involving celiac, superior mesenteric, and renal
    arteries, Ann Surg 188:404 (1978).

8. R. D. Weisel, K. W. Johnston, J. R. Baird, et al, Comparison of conduits for leg revascularization, Surgery 89:8 (1981).

9. L. R. Sauvage, M. W. Walker, K. Berger, et al, Current arterial prostheses, Arch Surg 114:687 (1979).

10. J. V. Robb and E. J. Wylie, Factors contributing to recurrent lower limb ischemia following bypass surgery for aoroiliac occlusive disease and their management, Ann Surg 193:346 (1981).

11. R. J. Sanders, R. F. Kempezinski, W. Hammond, et al, The significance of graft diameter, Surgery 88:856 (1980).

12. C. D. Campbell, D. H. Brooks, H. T. Bahnson, Expanded microporous polytetrafluorethylene (Gore-Tex) as a vascular conduit, in: "Vascular Grafts," P. N. Sawyer and M. J. Kaplitt, eds., Appleton-Century-Crofts, New York (1978).

13. H. Matsumoto, T. Hasegawa, K. Fuse, M. Yamamoto, M. Saigusa, A new vascular prosthesis for a small caliber artery, Surgery 74:519 (1973).

16. K. K. Unni, B. A. Kottke, J. L. Titus, R. L. Frye, R. B. Wallace, A. L. Brown, Pathologic changes in aortocoronary saphenous vein grafts, Am J Cardiol 34:526 (1974).

17. T. L. Spray and W. C. Roberts, Changes in saphenous veins used as aorto-coronary bypass grafts, Am Heart J 94:500 (1977).

18. B. H. Bulkley and G. M. Hutchins, Accelerated "atherosclerosis": a morphologic study of 97 saphenous vein coronary artery bypass grafts, Circulation 55:163 (1977).

19. T. Karino, Flow through venous valves: a mechanism for thrombogenesis, in: "Rheologic Contributions to Thrombosis and Hemostasis," NIH, Bethesda, Maryland, June 4-5 (1981).

20. C. C. Haudenschild, K. E. Gould, W. C. Quist, F. W. Logerfo, Protection of endothelium in vessel segments excised for grafting, Circulation 64 (Suppl 2):101, (1981).

21. J. C. A. Fuchs, J. S. Mitchener III, P. O. Hagen, Postoperative changes in autologous vein grafts, Ann Surg 188:1 (1978).

22. V. S. Sottiurai and R. C. Batson, Autogenous vein grafts: experimental studies, in: "Biologic and Synthetic Vascular Prostheses," J. C. Stanley, W. E. Burkel, S. M. Lindenauer, R. H. Bartlett, J. G. Turcotte, eds., Grune & Stratton, New York (1982).

23. S. M. Schwartz and D. M. Standaert, Endothelial cell turnover in rats: effects of angiotensin, age and chronic hypertension, in: "Atherosclerosis Revised," A. M. Gotto and R. Paoletti, eds., Raven Press, New York (1982).

24. M. B. Stemerman, Hemostasis, thrombosis, and atherogenesis, Atherosclerosis Revised," A. M. Gotto and R. Paoletti, eds., Raven Press, New York (1979).

25. T. Henriksen, E. M. Mahoney, Steinberg, Interaction of plasma lipoproteins with endothelial cells, in: "Endothelium," A. P. Fishman, ed., New York Academy of Science, New York (1982).

26. M. K. Dewanjee, N. Shapira, M. Tago, V. Fuster, M. P. Kaye, Endothelial permeability and C-14-and I-131-cholesterol uptake

on aortocoronary femora vein bypass grafts in dogs, Thromb Haemost 46:74 (abstr.) (1981).

27. M. K. Dewanjee, M. Tago, S. Rao, R. D. Ellefson, M. P. Kaye, Endothelial permeability and uptake of c-14cholesterol, I-125-HDL and I-131-LDL in aortocoronary femoral vein bypass graft in dogs, Atherosclerosis, 5:404a (abstr.)(1982).

28. M. K. Dewanjee, Experimental evaluation of venosclerosis of aortocoronary femoral vein bypass graft in control and aspirin-persantine-treated dogs: correlation with atherosclerosis, in: "Radiation and Cell Response," G. P. Scott and H. W. Wahner, eds., Iowa State University Press, Ames, Iowas, 2nd John Lawrence Interdisciplinary Symposium, Sioux Falls, South Dakota, June 3-4 (1981).

29. J. F. Mustard, M. A. Packham, R. L. Kinlough-Rathbone, Platelets, atherosclerosis, and clinical complications in vascular injury and atherosclerosis, in: "Biochemistry of Disease," S. Moore, ed., M. Dekker Inc., New York (1981).

30. A. J. Marcus, M. J. Brockman, B. B. Weksler, E. A. Jaffe, L. B. Safier, H. L. Ullman, N. Islam, K. Tack-Goldman, Arachidonic acid metabolism in endothelial cells and platelets, in: "Endothelium," A. P. Fishman, ed., N Y Acad Sci (1982).

31. Y. Stein and O. Stein, Interaction between serum Lipoproteins and cellular components of the arterial wall, in: "Biochemistry of Atherosclerosis," A. M. Seann, R. W. Wissler, G. S. Getz, eds., vol. 7, (1980).

32. M. B. Stemerman, Thrombogenesis of the rabbit arterial plaque--an electron microscopic study, Am J Pathol 73:7 (1973).

33. T. Hoving and H. Holmsen, Release of platelet aggregation substance (adenosine diphosphte) from rabbit blood platelets induced by saline "extracts" of tendons, Thromb Diath Haemorrh 9:264 (1963).

34. J. F. Mustard, Thrombosis and arterial disease, in: "Venous and Arterial Thrombosis: Pathogenesis, Diagnosis, Prevention and Therapy," J. H. Joist and L. A. Sherman, eds., Grune & Stratton, New York (1979).

35. V. Fuster and J. H. Chesebro, 10. Antithrombotic therapy: role of platelet-inhibitor drugs. I. Current concepts of thrombogenesis; role of platelets, Mayo Clin Proc 56:102 (1981), and III. Management of arterial thromboembolic and atherosclerotic disease, Mayo Clin Proc 56:265 (1981).

36. M. A. Packham and J. F. Mustard, Clinical pharmacology of platelets, Blood 50:555 (1977).

37. S. Moncada and J. R. Vane, Unstable metabolites of arachidonic acid and their role in haemostasis and thrombosis, Br Med Bull 34:129 (1978).

38. M. Tence, J. Polonsy, J. P. LeCovedic, J. Benveniste, Release purification and characterization of platelet aggregating factor, Biochimie 62:251 (1980).

39. R. L. Kinlough-Rathbone, M. A. Packham, H. J. Reimers, J. P Cazenave, J. F. Mustard, Mechanisms of platelet shape change,

aggregation and release induced by collagen, thrombin, or A23187, J Lab Clin Med 90:707 (1977).

40. W. G. Liekweg Jr. and L. J. Greenfield, Vascular prosthetic infections: Collected experience and results of treatment, Surgery 81:335 (1977).

41. C. A. Stevick and H. D. Fawcett, Aortoiliac-graft infection. Detection by leukocyte scan, Arch Surg 116:939 (1981).

42. P. P. McKeowen, D. C. Miller, S. W. Jamieson, R. S. Mithcell, B. A. Reitz, C. Olcatt, J. T. Mehigan, R. J. Silberstein, I. R. McDougall, Diagnosis of arterial prosthetic graft infection by Indium-111 oxine white blood cell scans, Circulation 66 (Part 2) I:130 (1982).

43. N. E. Owen and G. C. Le Breton, The involvement of calcium in epinephrine or ADP potentiation of human platelet aggregation, Throm Res 17:855 (1980).

44. P. Massini and E. F. Luscher, On the significance of the influx of calcium ions into stimulated human blood platelets, Biochim Biophys Acta 436:652 (1976).

45. S. Chierchia, F. Crea, W. Bemini, G. Gensini, O. Parodi, R. De Caterihna, A. Maseri, Antiplatelet effects of verapmil in man, Am J Cardiol 47:399 (Abstr.)(1981).

46. P. D. Henry, Comparative pharmacology of calcium antagonists: nifedipine, verapamil and diltiazem, Am J Cardiol 46:1046 (1980).

47. S. S. Adams, R. G. Bough, E. E. Cliffe, B. Lessel, R. F. N. Mills, Absorption, distribution and toxicity of ibuprofen, Toxicol Appl Pharmacol 15:310 (1969).

48. S. H. Ferreira and J. R. Vane, New aspects of the mode of action of nonsteroid anti-inflammatory drugs, Am Rev Pharmacol 14:57 (1974).

49. S. Moncada and R. Korbut, Dipyridamole and other phosphodiesterase inhibitors act as antithrombotic agents by potentiating endogenous prostacyclin, Lancet 1286 (1978).

50. M. B. Zucker and J. Peterson, Effect of acetylsalicyclic acid, other nonsteroidal anti-inflammatory agents and dipyridamole on human blood platelets, J Lab Clin Med 76:66 (1970).

51. J. J. Thebault, C. E. Blatrix, J. F. Blanchard, E. A. Panak, Effects of ticlopidine, a new platelet aggregation inhibitor in man, Clin Pharmacol Ther 18:485 (1975).

52. L. H. Rome, W. E. M. Lands, G. J. Roth, P. W. Majerus, Aspirin as a quantitative acetylating reagent for the fatty acid oxygenase that forms prostaglandins, Prostaglandins 11:23 (1976).

53. G. Masotti, L. Poggesi, G. Galanti, R. Abbate, G. G. Neri Serneri, Differential inhibition of prostacyclin production and platelet aggregation by aspirin, Lancet 2:1213 (1979).

54. V. D'Angelo, S. Villa, M. Mysliwiec, M. B. Donati, G. de Gaetano, Defective fibrinolytic and prostacyclin-like activity in human atheromatous plaques (letter to the editor), Thromb Haemost 39:535 (1978).

55. P. J. Fraker and J. C. Speck, Jr., Protein and cell membrane iodinations with a sparingly soluble chloroamide, 1, 3, 4, 6-tetrachloro-3a, 6a diphenylglycoluril, Biochem Biophys Res Commun 80:849 (1978).

56. M. L. Thakur, M. J. Welch, J. H. Joist, R. E. Coleman, Indium-111-labeled platelets: studies on preparation and evaluation of in vitro and in vivo functions, Thromb Res 9:345 (1976).

57. M. K. Dewanjee, S. A. Rao, P. Didisheim, Indium-111 tropolone, a new high-affinity platelet label: preparation and evaluation of labeling parameters, J Nucl Med 22:981 (1981).

58. W. A Heaton, H. H. Davis, M. J. Welch, C. J. Mathias, J. H. Joist, L. A. Sherman, B. A. Siegel, Indium-111: a new radionuclide label for studying human platelet kinetics, Br J Haematol 42:613 (1979).

59. H. W. Wahner, W. L. Dunn, M. D. Dewanjee, Distribution and survival of In-111-labeled platelets in normal persons, in: "Indium-111-labeled platelets and leukocytes," Proceedings of Second Annual Symposium, Mayo Clinic, Rochester, MN, H. W. Wahner and D. A. Goodwin, eds., October 24, (1981).

60. J. S. Robertson, M. K. Dewanjee, M. L. Brown, V. Fuster, J. H. Chesebro, Distribution and dosimetry of In-111-labeled platelets, Radiology 140:169 (1981).

61. M. K. Dewanjee, S. R. Vogel, K. A. Peterson, M. F. Lim, M. P. Kaye, Quantitation of platelet lysis, platelet consumption on oxygenator, and stabilization of platelet membrane with prostacyclin and ibuprofen during cardiopulmonary bypass surgery in dogs, Trans Am Soc Artif Intern Organs, 27:197 (1981).

62. M. K. Dewanjee, C. W. Pumphrey, K. P. Murphy, J. A. Rosemark J. H. Chesebro, V. D. Fuster, M. P. Kaye, Evaluation of platelet-inhibitor drugs in a canine bilateral femoral graft implant model, Trans Am Soc Artif Intern Organs 28:504 (1982).

63. M. K. Dewanjee, V. Fuster, F. Trastek, J. S. Rosemark, M. P. Kaye, Quantitation of lysis of cellular elements of blood during cardiopulmonary bypass surgery in dogs, J Nucl Med, P58 (abstr.)(1982).

64. L. A. Harker and S. R. Hanson, Graft thrombus formation, detection and resolution, in: "Biologic and Synthetic Vascular Prostheses," J. C. Stanley, W. E. Burke, S. M. Lindenauer, R. H. Bartlett, J. G. Turcotte, eds., Grune & Stratton, New York (1982).

65. R. W. Oblath, F. O. Buckley, R. M. Green, S. I. Schwartz, J. A. Deweese, Prevention of platelet aggregation and adherence to prosthetic vascular grafts to aspirin and persantine, Surgery 84:37 (1978).

66. M. K. Dewanjee, V. Fuster, M. P. Kaye, M. Josa, Imaging platelet deposition with In-111-labeled platelets in coronary bypass in dogs, Mayo Clin Proc 53:327 (1978).

67. V. Fuster, M. K. Dewanjee, M. P. Kaye, M. Josa, M. P. Metke, J. H. Chesebro, Noninvasive radioisotopic technique for detection

of platelet deposition in coronary artery bypass grafts in dogs and its reduction with platelet inhibitors, Circulation 60:1508 (1979).

68. K. M. Strathy, M. K. Dewanjee, M. P. Kaye, S. Rao, V. Fuster, Quantitation of platelet deposition in Gore-Tex femoral artery implants in the canine model and its reduction with dipyridamole and prostatcyclin, Am J Cardiol 45:424 (abstr.) (1980).

69. M. J. Welch, C. J. Mathias, B. A. Siegel, Clinical experience with Indium-111-labeled platelets, in: "Indium-Labeled Neutrophils, Platelets, and Lymphocytes," M. L. Thakur and A. Gottschalk, eds., Triverium Press, New York (1980).

70. D. A. Goodwin, J. T. Bushberg, P. W. Doherty, M. J. Lipton, F. K. Conley, C. I. Diamanti, C. F. Meares, In-111-labeled autologous platelets for location of vascular thrombi in humans, J Nucl Med 19:626 (1978).

71. J. L. Ritchie, J. R. Stratton, B. Thiele, G. W. Hamilton, L. N. Warrick, T. W. Huang, L. A. Harker, Indium-111 platelet imaging for detection of platelet deposition in abdominal aneurysms and prosthetic arterial grafts, Am J Cardiol 47:882 (1981).

72. E. F. Grabowski, P. Didisheim, J. C. Lewis, J. T. Franta, J. Q. Stropp, Platelet adhesion to foreign surfaces under controlled conditions of whole blood flow: human vs rabbit, dog, calf, sheep, pig, macaque, and baboon, Trans Am Soc Artif Intern Organs 23:141 (1977).

73. P. Didisheim, M. K. Dewanjee, C. S. Frisk, M. P. Kaye, D. N. Fass, Animal models useful for predicting clinical performance of biomaterial for cardio-vascular use, in: "Biological Performance of Materials in the Human Body," Concensus Development Conference on Clinical Applications of Biomaterials, NIH Campus, November 1-3 (1982).

74. C. W. Pumphrey, V. Fuster, M. K. Dewanjee, J. H. Chesebro, R. E. Vlietstra, M. P. Kaye, Comparison of the antithrombotic action of calcium antagonist drugs with dipyridamole in dogs, Amer J Cardiol 51:591 (1982).

75. S. R. Hanson, L. A. Harker, B. D. Ratner, A. S. Hoffman, In vivo evaluation of artificial surfaces with a non-human primate model or arterial thrombosis, J Lab Clin Med 95:289 (1980).

76. J. R. Stratton, B. L. Thiele, J. L. Ritchie, et al, Platelet deposition on aortic bifurcation grafts in man: quantitation with In-111-labeled platelet imaging, Circulation 66:1287 (1982).

77. J. R. Stratton and J. L. Ritchie, The natural history of platelet deposition on Dacron aortic bifurcation grafts in the first year following implantation in man, J Nulc Med 25:P73 (abstr.) (1983).

78. M. K. Dewanjee, P. Gloviczki, D. Kluge, M. P. Kaye, Inhibition of thrombotic occlusion of aortocoronary bypass graft with verapamil and papaverine in dogs, J Nucl Med 24:P60 (abstr.) (1983).

79. C. F. Pope, S. Rapaport, E. O. Smith, C. Balint, D. Sostman, M. Glickman, B. L. Zaret, M. D. Ezekowitz, A preliminary evaluation of Indium-111-labeled platelet scintigraphy in patients undergoing peripheral percutaneous transluminal angioplasty, J Nucl Med 24:P89 (abstr.) (1983).

80. B. T. Allen, C. J. Mathias, R. E. Clark, K. T. Hopkins, M. J. Welch, Use of Indium-labeled platelets to monitor platelet deposition on vascular grafts, J Nucl Med 24:P60 (abstr.) (1983).

81. W. M. Whitehouse, Jr., T. W. Wakefiedd, D. W. Vinter, J. W. Ford, D. P. Swanson, J. H. Thrall, J. W. Froelich, W. E. Burkel, L. M. Graham, J. C. Stanley, Indium-111 oxine-labeled platelet imaging of endothelial seeded Dacron thoracoabdominal vascular prostheses in a canine experimental model, ASAIO, vol. 9, (abstr. 12) (1983).

82. C. W. Pumphrey, J. H. Chesebro, M. K. Dewanjee, H. W. Wahner, L. H. Hollier, P. C. Pairolero, V. Fuster, In vivo quantitation of platelet deposition on human peripheral arterial bypass grafts using Indium-111-labeled platelets. Effect of dipyridamole and aspirin, Am J Cardiol 51:796 (1983).

83. P. Gloviczki, M. K. Dewanjee, V. F. Trastek, E. A. Hoffman, M. P. Kaye, Experimental replacement of the inferior vena cava, Surgery 95:160-168 (1983).

84. R. Rodvien, J. Robinson, R. R. Mitchell, P. Litwak, D. C. Price, A new model for in vivo platelet and thrombus kinetics, in: "Biomaterials: Interfacial Phenomena and Applications," S. L. Cooper and N. A. Peppas, eds., American Chem Society, Washington, D.C. (1982).

85. B. R. Young, L. K. Lambrecht, S. L. Cooper, Plasma proteins: their role in initiating platelet and fibrin deposition on biomaterials, in: "Biomaterials: Interfacial Phenomena and Applications," S. L. Cooper and N. A. Peppas, eds., American Chem. Society, Washington, D.C. (1982).

86. K. C. Agarwal, H. W. Wahner, M. K. Dewanjee, V. Fuster, F. J. Puga, G. K. Danielson, J. H. Chesebro, R. H. Feldt, Imaging of platelets in right-sided extracardiac conduits in humans, J Nucl Med 23:342 (1982).

87. M. Goldman, C. N. McCallum, R. J. Hawker, Z. Drole, G. Stanley, Dacron arterial grafts: the influence of porosity, velour, and maturity on thrombogenicity, Surgery 92:947 (1982).

88. A. D. Callow, R. Connolly, T. F. O'Donnell, R. Gembarowicz, E. Keough, K. Ramberg-Laskaris, C. R. Valeri, Platelet-arterial synthetic graft interaction and its modification, Arch Surg 117:1447 (1982).

89. B. L. Engelstad, R. J. Lusley, D. C. Price, M. J. Lipton, R. J. Stoney, A. S. Holly, Indium-111 platelet deposition following carotid endarterectomy in dogs and man, J Nucl Med 23:58 (1982).

90. J. T. Christenson, J. Megerman, K. C. Hanel, G. J. L'Italien, H. W. Strauss, W. M. Abbott, The effect of blood flow rates on

platelet deposition in PTFE arterial bypass grafts, Am Soc Artif Intern Organs 188 (1981).

91. J. T. Christenson, J. Megerman, K. C. Hanel, G. J. L'Italien, H. W. Strauss, W. M. Abbott, Prediction of early graft occlusion using Indium-111-labeled platelets, J Cardiovasc Surg 22:464 (1981).

92. M. E. Lovaas, P. Gloviczki, M. K. Dewanjee, M. P. Kaye, The role of anti-platelet therapy, J Surg Res 35:234-242 (1983).

93. T. Yui, T. Uchida, S. Matsuda, K. Iwaya, M. Umina, K. Ono, S. Muroi, Detection of platelet consumption in aortic graft with In-111-labeled platelets, Eur J Nucl Med 7:77 (1982).

94. L. M. Graham, W. E. Burkel, J. W. Ford, D. W. Vinter, R. H. Kahn, J. C. Stanley, Expanded polytetrafluoroethylene vascular prostheses seeded with enzymatically derived and cultured canine endothelial cells, Surgery 91:550 (1982).

95. M. Herring and A. Gardner, A single-staged technique for seeding vascular grafts with autogeneous endothelium, Surgery 84:498 (1978).

96. W. E. Burkel, J. W. Ford, D. W. Vinter, R. H. Kahn, L. M. Graham, J. C. Stanley, Fate of knitted Dacron velour vascular grafts seeded with enzymatically derived autologous canine endothelium, Trans Am Soc Artif Intern Organs 28:178 (1982).

97. M. K. Dewanjee, M. Tago, M. Josa, V. Fuster, and M. P. Kaye, Quantification of platelet retention in aortocoronary femoral vein bypass graft in dogs treated with dipyridamole and aspirin, Circulation 69:350 (1984).

98. M. K. Dewanjee, Carciac and Vascular imaging with labeled platelets and leukocytes, Seminars in Nuclear Medicine XIV, No. 3:154-187 (1984).

PRESENT TRENDS AND FUTURE DIRECTIONS

IN "LEUKOCYTE LABELING"

John G. McAfee, Gopal Subramanian, and George Gagne

Divisions of Radiological Sciences and Nuclear Medicine
Department of Radiology
Upstate Medical Center, S.U.N.Y.
Syracuse, New York 13210

INTRODUCTION

Imaging with In-111-labeled leukocytes has become an established clinical method for the detection of focal inflammatory lesions in many medical centers but has been rejected by others because of the technical complexities of the labeling procedure. Gaining knowledge of the in vivo migratory pattern of the different leukocyte populations and subtypes in health and disease remains an important goal(1), particularly in the field of immunology. Techniques for examining the in vivo distribution of lymphocytes, monocytes and the eosinophils are still under development, and differences in migratory patterns of mononuclear subtypes remain to be explored. This paper attempts to summarize recent progress in the techniques of harvesting and labeling leukocytes and suggests possible directions for future research.

LEUKOCYTE HARVESTING TECHNIQUES

The current radioactive agents capable of labeling cell suspensions with a high yield usually tag many cell types indiscriminately, and plasma proteins often compete in the labeling reaction. Hence, the cell population of interest, as a rule, has to be isolated before labeling. From the withdrawal of fresh anticoagulated blood to rein-

(This investigation was supported in part by PHS Grant No. CA-32853 awarded by the National Cancer Institute, DHHS)

265

jection of labeled cells, sterility and cell viability must be preserved, avoiding chemical toxicity and the mechanical trauma of centrifugation as much as possible.

ERYTHROCYTE SEDIMENTATION

    Gravity sedimentation remains the simplest and most popular
method for removing the bulk of erythrocytes from leukocyte-platelet
suspensions at room temperature. We prefer ACD solution to heparin as
anticoagulant because leukocytes are less adherent to plastic tubes or
syringes with the former agent. Boyum(2) believes the most potent
erythrocyte-aggregating agent is 2% methylcellulose in saline (1 part
to 10 parts blood). Six percent dextran (1 part to 5 parts blood) is
still widely used; however, 6% hydroxyethyl starch† (HES 1 part to 5
parts blood) is more effective(2) and is FDA approved. When the hematocrit is above 30%, Roy et al(3) recommend some dilution of the blood
before gravity sedimentation. About 70% of the leukocytes are
recovered in the supernatant plasma, but from 1 to 3 erythrocytes per
leukocyte still remain.

    For many in vitro studies, residual red cells in the leukocyte
suspension are removed by ammonium chloride lysis. This is
unsatisfactory, however, for in vivo studies because it adversely
affects the osmotic fragility of granulocytes(4) and their in vivo
survival, resulting in greater sequestration in the liver(5).

    The disadvantage of erythrocyte sedimentation is that it gener-
ally requires 1 hour. This time can be greatly shortened by low speed
centrifugation(6). In a modification of this technique, we have used
1% methylcellulose (1 part to 4 parts blood-ACD) spun at 20 G for 14
minutes and have recovered almost 100% of the leukocytes in the supernatant fraction.

DENSITY GRADIENT (ISOPYKNIC) CENTRIFUGATION

    Boyum's technique(7) is still widely used to isolate mononuclear
leukocytes (lymphocytes and monocytes). Anticoagulated whole blood is
layered above a solution of Ficoll-Hypaque (density 1.077 g/ml).
After centrifugation at 800 G for 15 minutes(4), the lighter platelets and mononuclears float above the heavier erythrocyte-granulocyte
layer. More recently, nonionic water-soluble iodinated contrast
media, including metrizamide or Nycodenz* have been used in place of
Hypaque. Platelets may be eliminated by resuspending and spinning
down the leukocytes and removing the supernatant.

---

†Volex, American Critical Care, McGaw Park, IL.
*Nyegaard & Co., Norway

In another one-step method, Ferrante and Thong(8) layer 5 ml of heparinized blood over 3 ml of Ficoll Hypaque (8.2% Ficoll density 1.114) in a 15 mm diameter tube, spinning at 300 G for 30 minutes at room temperature. Two bands appear above the red cell layer, the upper containing mononuclear cells and platelets and the lower, granulocytes. Aguado et al(9) use two 5 ml layers with densities of 1.097 and 1.075 g/ml in a 28 mm diameter tube. Heparinized blood (3.3 ml diluted to 10 ml) with phosphate-buffered saline are layered on top and centrifuged at 400 G for 40 minutes at room temperature. The upper layer contains mononuclear and platelets and the lower layer 98% granulocytes, with a recovery of 65%. The above methods seem to produce good leukocyte separations only for small volumes of blood. With 3 density gradients(10) (1.055, 1.062, and 1.095 g/ml, monocyte and lymphocyte layers can be separated with only a 10% cross contamination for up to 3 x 10^7 cells. However, meticulous loading of the gradients without interface disturbance is required.

Percoll**, a relatively new density gradient medium contains colloidal silica coated with PVP. It is nontoxic and has a low osmolality, low viscosity, and the sterile stock solution has a density of 1.13. With balanced salt solution or 0.25 M sucrose, the density can be varied over a wide range while controlling the osmolality(11). It can be used for multiple discontinuous gradients. A continuous gradient also can be generated by high-speed centrifugation (20,000 to 100,000 G). The densities obtained are monitored with calibrated density colored marker beads. After cell separation, the Percoll is removed by centrifuging the cells and washing.

Percoll (density 1.077) may substitute for Ficoll-Hypaque for flotation of mononuclears and platelets. Mononuclears with enriched monocytes (20%) can be separated in high yield from defibrinated blood(11). With a second centrifugation with Percoll (density 1.064) a band containing 90% monocytes may be obtained, with a low yield of 35%. With 7 discontinuous layers of Percoll of graduated densities, a better separation of leukocytes from monocytes is obtained(13). Many other variations in technique have been reported for lymphocyte-monocyte separation but it is difficult to achieve both a high purity and high yield simultaneously. Moreover, most techniques tend to work well for only small volumes of blood. Recently, three layers of Percoll (density 1.10, 1.087, and 1.070) have successfully separated large numbers of neutrophils from mononuclears (3 x 10⁸ cells) with high yield and purity after gravity sedimentation of a buffy coat(14).

Natural killer (NK) cells are among the most difficult cell populations to separate. One method(15) requires three steps. After flotation of mononuclears on a gradient of Ficoll-Hypaque, adherent monocyte B cells are eliminated by passage through a nylon wool

---

** Pharmacia Fire Chemicals AB, Uppsala, Sweden

column. The cells are then centrifuged on seven discontinuous
gradients of Percoll. The NK cells appear in the upper two layers,
but the purity is only about 40%.

Monocytes also have proven difficult to isolate. In one common
method, mononuclears are first harvested by density gradient
flotation. They are then added to glass dishes precoated with
autologous serum and incubated for one hour at 37° C. The lymphocytes
are rinsed off while the monocytes remain adherent. EDTA or lidocaine
solutions are added to detach some monocytes, and partially adherent
cells are scraped off with a rubber policeman. This method yields 50
to 70% of the monocytes, but some cell functions are altered by this
manipulation(16).

ELUTRIATION

Cell separation with the Beckman elutriator rotor has two major
advantages--cells of different diameter are separated, as well as
cells of different density. Furthermore, the cells in the chamber
remain in suspension and are not pelleted as in conventional
centrifugation(17). Cells in the suspending medium enter a rotating
4.5 ml conical chamber at the outer apex by continuous flow and are
exposed to two opposing forces--an outward centrifugal force and an
inward centripetal flow. If, for a particular cell, the flow force
exceeds the centrifugal force, the cell leaves the chamber for the
collecting vessel; if not, the cell remains in the chamber. Although
several cell populations can be harvested in high yield and high
purity with this method, the technique is exacting. Many variables
must be controlled, including the velocities of rotation and flow, and
the temperature, viscosity, osmolality, and pH of the medium.

Elutriation has been most widely employed to obtain granulocytes
in high purity from 5 to 120 ml of blood. From 120 ml, 3×10^8
granulocytes may be isolated with 96% purity and a recovery of at
least 70%(18). With special enlarged double chambers in the rotor,
10^9 or more cells can be harvested from larger volumes of blood(19).
In elutriating "buffy coat" blood diluted in suspending medium,
platelets promptly leave the chamber, followed by red cells and small
lymphocytes. In our experience, the cells remaining in the chamber
contain about 90% granulocytes, 3% monocytes, and 5 to 7%
erythrocytes. The suspending medium must contain some macromolecules
to prevent granulocyte aggregation--we use a special 0.5% hydroxyethyl
starch solution(20) instead of dilute serum albumin because the former
does not interfere with cell labeling.

Elutriation is successful in the difficult task of isolating
intact monocytes from lymphocytes. However, multiple steps are
required. The bulk of erythrocytes are removed by two buffycoat
centrifugations or red cell sedimentation. The mononuclear cells are

then harvested by flotation with Ficoll-Hypaque or Percoll gradients.
After three washes in Ca-Mg-free Hank's solution with EDTA (0.1
mg/ml), the cells are elutriated with the same medium in a special
Sanderson chamber(21, 22, 23) with three stepwise increases in flow
rates.

Lymphocytes are sometimes separated from monocytes and
granulocytes by elutriation(24) to avoid the reduced response to
stimulation with mitogens or antigens encountered after harvesting by
Ficoll-Hypaque flotation.

FLOW CYTOMETRY

Flow cytometry is an ingenious method for either analyzing or
sorting cell suspensions by differences in light scattering or natural
or induced fluorescence(25, 26). From a chamber containing cells in a
serum-free medium (5×10^4 to 5×10^6 cells/ml), cells exit one at a
time under controlled pressure in a thin jet through a UV-enhanced
argon-ion laser beam. The scatter or fluorescent emissions are
detected by photomultiplier tubes which convert the signals to
proportional voltage pulses. These pulses are then amplified and
converted to digital form for display in numerical or histogram
formats.

Both forward angle (1 to 19.5°) and 90° angle light scatter are
measured. Light scattered in the forward direction is generally
proportional to the cell size. Ninety-degree light scatter varies
with differences in internal structure of the cell, such as the
nuclear to cytoplasmic ratio or granularity of the cytoplasm. Light-
scatter measurements differentiate cells varying in size, refractive
index, or structural granularity and can readily distinguish between
live and dead cells. Fluorescence detection depends on differences in
staining properties between various cell types with fluorescent dyes.
Fluoroscein or rhodamine conjugates of monoclonal antibodies or
lectins may be used to distinguish between cells with different
surface receptors. On two- or three-dimensional displays, the
parameters of scatter, fluorescence, and cell number are depicted in
various combinations.

Cell separation (sorting) based on scatter or fluorescent
properties is achieved by vibrating the sample-sheath stream with a
piezoelectric crystal. The vibration causes the stream to break into
cell-containing droplets. When a cell satisfies the user's sort
criteria, charge is placed on the droplet. As the droplet descends,
it passes through a pair of deflection plates. Depending on the
polarity of the charge, the droplet deflects into different
containers.

The maximum rate of analysis is 10^4 positive cells/second, and
the maximum sorting rates 5,000 positive cells/second with 98% purity

(1.8 x 10^7 positive cell/hour). Cell recovery up to 90% may be achieved. Analysis and sorting of leukocyte populations is now one of the most important applications of flow cytometry.

NONSPECIFIC GAMMA-EMITTING AGENTS FOR LABELING LEUKOCYTES

Indium-111 Lipophilic Complexes

Two lipophilic chelating agents, 8-hydroxyquinoline (oxine) and tropolone, complexed with In-111 are popular for leukocyte labeling. Recently, another chelating agent, 2-mercaptopyridine 1-oxide, has been proposed by Thakur and Barry[28]. These agents form 3 to 1 complexes with trivalent indium or iron with a net charge of 0. Because they are liphophilic, they readily penetrate cellular or bacterial membranes. Once intracellular, the complexes apparently dissociate and the indium binds firmly to nuclear and cytoplasmic proteins whereas the chelating agents remain diffusible. The complexes indiscriminately label leukocytes, platelets, erythrocytes, tumor cells, or bacteria[29, 30].

The evolution of labeling techniques has been reviewed by Thakur[31]. With the older technique[32] In-111-oxine (50 μg/ml) was prepared by solvent extraction in chloroform or ethylene chloride, evaporated and dissolved in 50 μl of ethanol, and added to about 10^8 mixed leukocytes suspended in 5 ml of plasma-free medium. The final concentration of ethanol therefore was 1%. However, structural changes in neutrophils have been observed with concentrations as low as 0.125%[33]. In subsequent methods, ethanol was avoided entirely by In-111-oxine in acetate-TRIS buffer[34] or by using the soluble salt, oxine sulphate, in TRIS or HEPES buffer[20, 35]. In one commercial preparation free of ethanol, 50 μg oxine/ml are dissolved in HEPES-saline buffer with polysorbate 80.

In some preparations of In-111 in HCl used to form these complexes, the trace metals Zn, Cd, Cu, or Fe are present up to 3 μg/ml and compete with In-111 in forming 2:1 or 3:1 complexes with the microgram quantities of chelating agent. Moreover, concentrations of zinc as low as 0.3 μg/ml may be harmful to neutrophils[35].

The virtues of tropolone compared with oxine for leukocyte labeling remain controversial[31]. There is general agreement that both agents are toxic at levels above 50 μg/10^8 cells. Burke recommends this amount of tropolone for labeling in 10 ml of Hank's solution for a labeling efficiency comparable to 15μg of oxine. In vitro tests of cell function, including chemotaxis, comparing tropolone with oxine has produced variable results. Workers from Hammersmith[39, 40] recommend 100μl of tropolone (approximately 54 μg in HEPES-saline buffer and less than 50 μl or 300 μCi In-111 in 0.04 M HCL) for leukocytes obtained from 80 ml of blood but suspended in only

1 ml of cell-free plasma. They obtained yields of 50 to 80%. The
early transient accumulation of labeled neutrophils in the lungs was
less with tropolone than with oxine.

We prefer tropolone for clinical leukocyte imaging because of its
ability to label cells in small volumes of plasma. Keeping them in
their natural environment should avoid any loss of viability from
preliminary washing in plasma-free media. Moreover, the procedure
time is reduced. Thirty-six ml of fresh blood are drawn into a 60 ml
syringe containing 4 ml of ACD solution. Seven ml of Volex solution
(6% HES) are added from a sterile vial and the blood gently mixed.
After gravity settling of the red cells for 1 hour with the syringe
vertical, needle pointing up, the cell-rich plasma is expelled through
a butterfly infusion set into a sterile plastic 30 ml capped and
vented tube. The tube is spun at 450 g for 5 minutes and all plasma
removed with a spinal needle except the last 0.5 ml.

A minute volume containing 0.5mCi of high specific activity
In-111-chloride (5 mCi/100 µl) in 0.03 M HCL is added to 50 µl of
tropolone (50 µg) in HEPES-saline in a sterile vacutainer tube. The
mixture is added to the resuspended cells in the laminar flow hood and
incubated for 15 minutes at room temperature with gentle agitation.
With this "minimal volume" technique, about 10^8 leukocytes are labeled
in a total volume of 0.51 ml with an average yield of 90%, so that
washing the cells with plasma after labeling is usually unnecessary.
Five ml of plasma supernatant are then added for injection into the
patient. When a similar "minimal volume" technique is tried for
oxine, the labeling yield is only 36%.

In evaluating any leukocyte labeling technique, the most
important criteria are satisfactory migration and recovery of the
labeled cells in the circulation after reinjection. None of the
techniques to date, using gamma emitters have approached the 50%
recovery in the circulating granulocyte pool obtained with diisopropyl
fluorophosphonate (DFP-32). With In-111-oxine we obtained a recovery
of 34% in the dog(20), and 30% was obtained in man by Weiblen et
al(41). With In-111-tropolone and leukocyte labeling in 0.5 ml of
plasma, the cell recovery in the circulation was 36%, as determined by
elutriation of blood samples obtained 1 hour after injection.

With any "indiscriminate" agent, the labeling efficiency
increases with a greater number of cells in the smallest possible
volume. For a given agent and given labeling efficiency, the
intrinsic cell radiation dose is obviously independent of volume.
From Silvester's data(42), 10^8 neutrophils receive 8,750 rads to
complete decay when labeled with 0.5 mCi of In-111. From the data of
Bassano and McAfee(43), under the same conditions, the estimated
radiation dose to neutrophils while circulating (effective T 1/2 - 5.5
hours) is 984 rads. Fortunately, neutrophils are very resistant to
ionizing radiation. Little or no impairment of chemotaxis, bacterial

killing, superoxide production, phagocytosis, or morphological changes
are detected up to 5,000 rads(44, 45), and only a 50% reduction in
response is induced with 50,000 rads(44).

On the contrary, lymphocytes are notoriously sensitive to
radiation effects. Sparshott et al(46) found that 4 to 10 μCi In-111
10^8 cells was the optimal dose for rat lymphocytes, and the highest
level compatible with cell survival up to 24 hours was only 20 μCi/10^8
cells. The estimated radiation dose to the cells from 20 μCi/10^8 was
initially 7.6 rads per hour, 160 rads by 24 hours and the total
cumulative cell dose 740 rads, assuming an 80% labeling efficiency.
Unfortunately, the lowest activity for gamma camera imaging in humans
is about 100 μCi of In-111.

Tc-99m LEUKOCYTE LABELING METHODS

Leukocytes were first labeled with Tc-99m pyrophosphate for
clinical imaging by Uchida et al(47), but elution of the label and
diffuse abdominal activity were encountered. In another method(48)
cells obtained after red cell sedimentation, ammonium chloride lysis
and washing in saline were incubated at 37° C for 15 minutes with cold
stannous pyrophosphate (59.4 mg pyrophosphate, 1.3 mg stannous
chloride). These cells were rewashed, high specific activity Tc-99m
pertechnetate added for 5 minutes, and rewashed. In vitro tests of
cell function were satisfactory(49). In dogs(48), however, only 10%
of the injected activity was recovered in the white cell fraction from
1 to 3 hours after injection. Moreover, the abscess to blood
concentration ratio at 6 hours was only 1.5 to 3:1.

We previously evaluated Tc-99m-Sn oxine for labeling leukocytes
and abscess localization in dogs(50). Although abscess concentrations
were higher than those with simultaneously injected Ga-67, the
recovery of labeled cells from the blood and abscess localization were
inferior to those with In-111-oxine. Moreover, images showed
considerable gastrointestinal activity. In more recent experiments,
Tc-99m tropolone produced no better results.

Jones et al(51) labeled Chinese hamster B79 lung fibroblasts with
a univalent cationic complex (hexakis-t-butyl isonitrile) Tc-99m (I).
We incubated 2 x 10^8 mixed leukocytes in a 5 ml suspension with 20 μl
of this complex at room temperature at 30 minutes. When suspended in
plasma, we obtained a labeling yield of 80% and suspended in saline,
96%. On elutriating the cells, however, most of the radioactivity
washed off, indicating that the label was promptly reversible. With
methyl and ethyl iosonitrile complexes, we observed a cell labeling
yield of only 5%.

Labeling by Phagocytosis

From numerous experiments in bacteriology, phagocytosis by
neutrophils and macrophages is optimal at 37° C with gentle agitation

over about 1 hour, and plasma in the medium is essential for
opsonization. Particles must first adhere to the cell membrane
(a reversible phenomenon) before irreversible engulfment can
occur(52). As a rule, the optimum labeling yield is only 30 to 40%.
With carefully sized polystyrene latex spheres, the optimal size for
phagocytosis by neutrophils is 0.25 to 3 μm(53). Particles less than
0.1 μm in diameter, like colloidal Au-198, are phagocytosed by
macrophages but poorly by neutrophils. Tc-99m sulfur colloid (0.1 to
1 μm) has been tried for labeling neutrophils(54). The unbound
particles were removed by spinning down the cells and washing. In our
in vivo experiments in dogs, however, the recovery of activity in the
circulation was very low, and activity in the lungs and liver was
high. On elutriation, most of the activity washed away from the
cells, indicating that most of the binding was due to surface
adherence. With Tc-99m albumin millimicrospheres (0.5 to 2 μm)(18),
there was only minimal nonspecific adherence to erythrocytes. By
elutriation, unengulfed particles were removed and considerable
activity remained cell bound. On reinjection of the cells in the dog,
however, the 1 hour recovery in the circulation was only 10%, with
marked sequestration in the liver and lungs.

Schroth et al(55) revived the approach of labeling by
phagocytosis using Tc-99m-Sn colloid. Five ml of saline containing
pertechnetate eluate were added to a vial of stannous fluoride colloid
(125 μg without stabilizing agent) and rotated at room temperature for
1 hour to grow larger size colloidal particles. Ten ml of whole blood
with 50 IU of heparin were added and rotated for 1 hour. Then 1 ml of
3.8% sodium citrate was added and the blood rotated another hour, to
dissolve the colloidal particles which had not been engulfed. When
the cell binding was less than 80%, the plasma was replaced with
saline before reinjection. This method required no leukocyte
separation steps but still consumed 3 hours. For human and canine
blood, we obtained a high labeling efficiency of 87 to 90% with this
method and most of the label was not removed by elutriation. The in
vivo survival of reinjected cells in the dog, however, was poor. The
activity in the circulating blood volume was 28% at 2 minutes but fell
to 2 to 5% by 2 hours. Camera images showed high activity in the
liver and spleen and some activity in the marrow, bladder, and
thyroid.

So far, therefore it must be concluded that none of the methods
for labeling leukocytes with Tc-99m have produced in vivo results
comparable to those achieved with the In-111-lipophilic chelates.

LABELING SURFACE MEMBRANE PROTEINS

Like other cells, leukocytes contain "exposed" surface proteins
in their cell membranes which are capable of being labeled directly.
Tyrosyl groups in the surface membrane proteins of lymphocytes have

been directly radioiodinated with lactoperoxidase(56) but the labeling
yield was only 2 to 12%. Cells may be labeled with compounds
containing reactive groups that bind covalently to the proteins, such
as diazonium compounds. Diazotized radioiodinated diiodosulphanilic
acid has been used for labeling erythrocytes or platelets(57). The
advantage of this approach is a relatively low cell radiation dose
when the labeling agent is not internalized by the cells. The
disadvantage is a low labeling yield (3 to 10%), even with high
specific activity material. The label tends to be shed in vivo. We
radioiodinated 5×10^7 canine leukocytes with Bolton-Hunter
reagent(58), with a yield of 20%. After reinjection in the dog,
however, most of the radioactivity was eluted from the cells by 1
hour. To date, the idea of labeling cells by iodination of surface
proteins does not look promising for in vivo studies.

GAMMA-EMITTING AGENTS FOR SELECTIVE LEUKOCYTE LABELING

    The agents discussed so far label all cell types
indiscriminately. Certain peptides and proteins, however, have the
potential for labeling different types or subpopulations of leukocytes
selectively, by binding to specific cell surface receptors.

Lactoferrin

    Bennett et al(59) have used radioiodinated lactoferrin as an in
vitro label for B cells and monocytes because they possess specific
membrane surface receptors for this protein. In a medium containing
calcium and serum albumin, 10 to 20 µg of radioiodinated protein
labeled isolated suspensions of 2×10^6 B cells or monocytes with a
yield of 50 to 66%. The label did not readily dissociate, except in
the presence of cold lactoferrin. However, in mixed leukocyte
suspensions about half the radioactivity nonspecifically bound to
neutrophils.

Chemotactic Oligopeptides

    Synthetic N-formyl oligopeptides of l-amino acids are powerful
chemotactic factors for neutrophils and monocytes(60). The specific
membrane-binding sites for these chemoattractants are distinct from
those for Fc, C_3a, and C_5a. The chemotactic effects of over 100 of
these sythetic peptides have been studied.

    One of the most potent chemotactic factors is N-formyl-methionyl-
leucyl-phenylalanine (CHO-met-leu-phe). Zoghbi et al(61) covalently
coupled transferrin to this agent with carbodiimide for labeling with
In-111 and obtained a maximum labeling yield of 60% in plasma.
However, less than nanomolar quantities of this free tripeptide have
activated neutrophils and released lysosomal enzymes. Injected

intravenously, about 1 nanomol caused an immediate and profound
leukopenia in rabbits(62) with recovery in about 30 minutes. We also
found this agent very toxic to the dog and rabbit when injected
intravenously in doses of less than 1µg.

Several antagonists(63) competitively bind to neutrophils,
thereby blocking the action of chemotactic peptides without
stimulating chemotaxis or subsequent metabolic events. Theoretically,
therefore, an antagonist would be better for cell labeling than an
active chemoattractant. Unfortunately, at least one well known
antagonist, carbobenzoxy-phenylalanyl-methionine (CBZ-phe-met),
promptly releases from cells on washing(64).

An even more effective chemoattractant for human neutrophils is
the hexapeptide N-formyl-nle-leu-phe-nle-tyr-lys(65). Because this
contains a tyrosyl group, it can be readily labeled with radioiodine
with high specific activity. With 0.1 nanomols of this I-131 labeled
compound, we obtained a leukocyte-labeling yield of about 20% in
either plasma or saline. Nanomolar quantities bind rapidly and
irreversibly to neutrophils but induce a chemotactic response. The
peptide is promptly internalized and proteolytic fragments are
released.

From the available evidence, labeled oligopeptide chemotactic
factors do not look promising as leukocyte-labeling agents because
minute quantities induce leukopenia, and, when internalized by the
cells, they trigger profound metabolic changes.

MONOCLONAL ANTIBODIES AGAINST SURFACE ANTIGENS OF LEUKOCYTES

Over 300 murine monoclonal antibodies (Mab) (IgG or IgM) already
have been produced from hybridomas which bind to different specific
cell surface antigens of human leukocytes. Some bind to all
leukocytes, whereas others bind only certain cell types or
subpopulations. Some bind almost exclusively to T lymphocytes (66,
67), B lymphocytes, the T helper-inducer subset(68), monocytes (69,
70), granulocytes (71,72,73), or K and NK cells(74,75). Representative examples are listed in Table 1. Frequently, the binding is
different between normal and activated T cells or monocytes, between
peripheral blood cells and their marrow or thymic precursors or
between normal and neoplastic cell lines.

These monoclonal reagents are now widely employed for cell
identification and cell separation(76). For fluorescence microscopy
or flow cytometry(26), cell staining is obtained by a two-step
indirect fluorescence technique. The cell suspension is first
incubated with the primary antibody in serum-free buffered solution
containing 0.02% sodium azide to prevent capping and internalization.
Fluorescein or rhodamine-conjugated anti-mouse immunoglobulin is added
to bind to the primary antibody. Radioiodinated secondary antibody or

Table I. Selected Monoclonal Antibodies for Human Leukocyte Surface
          Antigens In "Normal" Peripheral Blood Cells

Clone Designation	Mouse Antibody Class	Antigen Distribution	Author	Commercial Source
HLe-1	IgG_1	all leukocytes	Beverley	BD
IG10	IgM	all granulocytes; weak monocytes	Bernstein	NEN
G7E(LeuM4)	IgM	95% granulocytes; weak monocytes	Thompson	BD
FMC-11	IgG_1	granulocytes	Zola	Sera-Lab
63D3	IgG_1	84% monocytes weak granulocytes	Ugolini	BRL
MOP9(LeuM3)	IgG_{2b}	86% monocytes; weak granulocytes	Dimitriu-Bona	BD
5F1	IgM	monocytes; platelets	Bernstein	--
VIB-C5	IgM	B cells, granulocytes	Knapp	NEN
9.6(Lyt-3)	IgG_{2b}	E receptor lymphocytes weak monocytes	Kamoun	NEN
OKT4+	IgG_{2b}	T helper-inducer subset	Reinherz	Ortho
HNK-1(Leu7)	IgM	K + NK cells	Abo	BD

staphylococcal protein A may be used for the quantitation of the
antigenic sites.

     For cell separation, unwanted cell types may be eliminated by
indirect fluorescence labeling in a cell sorter, leaving the cell type
of interest unstained and isolated in high purity. Alternatively,
unwanted cell types may be lysed with specific cytotoxic Mab in the
presence of complement, followed by washing.

Mab labeled with gamma-emitters may prove useful in the future
for in vivo cell migration studies. Because of their great
specificity, they would eliminate the need for stringent cell
separation techniques before labeling, and leukocytes could remain in
their normal medium, plasma. Many practical problems confront this
approach. Commercial monoclonals are expensive. Those obtained from
mouse ascitic fluid remain contaminated with other proteins even when
purified by salt precipitation and affinity chromatography. Those
obtained from hybridoma culture supernatants usually contain fetal
calf serum. For this application, they should be cultured in media
containing serum without immunoglobulins. Sodium azide, a common
preservative, must be removed by dialysis because it is cytotoxic.
Aggregation of immunoglobulins must be avoided.

Complement-mediated lysis varies greatly from one antibody to
another, depending on the subclass of immunoglobulin. Most IGM,
IgG2a, and IgG2b Mabs are cytotoxic, while IgG1 is not.use cytotoxic
antibodies, $F(ab')_2$ or $F(ab')_2\mu$ fragments may be isolated. This would
avoid the problems of aggregation and nonspecific binding to Fc
neutrophil receptors.

Radioiodinated Mab and antibody fragments are employed to study
cell-binding kinetics in vitro(81). Monovalent Fab fragments have
relatively weak binding, but bivalent IgG of $F(ab')_2$ fragments do not
readily dissociate, and polyvalent IgM binding is strongest. For cell
labeling, Mabs are generally used in micro to nanomolar concentrations,
and the number of binding sites per cell probably ranges from 100,000
to 200,000. If 10^8 cells were totally labeled with 1 microgram of
IgG, about 80,000 binding sites per cell would be occupied. From the
experience with fluorescein-tagged antibodies, up to 5 sites on the
antibody molecule could be conjugated without blocking antigen
binding.

To investigate the feasibility of using labeled monoclonals for
tagging leukocytes for in vivo migration studies, we performed limited
preliminary experiments to assess the stability of the label. The
monoclonal G7E11 (anti-leu M_4), an IgM specific for granulocytes,
weakly reactive with monocytes, was radioiodinated. One point five
micrograms were added to 10^8 mixed leukocytes in 0.5 ml plasma, and a
labeling yield of 20% was obtained. Elutriation of the cells for 1
hour showed that most of the label remained cell bound.

Several techniques are being developed for conjugating chelating
groups to monoclonal antibodies, for labeling with In-111 or Tc-99m.
A cyclic dianhydride of DTPA(78) has been successfully coupled to
amino groups of monoclonal antibodies(79) forming amide bonds, using
300 µg of protein. The In-111-labeled protein has been more stable in
vivo than radioiodinated protein. P-isothiocyanatobenzyl-EDTA has
been developed for coupling to amino groups of proteins at alkaline
pH, forming a thiourea linkage(80). This is the same coupling

reaction as that of fluorescein isothiocyanate, so widely used for antibody fluorescence labeling.

Some cell membrane receptor proteins are "resident" proteins which remain at the cell surface. Others, however, recycle regularly (81) from surface to interior by endocytosis and resurface again within minutes. The half-time of internalization of some of these cycling receptors is about 5 minutes. The internalization of ligands, such as Mab, therefore, will depend upon whether the antigen is a resident or a migrant receptor.

CONCLUSIONS

Much progress has been made in improving and simplifing the techniques of harvesting and labeling leukocytes with gamma emitters. Clinical imaging studies for the detection of focal inflammatory foci can now be conveniently performed with the In-111-lipophilic chelates for efficient labeling of mixed leukocytes. The ideal radioactive agent should label cells irreversibly with high yield in plasma, rather than in an artificial media, with as little in vitro manipulation as possible, and be nontoxic. Indium-111 tropolone appears the closest agent to this ideal at present.

Efficient labeling has not yet been achieved for less than 10^7 cells, to permit in vivo migration studies of smaller cell populations such as eosinophils or lymphocytes subsets. Tc-99m would be the best and most convenient radionuclide for imaging studies up to 24 hours. However, there is still no technique for tagging it irreversibly to leukocytes without activating or damaging them. Because lymphocytes are so exquisitely sensitive to radiation, they need an agent which is not internalized, or at least confined to the cytoplasm.

The most fruitful approach for the future appears to be the development of labeled ligands for specific cell surface receptors. Monoclonal antibodies and their $F(ab')_2$ fragments for leukocyte surface antigens appear particularly promising. They already have labeled leukocytes successfully for in vitro studies. Selective irreversible labeling of individual cell types and subsets without tedious isolation procedures should be attainable.

References

1. J. G. McAfee, Importance of cell labeling techniques: Indium-111-labeled neutrophils, platelets and lymphocytes, in: "Proceedings of the Yale symposium," M. L. Thakur and A. Gottschalk, eds., Trivirum, New York (1980).
2. A. Boyum, Separation of blood leucocytes, granulocytes and lymphocytes, Tissue Antigens 4:269 (1974).

3. A. J. Roy, A. Franklin, W. B. Simmons, et al, A method for separation of granulocytes from normal human blood using hydroxyethyl starch, Prep Biochem 1:197 (1971).

4. D. C. Dooley, T. Takahashi, The effect of osmotic stress on the function of the human granulocyte, Exp Hematol 9:731 (1981).

5. M. L. Thakur, J. P. Lavender, R. N. Arnot, et al, Indium-111-labeled autologous leukocytes in man, J Nucl Med 18:1012 (1977).

6. P. Madyastja, K. R. Madyastha, T. Wade, et al, An improved method for rapid layering of Ficoll-Hypaque double density gradients suitable for granulocyte separation, J Immunol Methods 48:281 (1982).

7. A. Boyum, Isolation of mononuclear cells and granulocytes from human blood, Scand J Clin Lab Invest 21:(Suppl. 97) 1 (1968).

8. A. Ferrante, Y. H. Thong, Optimal conditions for simultaneous purification of mononuclear and polymorphonuclear leukocytes from human blood by the Hypaque-Ficoll method, J Immunol Methods 36:109 (1980).

9. M. T. Aguado, N. Pujol, E. Rubiol, et al, Separation of granulocytes from peripheral blood in a single step using discontinuous density grandients of Ficoll-Urografin: a comparative study with separation by dextran, J Immunol Methods 32:41 (1980).

10. H. Loos, B. Blok-Schut, R. Van Doorn, et al, A method for the recognition and separation of human blood monocytes on density gradients, Blood 48:731 (1976).

11. H. Pertoft, A. Johnson, B. Warmegard, et al, Separation of human monocytes on density gradients of Percoll, J Immunol Methods 33:221 (1980).

12. F. Gmelig-Meyling, T. A. Waldmann, Separation of human blood monocytes and lymphocytes on a continuous Percoll gradient, J Immunol Methods 33:1 (1980).

13. A. J. Ulmer, H-D Flad, Discontinuous density gradient separation of human mononuclear leucocytes using Percoll as gradient medium, J Immunol Methods, 30:1 (1979).

14. D. C. Dooley, J. F. Simpson, H. T. Merryman, Isolation of large numbers of fully viable human neutroqhils: a preparative technique using Percoll density gradient centrifugation, Exp Hematol 10:591 (1982).

15. T. Timonen, C. W. Reynolds, J. R. Ortaldo, Isolation of human and rat natural killer cells, J Immunol Methods, 51:269 (1982).

16. D. G. Fischer, H. S. Koren, Isolation of human monocytes in Methods for studying mononuclear phagocytes, Adams DO, Edelson PJ and Kosen HS, editors Academic Press, New York (1981).

17. F. J. Lionetti, S. M. Hunt, R. Valeri, Isolation of human blood phagocytes by counter flow centrifugation elutriation, in: "Methods of Cell Separation," vol. 3, N. Catsimpoolas, ed., Plenum Press, New York (1980).

18. J. F. Jemionek, T. J. Contreras, J. E. French, et al, Technique for increased granulocyte recovery from human whole blood by

counterflow centrifugation-elutriation. I. In vivo analysis,
Transfusion 19:120 (1979).

19. J. F. Jemionek, T. J. Contreras, D. N. Stevens, et al, Use of a
    modified rotor and enlarged separation chamber for isolation
    of human granulocytes by counterflow
    centrifugation-elutriation, Cryobiology 17:230 (1980).

20. J. G. McAfee, G. M. Gagne, G. Subramanian, et al, Distribution of
    leukocytes labeled with In-111 oxine in dogs with acute
    inflammatory lesions, J Nucl Med 21:1059 (1980).

21. R. J. Sanderson and K. E. Bird, Cell separation by counter flow
    centrifugation, in: "Methods in Cell Biology", vol. 15,
    D. M. Prescott, ed., Academic Press, New York (1977).

22. R. J. Sanderson, F. T. Shepperdson, A. E. Vatter, et al,
    Isolation and enumeration of peripheral blood monocytes, J
    Immunol 118:1409 (1977).

23. D. A. Norris, R. M. Morris, R. J. Sanderson, et al, Isolation of
    functional subsets of human peripheral blood monocytes, J
    Immunol 123:166 (1979).

24. T. G. Pretlow II, T. P. Pretlow Centrifugal elutriation
    (counterstreaming centrifugation) of cells, Cell Biophys
    1:195 (1979).

25. C. L. Berger, R. L. Edelson, Comparison of lymphocyte function
    after isolation by Ficoll-Hypaque flotation or elutriation, J
    Invest Dermatol 73:231 (1979).

26. L. A. Herzenberg and L. A. Herzenberg, Analysis and separation
    using the fluorescence activated cell sorter. (FACS), in:
    "Handbook of Experimental Immunology," vol. 2, Cellular
    Immunology, 3rd edition, D. M. Weir, ed., Blackwell, Oxford
    (1978).

27. M. R. Loken and A. M. Stall, Flow cytometry as an analytical and
    preparative tool in immunology, J Immunol Methods 50:R85
    (1982).

28. M. L. Thakur and M. J. Barry, Preparation and evaluation of a new
    In-111 agent for efficient labeling of human platelets in
    plasma (Abstr.), Fourth International Symposium on
    Radiopharmaceutical Chemistry, August 23-27, 1982,
    Kernforschungsanlage Julich GmbH (1982).

29. J. G. McAfee and M. L. Thakur, Survey of radioactive agents for
    in vitro labeling of phagocytic leukocytes. I. Soluble
    agents, J Nucl Med 17:480 (1976).

30. R. H. Wiltrout, E. Gorelik, M. J. Brunda, et al, Assessment of in
    vivo natural antitumor resistance and lymphocyte migration in
    mice: comparison of I-125-iododeoxyuridine with
    Indium-111-oxine and Chromium-51 as cell labels, Cancer
    Immunol Immunother 14:172 (1983).

31. M. L. Thakur, Cell labeling: achievements, challenges and
    prospects, J Nucl Med 22:1011 (1981).

32. J. G. McAfee, Techniques of harvesting platelets and neutrophils
    and labeling with In-111 oxine, in: "Proceedings of the Yale
    Symposium," M. L. Thakur and A. Gottschalk, eds,. Trivirum,
    New York (1980).

33. M. Lichtman, P. A. Santillo, E. A. Kearney, et al, The shape and
     surface morphology of human leukocytes: the in vitro effect
     of temperature, metabolic inhibitions and agents that
     influence membrane structure, Blood Cells 2:507 (1976).
34. W. T. H. Goedemens, Simplified cell labeling with In-111
     acetylacetonate and In-111 oxinate, Br J Rad 54:636 (1981).
35. D. Ducassou, J. P. Nouel, A. Brendel, Le marquages des elements
     figures du sang par l'indium radioactif-methodology-resultats-
     indications, Rad Isot in Klinik und Forschung 13:91 (1978).
36. M. Chvapil, L. Stankova, C. Zukoski, et al, Inhibition of some
     functions of polymorphonuclear leukocytes by in vitro zinc J
     Lab Clin Med 89:135 (1977).
37. M. R. Hardeman, Tropolone, the favourite ligand for cell
     labeling? (letter to the editor), Eur J Nucl Med 7:528
     (1982).
38. J. E. T. Burke, S. Roath, D. Ackery, P. Wyeth, The comparison of
     8-hydroxyquinoline, tropolone, and acetylacetone as mediators
     in the labelling of polymorphonuclear leukocytes with In-111:
     a functional study, Eur J Nucl Med 7:73 (1982).
39. H. J. Danpure, S. Osman, F. Brady, The labeling of blood cells in
     plasma with In-111-tropolonate, Brit J Radiol 55:247 (1982).
40. A. M. Peters, S. Saverymuttu, H. J. Reavy, et al, Imaging of
     inflammation with In-111 tropolonate labeled leukocytes, J
     Nucl Med 24:39 (1983).
41. B. J. Weiblen, L. Forstrum, J. McCullough, Studies of the
     kinetics of In-111-labeled granulocytes J Lab Clin Med 94:246
     (1979).
42. D. J. Silvester, Consequences of In-111 decay in vivo: calculated
     absorbed radiation dose to cells labeled by In-111 oxine, J
     Label Comp Radiopharm 16:193 (1979).
43. D. A. Bassano and J. G. McAfee, Cellular radiation doses of
     labeled neutrophils and platelets, J Nucl Med 20:255 (1979).
44. T. R. Holley, D. E. Van Epps, R. L. Harvey, et al, Effect of high
     doses of radiation on human neutrophil chemotaxis,
     phagocytosis and morphology, Am J Path 75:61 (1974).
45. L. N. Button, W. C. DeWolf, P. E. Newburger, et al, The effects
     of irradiation on blood components, Transfusion 21:419 (1981).
46. S. M. Sparshott, H. Sharma, J. D. Kelly, et al, Factors
     influencing the fate of In-111-labelled lymphocytes after
     transfer to syngeneic rats, J Immunol Methods, 41:303 (1981).
47. T. Uchida, T. Nemoto, T. Yui, et al, Use of Technetium-99m as a
     radioactive label to study migratory patterns of leukocytes, J
     Nucl Med 20:1197 (1979).
48. N. Linhart-Colas, M. Meignan, B. Bok, et al, "In vivo" kinetics
     of Technetium-99m-labeled leukocytes in dogs and the effects
     of an abscess Biomedicine 32:133 (1980).
49. N. Colas-Linhart, M. Barbu, M. A. Gougerot, et al, Five
     leukocyte-labeling techniques: a comparative in-vitro study,
     Br J Haematol 53:31 (1983).

50. G. Subramanian, J. G. McAfee, G. M. Gagne, R. W. Henderson, M. Rosenstreich, Tc-99m-oxine: a new lipophilic radiopharmaceutical for labeling leukocytes and platelets, in: Nuklearmedizin 15th International Annual Meeting of Society of Nuclear Medicine, Gronigen, Sept. 13-16, 1977, H. A. E. Schmidt and M. F. K. Woldring, eds., Schattauer Verlag, Stuttgart-New York, 1978.

51. A. G. Jones, A. Davison, M. J. Abrams, et al, A new class of water soluble low valent technetium unipositive cations, (abstr.), Fourth International Symposium on Radiopharmaceutical Chemistry, August 23-27, 1982, Kernforschungsanlag Julich GmbH (1982).

52. J. G. McAfee and M. L. Thakur, Survey of radioactive agents for in vitro labeling of phagocytic leukocytes. II. Particles, J Nucl Med 17:488 (1976).

53. J. Roberts and J. H. Quastel, Particle uptake by polymorpho-nuclear leukocytes and Ehrlich ascites-carcinoma cells, Biochem J 89:150 (1963).

54. D. K. English and B. R. Andersen, Labeling of phagocytes from human blood with Tc-99m-sulphur colloid, J Nucl Med 16:5 (1975).

55. H. J. Schroth, E. Oberhausen, R. Berberich, Cell labeling with colloidal substances in whole blood, Eur J Nucl Med 6:469 (1981).

56. J. J. Marchalonis, R. E. Cone, V. Santer, Enzymatic iodination: a probe for accessible surface proteins of normal and neoplastic lymphocytes, Biochem J 124:921 (1971).

57. J. N. George, P. C. Lewis, D. A. Sears, Studies on platelet plasma membranes. II. Characterization of surface proteins of rabbit platelets in vitro and during circulation in vivo using diazotized (I-125)-diiodosulfanilic acid as a label, J Lab Clin Med 88:247 (1976).

58. A. E. Bolton, W. M. Hunter, The labeling of proteins to high specific radioactivities by conjugation to a I-125-containing acylating agent, Biochem J 133:529 (1973).

59. R. M. Bennett, J. Davis, Lactoferrin binding to human peripheral blood cells: an interaction with a B-enriched population of lymphocytes and a subpopulation of adherent mononuclear cells, J Immunol 127:1211 (1981).

60. H. J. Showell, R. J. Freer, S. H. Zigmond, et al, The structure-activity relations of synthetic peptides as chemotactic factors and inducers of lysosomal enzymes secretion for neutrophils, J Exp Med 143:1154 (1976).

61. S. S. Zoghbi, M. L. Thakur, A. Gottschalk, et al, A potential radioactive agent for the selective labeling of human neutrophils, (To be published).

62. J. T. O'Flaherty, H. J. Showell, P. A. Ward, Neutropenia induced by systemic infusion of chemotactic factors, J Immunol 118:1586 (1977).

63. R. J. Freer, A. R. Day, N. Muthukumaraswamy, et al, Antagonists of the formylated peptide chemoattractants: structure-

activity comparison with formyl-methionyl-leucyl-phenyl-
alamine-OH in: " Biochemistry of the Acute Allergic
Reactions," Alan R. Liss, NY, (1981).

64. J. T. O'Flaherty, H."J. Showell, D. L. Kreutzer, et al,
     Inhibition of in vivo and in vitro neutrophil responses to
     chemotactic factors by a competitive antagonist, J Immunol
     120:1326 (1978).

65. J. Niedel, S. Wilkinson, P. Cuatrecasas, Receptor-mediated uptake
     and degradation of I-125-chemotactic peptide by human
     neutrophils, J Biol Chem 254:10700 (1979).

66. B. F. Haynes, Human T-lymphocyte antigens as defined by
     monoclonal antibodies, Immunol Rev 57:127 (1981).

67. M. Kamoun, J. Martin, J. A. Hansen, et al, Identification of a
     human T-lymphocyte surface protein associated with the
     E-rosette receptor, J Exp Med 153:207 (1981).

68. E. L. Reinherz, P. C. Kung, G. Goldstein, et al, Further
     characterization of the human inducer T-cell subset defined by
     monoclonal antibody, J Immunol 123:2894 (1979).

69. A. Dimitriu-Bona, G. R. Burmester, S. J. Waters, et al, Human
     mononuclear phagocyte differentiation antigens. I. Patterns
     of antigenic expression on the surface of human monocytes and
     macrophages defined by monoclonal antibodies, J Immunol
     130:145 (1983).

70. V. Ugolini, G. Nunez, R. G. Smith, et al, Initial characteri-
     zation of monoclonal antibodies against human monocytes, Proc
     Natl Acad Sci, USA, 77:6764 (1980).

71. H. Zola, P. McNamara, M. Thomas, et al, The preparation and
     properties of monoclonal antibodies against human granulocyte
     membrane antigens, Br J Haemat 48:481 (1981).

72. W. Knapp, Monoclonal antibodies against differentiation antigens
     of myelopoiesis, Blut 45:301 (1982).

73. I. D. Bernstein, R. G. Andrews, S. F. Colen, et al, Normal and
     malignant human myelocytic and monocytic cells identified by
     monoclonal antibodies, J Immunol 128:876 (1982).

74. J. R. Ortaldo, S. O. Sharrow, T. Timonen, et al, Determination of
     surface antigens in highly purified human NK cells by flow
     cytometry with monoclonal antibodies, J Immunol 127:2401
     (1981).

75. T. Abo and C. M. Balch, A differentiation antigen of human NK and
     K cells identified by a monoclonal antibody (HNK-1), J Immunol
     127:1024 (1981).

76. S. Hirsch and S. Gordon, The use and limitations of monoclonal
     antibodies against mononuclear phagocytes, Immunobiology
     161:298 (1982).

77. D. W. Mason and A. F. Williams, The kinetics of antibody binding
     to membrane antigens in solution and at the cell surface,
     Biochem J 187:1 (1980).

78. W. C. Eckelman, S. M. Karesh, R. C. Reba, New compounds; fatty
     acid and long chain hydrocarbon derivatives containing a
     strong chelating agent, J Pharm Sci 64:704 (1975).

79. D. J. Hnatowich, W. W. Layne, R. L. Childs, et al, Radioactive
     labeling of antibody: a simple and efficient method, Science
     220:613 (1983).
80. C. F. Meares, L. H. DeRiemer, CS-H Leung, et al, Properties in
     vivo of chelate-tagged proteins and polypeptides, in:
     Modification of Proteins: Food, Nutritional and
     Pharmacological Aspects, R. E. Feeney and J. R. Whitaker,
     eds., Advances in Chemistry Series 198, American Chemical
     Society, Washington, D.C. (1982).
81. M. S. Brown, R. G. W. Anderson, J. L. Goldstein, Recycling
     receptors: the round-trip itinerary of migrant membrane
     proteins, Cell 32:663 (1983).

GRANULOCYTE KINETICS

A. M. Peters*, S. H. Saverymuttu†, and J. P. Lavender+

Departments of Diagnostic Radiology+ and Medicine†
Hammersmith Hospital
London, England
Glaxo Group Research Ltd.
Ware, Hertfordshire*, England

INTRODUCTION

When DFP-32 (1-4) and later Cr-51 (5,6) were used to label granulocytes, much interest was generated in the field of granulocyte kinetics because the means became available to noninvasively quantify earlier direct observations made on living tissues. Such observations had revealed how granulocytes adhered to the endothelial surface of blood vessels, displaying a "rolling" motion along the surface (7,8).

Following injection of autologous radiolabeled granulocytes in intact normal subjects, the size of this "marginating" granulocyte pool (MGP) as a fraction of the total blood granulocyte pool (TBGP) was measured and its response to various interventions studied (2-4). Since P-32 is not a gamma emitter and Cr-51 only a weak one, there was little interest in the use of radiolabeled granulocytes to image abscesses, although some investigators attempted to do so with Cr-51 (9-11).

With the introduction of other, gamma-emitting, leukocyte labeling radionuclides, such as Ga-67 (12,13), Tc-99m (14-17) and particularly In-111 (18-21), this situation has reversed, and most studies of labeled granulocytes have been applied to abscess imaging. This emphasis can partly be explained by the necessity, when labeling with In-111, to isolate cells from plasma, which otherwise preferentially binds In-111. Following such in vitro manipulation, the behavior of In-111-leukocytes upon injection, typified by prolonged lung sequestration, led to uncertainties about the physiological integrity of In-111-granulocytes, which, nonetheless, were still sufficiently functional to localize in sites of inflammation. In contrast to In-111, leukocytes can be labeled with DFP-32 in whole blood, or with Cr-51 in plasma. Although other cells are also labeled by these isotopes, erythrocytes with DFP-32 and lymphocytes with Cr-51, meaningful kinetic data can be obtained by counting the activity in granulocytes isolated from blood samples obtained after injection of the labeled cells. Notwithstanding the problem of plasma affinity for In-111, the technique of post-injection activity counting in isolated cells is also available with In-111, although the advantages of gamma emission offered by this isotope would be lost.

During the past two years, our group has extensively used an In-111-chelate, tropolone, for cell labeling (22-24). This ligand achieves a usefully high labeling efficiency in plasma. By using density gradient materials enriched with autologous plasma, we have been able to isolate granulocytes from other cellular elements and label them with In-111 without separation from a plasma environment. The kinetic behavior of these cells suggests that phenomena attributable to granulocyte activation are greatly reduced by this labeling technique. We therefore decided to study granulocyte kinetics in health and disease using In-111 in the hope of quantifying sites of margination and identifying principal sites of destruction. This work will be reviewed, like platelet kinetics, under the three headings of distribution, life-span, and destruction, making comparisons with the kinetic data acquired on the basis of labeling with other isotopes.

DISTRIBUTION

The distribution of granulocytes is more complex than that of platelets in that there appears to be multiple vascular beds within which the former pool, i.e., are present in excess of plasma with respect to peripheral venous blood. These pools collectively constitute the MGP, evidence for the existance of which was originally obtained from direct microscopic observations on the relationship of granulocytes with the endothelial surface of post-capillary venules (7,8).

Estimates of the MGP based on blood clearance kinetics

DFP-32. Athens et al (2-4) were the first to attempt to measure the size of the MGP using radiolabeled granulocytes. They subtracted the activity present in circulating blood 3 minutes after the completion of a 20-minute infusion of DFP-32 labeled granulocytes from the activity anticipated on the basis of the dose and the recipient's total blood volume, making the assumption that the difference represented granulocytes that had entered the MGP (Fig. 1). Evidence that this difference was available to the circulation and therefore that the MGP was physiological was that it could be substantially reduced by exercise or epinephrine infusion. Furthermore, cell function, determined in vitro, appeared to be normal. This group gave

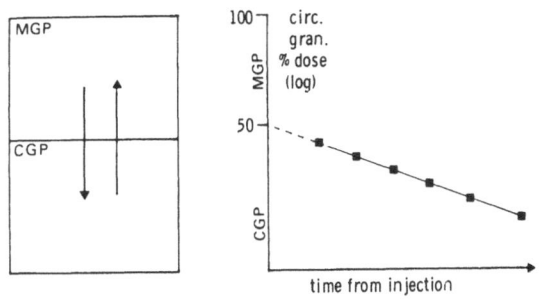

Fig. 1. Method of calculating the relative sizes of the CGP and MGP by extrapolation of the granulocyte survival curve (semilog replot) to zero time.

the normal values of the circulating granulocyte pool (CGP), i.e., the TBGP minus the MGP, as 46% of the TBGP, rising to 79% in response to epinephrine or exercise. These findings were largely confirmed in the dog by Raab et al (25). Other workers obtained similar values of the CGP using DFP-32 (5,6,26-28). Dresch et al (6) took more frequent samples after bolus injection of DFP-32 labeled granulocytes and obtained bi-exponential clearance curves which they interpreted to respectively represent 1) equilibration of granulocytes between the CGP and MGP and 2) subsequent granulocyte loss from the TBGP. Extrapolation of the clearance curve to zero time suggested a granulocyte recovery of 100%, and analysis of the respective zero time intercepts of the 2 exponentials gave an MGP of 44%.

Cr-51. McMillan and Scott's recovery of Cr-51 labeled granulocytes, 20%, was much less than that, 56%, of their DFP-32 labeled cells (5). For labeling with Cr-51, these workers prepared a leukocyte-rich cell suspension in phosphate buffer following red cell sedimentation. Granulocyte-associated counts were then determined following granulocyte isolation from post-injection blood samples, as with DFP-32 cells. On the other hand, Dresch et al (6) obtained very similar disappearance curves of cells, labeled as leukocyte concentrates in plasma, with Cr-51 and DFP-32.

H-3-Thymidine. Dancey et al (27), measured the recovery and survival time in normal subjects of H-3-thymidine granulocytes previously labeled in vivo in a donor. These cells had not, therefore, been manipulated or exposed to potential toxins ex vivo. Recovery was 59% and t1/2 7.6 hours. DFP-32-labeled granulocytes, studied simultaneously, gave a recovery of 48% and t1/2 of 5.4 hours. These workers derived recovery by extrapolation of the clearance curve based on samples taken from 1 hour, and it is worth noting that the 5-minute values were higher than the extrapolated 5-minute values by some 5-10% for the thymidine-labeled cells and somewhat less for the DFP-32 cells. Vincent et al (29) used a similar labeling technique in chimeric calves and obtained a t1/2 of 6.4 to 7.5 hours.

In-111. Weiblen et al (30), labeled "pure" granulocytes,
isolated on Ficoll-Hypaque double-density gradients. They obtained a
recovery of 30%, but this was a value achieved not initially but
between 10 minutes and 2 hours after injection. In dogs, McAfee et al
(31) also noted a progressive increase in the level of granulocyte
associated In-111 activity, reaching a maximum of 32% 4 hours after
injection. These authors concluded that since this delayed rise was
clearly a labeling artifact, In-111 could not be used to study granulocyte kinetics.

The comparison of recovery values between different groups is
made difficult by the differing methods of expressing them, i.e.,
sometimes as an extrapolated value, sometimes at a specified time, and
sometimes as a maximum. The biexponential curves obtained by some
workers deserve further discussion. Dresch et al (6) found that their
DFP-32 and Cr-51 curves became monoexponential between 30 minutes and
1.5 hours after injection. They concluded that the first exponential
represented equilibration between the CGP and MGP because it could be
abolished by epinephrine. McMillan and Scott (5) recorded biexponential curves for Cr-51 but not with DFP-32, concluding that the first
exponential, a relatively rapid one, of the Cr-51 curves was due to
clearance of damaged cells or to Cr-51 elution.

In our own studies with In-111-tropolone-labeled granulocytes,
recovery has been calculated on the basis of estimated total blood
volumes and on comparison with the circulating activity of In-111-
labeled red cells given prior to the granulocytes. The granulocyte
recovery values were similar with the two approaches, being about 45%
at 5 minutes and 35% at 40 minutes. Thus the shape of the clearance
curves were similar to those of Dresch et al (6), but unlike those of
Weiblen et al (30) and McAfee et al (31) in that the highest circula-

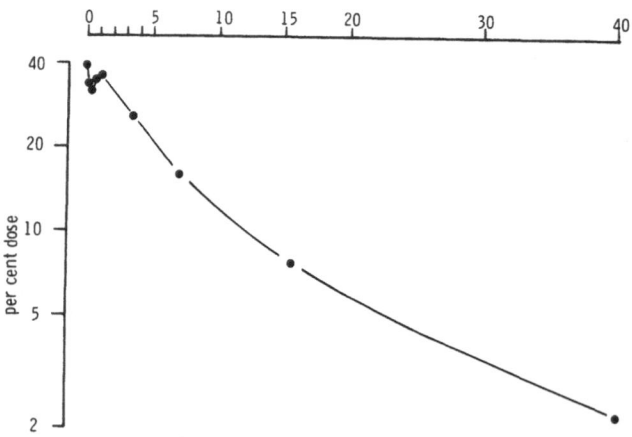

Fig. 2. Peripheral
blood plasmas labeled
In-111-granulocyte
time activity curve
in an asplenic pat-
ient showing a prom-
inent recirculation
"blip" at about 45
minutes. The source
of this "blip" is
probably the liver.

ting level was always the earliest one determined. However, we have
occasionally recorded curves showing a small secondary rise between 30
minutes and 3 hours (Fig. 2), as is also apparent from inspection of
Dresch's curves. We think this is due to release from the liver.

Lung Sequestration of Granulocytes Labeled with Gamma Emitters

Since gamma emitters became available to label cells, it has
become evident that there is prominent retention of granulocytes in
the lungs immediately following injection of the labeled cells. It
has been most frequently observed with In-111 (19,20,30,31), but also
with Tc-99m (14-16) and Cr-51 (11). Clearance of the activity has
been reported to occur over the ensuing 1-4 hours. This early lung
activity might be regarded as evidence in favor of the lung as a major
site of granulocyte margination. Thus Bierman et al (32) showed that
following epinephrine injection the arterial granulocyte level rose
before the venous level, suggesting release of granulocytes from the
lung, and that following infusion into nonleukemic recipients,
leukemic granulocytes were sequestered in the lungs (33). Furthermore, in more recent studies, based on unlabeled as well as Cr-51-
labeled granulocytes, Martin et al (34) concluded that the lung pooled
granulocytes to an extent that was inversely related to cardiac
output. On the other hand, heat-damaged In-111-granulocytes undergo
very prominent lung sequestration (20), suggesting that the latter is
not a physiological event. In addition, in crossed-circulation experiments in dogs, English and Andersen (15) showed that the Tc-99m-gran-
ulocyte lung retention was much higher in the primary than in the secondary recipient. Few would now doubt that extensive lung sequestration is indicative of cell damage and/or "activation" and is artifactual. Indeed, the progressive rise in cell-bound activity in peripheral blood for up to 4 hours after injection of In-111-granulocytes
(30,31) is related to the release of cells from the lungs (23). Thus
in English and Andersen's studies (15), labeled granulocyte blood
levels increased between 0 and 2 hours in the primary but not in the
secondary recipients. Leukocytes harvested from the peritoneum and
then labeled appear to undergo even more prominent lung retention
(12,35) possibly because, having already become extravascular in the
donor, they are already activated before labeling.

The role of isolation from plasma in lung retention has been
examined by Saverymuttu et al (23) who showed that granulocytes
separated in plasma-enriched media and labeled in plasma with In-111-
tropolone were sequestered in the lung to a much less extent than
granulocytes isolated and labeled in saline (Fig. 3). The optimal
functional integrity of plasma-labeled cells was obvious from their
very rapid rate of localization in abscesses -- measurable in minutes
(36) instead of the hours that were the case with saline-labeled
cells.

What influence immediate lung sequestration has had on estimates
of the size of the MGP based on recovery alone is difficult to assess.

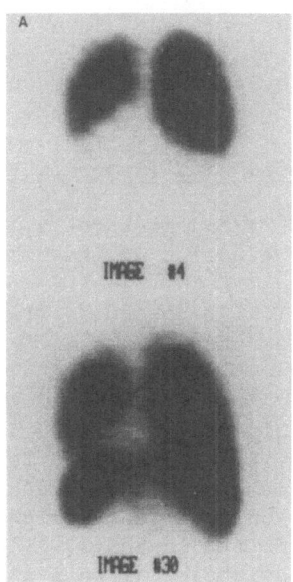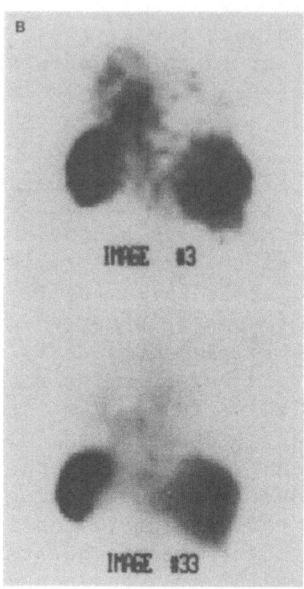

Fig. 3.   Lung sequestration of In-111-granulocytes immediately
          following injection. Upper images, 3 or 4 minutes after
          injection; lower images, 30 or 33 minutes after injection.
          A, Cells labeled in saline (with acetylacetone) following
          separation on a Percoll double-density gradient; B, cells
          labeled in plasma (with tropolone) following separation on a
          Percoll double-density gradient enriched with plasma.

In view of their minimal ex vivo manipulation, it seems doubtful that
the H-3-thymidine-labeled granulocytes of Dancey et al (27) could have
undergone much lung hold-up unless the recipients had unidentified
antigranulocyte antibodies. DFP-32-labeled cells, although maintained
in whole blood, may have undergone lung hold-up because DFP-32 a) may
be damaging to granulocytes (6) and b) labels the cell membrane (1)
and may therefore lead to activation insofar as this is manifested as
a surface- (granulocyte) to-surface (pulmonary endothelium)
phenomenon.

It should be emphasized that lung retention is probably a very
sensitive indicator of granulocyte activation and is not merely the
result of entrapment of granulocyte aggregates, because these have not
been visible in our suspensions of saline-labeled cells undergoing
sequestration. In vitro functional studies in our laboratory (A.J.
Pinching, unpublished) have failed to differentiate between plasma-
labeled granulocytes traversing the lung rapidly and saline-labeled
cells undergoing retention. Although isolation from plasma appears to
be detrimental to kinetic integrity, other manipulative procedures

used in granulocyte isolation, such as hypotonic lysis of contaminating red cells (37,38), pelletting (38) and exposure to Ficoll-Triosil (39), have been shown to be damaging.

Sites of Granulocyte Margination

The likeliest sites for granulocyte margination would appear to be the lungs, liver, spleen, and regions of the body such as the G.I. tract, from which extravasated granulocytes come into contact with the external environment.

Lungs. The evidence in favor of the lungs is presented above. We have attempted to quantitate intrapulmonary granulocyte margination by comparing the signal ratio recorded over the chest by the gamma camera with the corresponding signal ratio recorded simultaneously in p-eripheral blood samples following the sequential injection of In-111-labeled red cells and In-111-labeled granulocytes, as has been described for platelets. Soon after granulocyte injection this factor was about 5 but fell to about 2.5 at 40 minutes. Beyond 40 minutes, activity recorded over the chest starts to be influenced by bone marrow activity. The true value for the transit time of granulocytes (relative to red cells) is probably less than 2.5 as any ex vivo manipulation, however minimal, could be expected to prolong it.

Spleen. The importance of the spleen in granulocyte margination was previously assessed by measuring the size of the MGP before and after splenectomy in animals, and in humans without spleens, with normal-size spleens, and with splenomegaly.

Using DFP-32, Raab et al (25) showed that although the MGP in dogs is a similar fraction of the TBGP as it is in man, its size did not change when measured 6 weeks after splenectomy. In man Uchida and Kariyone (40) found no correlation between spleen size and DFP-32-labeled granulocyte recovery. However, others (26,41) have recorded an enlarged MGP in splenomegaly, and Brubaker and Johnson (42) found, in addition to a positive correlation between the size of the spleen and the MGP, a significant decrease in the MGP following splenectomy. Epinephrine injection also appears to mobilize granulocytes from the spleen, as recorded by McMillan and Scott (5) by surface counting over the spleen following Cr-51-labeled granulocyte injection, and by Bierman et al (43), who noted that, following epinephrine injection directly into the splenic artery, there was an early rise in the peripheral arterial granulocyte level followed by a positive hepatic venous-arterial gradient. Studies on granulocyte recovery in asplenic subjects appear to be fewer than those in splenomegaly. Dresch et al (6) found a reduced MGP in one asplenic but otherwise normal subject, and in a series of 8 patients Spivak and Perry (44) obtained their highest recovery value in an asplenic subject.

The possibility of an effect of lung sequestration in the above studies has to be remembered because it may result in paradoxical information with respect to quantitation of splenic margination based on recovery. Thus English and Andersen (15) in their crossed-circulation experiments in dogs showed reduced liver and lung uptake in the secondary recipient but greater splenic uptake. Furthermore, Saverymuttu et al (23) demonstrated that when lung sequestration was marked, the rate of uptake of activity by the spleen was reduced, and vice versa when transit was rapid. Thus recovery might be expected to vary directly with splenic uptake.

We have approached the spleen as a site of granulocyte margination in three ways: first, by recording peripheral blood clearance and splenic time activity curves following bolus injection of In-111-tropolone plasma labeled granulocytes and then applying deconvolution (45) and compartmental analysis (46); second, by measuring splenic granulocyte transit time as a factor of red cell transit time by the method described above for lung transit time; and third, by studying the decrease in splenic activity in patients with inflammatory disease.

The rate constant of uptake of In-111 granulocytes into the spleen is very similar to that of platelets (46). The latter pool in the spleen with a rate constant of equilibration between the spleen and blood which is equal to the sum of splenic input and output rate constants (46). The rate constant of granulocyte uptake could be this rapid only if 1) granulocytes also pooled in the spleen or 2) granulocytes were removed irreversibly by the spleen and simultaneously by another organ such that the combined clearance was greater than that afforded by splenic blood flow alone. However, when the distribution of In-111-granulocytes was quantitated shortly after injection, the liver, spleen, and CGP accounted for almost 100% of the dose, and since the hepatic rate constant is much more rapid than that of the

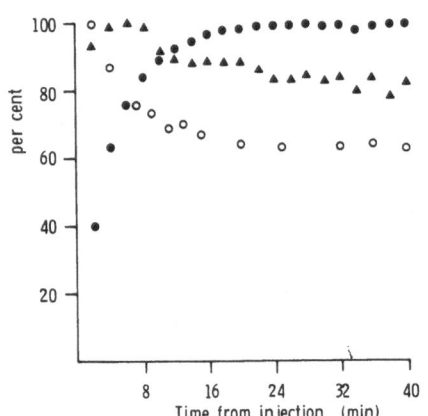

Fig. 4. Time activity curves recorded in peripheral venous blood, and by the gamma camera over the liver and spleen following injection of In-111-granulocytes, isolated and labeled in plasma. Liver activity reaches an early plateau and then falls slightly. The blood and spleen curves reach steady values after the liver plateau and appear as mirror images of each other. These contrasting time courses of activity are seen in the liver and spleen irrespective of whether the camera is positioned arteriorly or posteriorly.

spleen (Fig. 4), splenic uptake must represent pooling. An influence
of granulocyte accumulation in other organs on the splenic equilibration rate constant can also be discounted since little activity is
available for margination or uptake in other organs.

Deconvolution analysis (45) is a technique which, on the basis of
blood and splenic time activity curves, predicts the "washout" curve
of cells from the spleen that would take place if the cells were
injected as a bolus directly into the splenic artery and did not
recirculate. It assumes that the splenic activity and blood
activities are represented by a homogeneous cell population. Deconvolution analysis generated monoexponential splenic clearance curves
with rate constants indicating a mean cell transit time of about 10
minutes, similar to platelets. This was close to the value given by
the technique of comparing the granulocyte to red cell In-111 signal
ratios recorded by the gamma camera with those measured in peripheral
blood.

Further confirmation that the splenic activity is "in transit" is
its dramatic decrease in patients with extensive sepsis. Indeed, a
significant correlation was observed between the fall in splenic
activity (between equilibration, early after injection, and the value
at 24 hours) and the fecal In-111 recovery in a series of patients
with inflammatory bowel disease, a condition in which fecal granulocyte excretion has been extensively studied by Saverymuttu et al
(47,48).

Liver. The role of the liver in granulocyte margination is
difficult to evaluate because of the unknown contribution of cell
damage and/or activation to the hepatic activity following In-111-
granulocyte injection. Liver activity is greater when cells have been
extensively manipulated and tends to parallel the intensity of lung
sequestration. In our studies, deconvolution analysis and comparison
with In-111 red cells indicated an initial transit time of granulocytes through the liver of about 3 minutes, assuming all the activity
to be in transit.

Using our In-111-tropolone labeled granulocyte kinetic data, we
have attempted to quantitate the distribution of the MGP in man. If
the MGP and CGP are of equal size, an intrasplenic granulocyte transit
time(t) of 10 minutes and a splenic blood flow of 5% total blood
volume per minute (46) would indicate (because t = volume/flow) that
25% of the TBGP and therefore 50% of the MGP is in the spleen. If the
blood volume of the lungs is 10% of total blood volume (49), then a
granulocyte transit time through the lung of 2.5 that of red cells
would indicate that the lung MGP is 15% of the total MGP. Absolute
quantitation of the In-111 activity present in the liver (25%), spleen
(36%), and CGP (37%) about 1 hour after injection accounted for almost
100% of the injected dose, suggesting that most of the remaining 35%
of the MGP is located in the liver, although it is uncertain as to how
much of the liver activity is physiological.

The idea that in normal individuals granulocyte margination is not universal throughout all vessels (but confined, as we suggest, to the liver, spleen, and lung) is consistent with many of the direct observations made on living vessels. Thus Atherton and Born (50,51) and Gorog et al (52) have commented on the high "rolling" granulocyte counts seen in venules immediately following preparation for micro- scopy but which fall to low levels subsequently. Mayrovitz et al (53) developed a bat's wing model in order to observe granulocyte adhesive- ness to venular endothelium entirely free of preparation artifacts. They observed only the very rare granulocyte sticking to endothelium. Grant (54) has also suggested that the true incidence of granulocyte adherence may not be known as any preparation may lead to artifactual promotion of such adherence. It is of interest to compare these observations, which essentially indicate the sensitivity of endothe- lial "stickiness" to manipulations, with those described above which indicate sensitivity of granulocyte "stickiness" to manipulations.

In Vitro Models for Granulocyte Adhesiveness

Various systems have been developed to measure granulocyte adher- ence in vitro, which has then been extrapolated to margination in vivo, and correlated with the magnitude of MGP. Such systems include the nylon fiber column developed by MacGregor et al (55) and the glass bead column developed by Garvin (56), in both of which the effluent granulocyte count is compared with that in whole blood applied to the column. The relevance of in vitro adherence to margination in vivo is uncertain, but some interesting findings have emerged from these studies. Thus granulocyte adherence to nylon fiber is reduced in the presence of alcohol levels typical of those found in inebriated man (55), a finding that may be related to the known association of infec- tious complications with alcohol ingestion. Furthermore, 12 to 48 hours after alcohol consumption, granulocyte adhesiveness measured in whole blood from chronic alcoholics is increased, and this may be related to the temporary neutropenia seen in some alcoholics at the time of hospital admission, if this neutropenia is due to enlargement of the MGP (57). Recent elaborations of these techniques include measurement of granulocyte adherence to endothelial monolayers in cul- ture (58-60).

Divalent cations are essential for granulocyte adherence to nylon fiber and glass bead columns. It is of interest, therefore, that calcium chelating agents such as EDTA and sialic acid have been shown to result in inhibition of granulocyte adherence to directly observed microvessels in response to trauma and to result in granulocytosis within 30 minutes of administration to experimental animals (50-52). Such observations may have an interesting role to play in the study of mechanisms controlling granulocyte margination in vivo.

LIFE-SPAN

There is general agreement that, following equilibration between the CGP and MGP, intravascular granulocyte survival is exponential, indicating random granulocyte utilization and/or destruction. Areas of disagreement are 1) the half-time of survival and 2) the presence or otherwise of additional slower exponentials and their interpretation when present.

Disregarding the early rapid exponentials seen by some investigators, DFP-32 granulocyte survivals have been fairly consistently monoexponential with mean half-times ranging from 5.3 to 10.3 hours. Survivals based on Cr-51 labeling have been longer with mean half-times ranging from 7.4 hours (44) to 16.1 hours (6). Some workers have described Cr-51 survivals as monoexponential, whereas others have noted a departure from linearity (i.e., on semi-log replot) after about 6 hours. Dresch et al (6) pointed out that such "tails" may be due to labeling of mononuclears with Cr-51. The latter cells are not labeled to any significant extent by DFP-32. In their comparison between Cr-51- and DFP-32-labeled granulocytes, McMillan and Scott (5) found similar t1/2 values (10.5 and 10.3 days, respectively), whereas Dresch et al (6) obtained corresponding values of 16.1 and 5.4 hours Spivak and Perry (44) found them to be similar up to 6 hours, but the Cr-51 counts were relatively higher thereafter. DFP-32 probably elutes in vivo and may be cell damaging. Thus Dresch et al (6) compared the simultaneous survivals of DFP-32 and Cr-51 labeled cells after the same population had been 1) separately and 2) simultaneously labeled with the two isotopes. With simultaneous labeling the t1/2 values were 5.5 and 9.4 hours, respectively. The difference suggests elution, and the reduction of the Cr-51 t1/2 to 9.4 hours (from 16.1 hours) suggests damage.

The t1/2 of H-3-thymidine labeled granulocytes, not manipulated in vitro, was 7.6 hours, longer than the DFP-32 survival of 5.4 hours recorded in the same subjects (27). Weiblen et al (30) reported an intravascular t1/2 of In-111-granulocytes, isolated on double Ficoll-Hypaque gradients, as 5 hours. This was based on samples up to 6 hours after injection, although at 24 hours the counts were some five times higher than expected from the first 6 hour data points. We, too, find that the In-111-granulocyte survival is not monoexponential but has a tail from about 12 hours (Fig. 5.), although this is not invariably seen in the individual subject. Weiblen et al (30) suggested that this tail might be the result of labeling of reticulocytes in bone marrow following uptake by the latter of In-111-transferrin.

Granulocyte life-span in disease has received much less attention than platelet survival. In the acute stage of infection in dogs, DFP-32-granulocyte survival was reduced but became normal during the subacute (or established) phase and prolonged during convalescence (61). In three patients with chronic infection studied by McMillan

and Scott (5), Cr-51-granulocyte survival was similar to that in their
normal subjects. We also found no significant difference between
In-111-granulocyte survival in normal subjects and in patients with
positive white cell scans. These clinical data are consistent with
the canine data of Marsh et al (61) described above and seem to
suggest that in inflammation the rate constant governing granulocyte
destruction at sites encountered in the normal subject is "turned
down" so that the sum of the rate contants of losses in inflammatory
disease remains the same as in the normal. McCullough et al (62)
studied the effect of antileukocyte antibodies on the recovery and
survival of In-111-labeled "pure" granulocytes separated on double
Ficoll-Hypaque gradients. Granulocyte agglutinating, but not
granulocytotoxic or lymphocytotoxic, antibodies caused a reduced
recovery and t1/2.

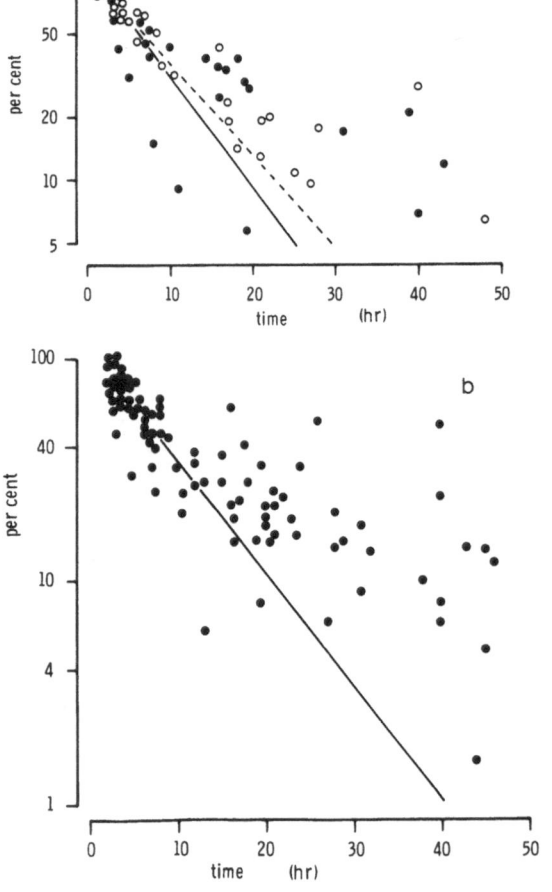

Fig. 5. In-111-granulo-
cyte survival curves. A,
Patients with negative
granulocyte scans; B,
patients with positive
scans for inflammatory patients with suspected
inflammatory disease
(closed circles and
continuous regression
line) and normal subjects
(open circles and discon-
tinuous regression line).
The latter comprised
normal volunteers and
patients suspected of
inflammatory bowel disease
but finally diagnosed as
having irritable bowel
syndrome and who were considered, therefore, to be
normal with respect to
granulocyte kinetics.

DESTRUCTION

The fate of granulocytes, like their life-span, is a rather obscure field.

Nonisotopic Data

There are very little data on the fate of unlabeled granulocytes, and the reader is referred to Murphy's review (63) for what little there are.

Isotopic Data

In order to obtain useful quantitative isotope data on sites of granulocyte loss, it is clearly essential to incorporate most of the dose of the radiolabel into granulocytes before injection. The isotope to be used, therefore, must have an efficiency of labeling sufficiently great to allow its deposition into the relatively small numbers of cells available in pure granulocyte suspensions.

Two approaches are available to study granulocyte kinetics with radioisotopes. One is to count activity in tissues isolated from the experimental animal at various times after the injection of labeled cells. Here, provided the blood cells that are labeled other than granulocytes are relatively long-lived in the circulation, fixed tissue activity can be assumed to be derived from granulocytes, and specific labeling of granulocytes is not absolutely essential. The other approach, which is to monitor granulocyte destruction in intact subjects by external scintillation counting and regional absolute quantitation of activity, does require specific labeling of granulocytes. Clearly, if the technique used to isolate granulocytes from blood prior to labeling results in their damage, then the quantitative data is unreliable and probably inferior to data based on mixed leukocyte labeling. From the foregoing discussion, therefore, it should come as no surprise to the reader that reliable data on sites of granulocyte destruction are very sparse.

There do not appear to be any experimental data on granulocyte destruction based on either DFP-32 or H-3-thymidine labeling. In the rabbit study of Winkelman et al (9), mixed leukocytes labeled with Cr-51 were principally deposited in liver and spleen, but abscess levels were less, suggesting poor granulocyte viability. The lung, liver, and spleen were the main organs within which Ga-67 was recorded by Burleson et al (12) 24 hour after injection of Ga-67-labeled leukocytes. Deposition of activity in these sites was even more prominent when peritoneal leukocytes were harvested and labeled. In cross-circulated dogs, the distribution at 12 hours (11 hours after cross-circulation) of leukocytes labeled with Tc-99m sulfur colloid was markedly different in the primary and secondary recipients (15). In the secondary recipient (assumed to more accurately reflect the

physiological situation), the spleen accumulated activity to the
greatest extent, followed by the liver. Bone marrow was not assessed.
In none of these studies could the precise contribution of
granulocytes to the tissue activity levels be evaluated.

Absolute quantitation of the distribution of In-111-leukocytes in
man was first attempted by Thakur et al (19) using In-111-labeled
mixed leukocytes and, in one patient, pure granulocytes separated on
Ficoll-Hypaque. The mixed leukocyte distribution at 24 hours is
difficult to evaluate because of contaminating labeled red cells, but
the splenic activity ranged from 8 to 19% and the liver activity 7 to
15% of the dose. With prior hypotonic lysis of red cells in mixed
suspensions, the liver activity was 37% and the splenic activity 13%
of the dose at 24 hours. Similar, relatively high, liver levels were
seen with the pure granulocytes, with 34% in the liver and only 6% in
the spleen.

Weiblen et al (30) studied the distribution of "pure" In-111-
granulocytes separated on double-density Ficoll-Hypaque gradients. In
one subject they calculated liver and spleen activity levels 24 hours
after injection as 12 and 19%, respectively. There was also visible
uptake in bone marrow but no activity readily apparent elsewhere.

It is difficult to evaluate these studies with regard to the fate
of granulocytes under physiological circumstances, since it is uncertain as to whether this predominantly reticuloendothelial distribution
of activity, at a time, 24 hours, at which the radiolabeled granulocytes have almost all completed their intravascular life-spans, represents physiological destruction, "artifactual" destruction of cells
damaged by labeling, or just uptake of non-cell-bound In-111. We have
attempted this evaluation by examining liver, spleen, and bone marrow
time activity curves over the duration of the granulocyte survival
time and beyond. Within each organ, changes occurred only during the
time interval in which In-111-granulocytes were circulating and
remained constant between 24 and 48 hours. Since plasma In-111 levels
tend to remain constant and furthermore persist after essentially all
labeled granulocytes have disappeared from the blood, it would appear
that organ activity levels at 24 hours reflect uptake of cell-bound
In-111 and not plasma In-111. Since bone marrow activity at 24 hours
is, if anything, greater in those studies showing less initial lung
sequestration and rapid and heavy uptake of In-111-granulocytes in
abscesses, we conclude that bone marrow uptake of granulocytes is
physiological. In studies showing early marked liver uptake, bone
marrow activity was minimal, suggesting that if cells are damaged,
they preferentially accumulate in the liver. Furthermore, in patients
with inflammatory bowel disease the final levels of activity in bone
marrow, liver, and spleen, as fractions of the initial levels, correlated inversely with the fecal In-111 excretion (Fig. 6.), further
suggesting the normal viability of cells undergoing ultimate reticuloendothelial destruction. It is instructive and interesting to compare

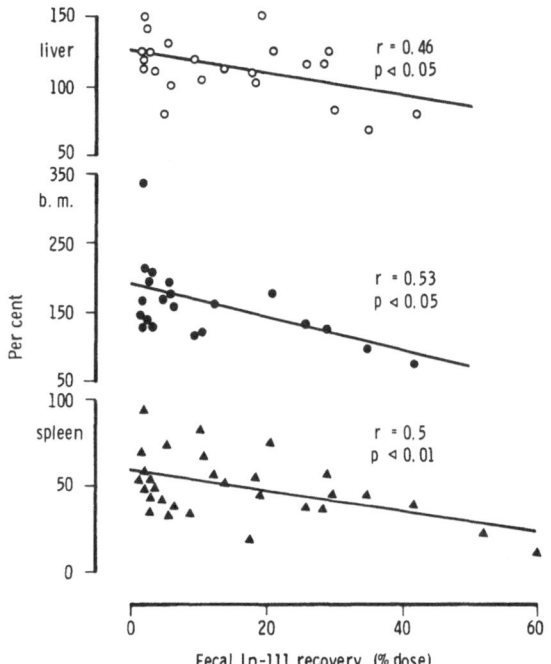

Fig. 6. Relationship between hepatic (top), bone marrow (middle), and splenic (bottom) In-111 levels at 24 hours after injection of 111-In-granulocytes and the fecal recovery of In-111 in patients with inflammatory bowel disease. Each value for the spleen is expressed as a percentage of the corresponding splenic value recorded at 40 minutes, or, if it was higher than the 40-minute value, the 3-hour value. Each value for the liver and marrow is expressed as a percentage of the corresponding value recorded at 3 hours.

granulocytes with platelets in terms of their normal destruction sites. Since the normal function of granulocytes takes them to extra-vascular locations, it is expected that their normal destruction sites are extravascular. The normal function of platelets is hemostasis, and it was previously thought that the vascular endothelium was the site for platelet loss. However, it is now known that even in conditions associated with thrombosis formation, the great majority of platelets are destroyed in the reticuloendothelial system, almost certainly including the bone marrow. It seems quite plausible, there-fore, that in the abscence of extensive inflammation, granulocytes are destroyed in bone marrow even though this is at random. The probability of the granulocyte being destroyed in the reticuloen-dothelial system is inversely related to the amount of inflamed tissue present in the body, and, similarly, for a platelet, inversely related to the extent of thrombosis.

ACKNOWLEDGMENT

The authors' work described in this review was supported by grants from the Cancer Research Campaign, Wellcome Trust and Nuffield Foundation. We are grateful to Mrs. Helen Reavy for expert technical assistance and to Amersham International for supplies of In-111-chloride. The manuscript was cheerfully typed by Miss Beverley Shambrook of Glaxo Group Research Ltd, Ware.

REFERENCES

1. J. W. Athens, A. M. Mauer, H. Ashenbrucker, G. E. Cartwright, M.
    M. Wintrobe, Leukokinetic studies. I. A method for labeling
    leukocytes with diisopropylfluorophosphate (DFP-32), Blood 14:
    303 (1959).
2. A. M. Mauer, J. W. Athens, H. Ashenbrucker, G. E. Cartwright, M.
    M. Wintrobe, Leukokinetic studies. II. A method for labeling
    granulocytes in-vitro with radioactive diisopropyl-
    fluorophosphate, J Clin Invest 39:1481 (1960).
3. J. W. Athnes, A. M. Mauer, H. Ashenbrucker, G. E. Cartwright, M.
    M. Wintrobe. Leukokinetic studies. III. The distribution of
    granulocytes in the blood of normal subjects, J Clin Invest
    40:159 (1961).
4. J. W. Athens, O. P. Haab, S. O. Raab, A. M. Mauer, H.
    Ashenbrucker, G. E. Cartwright, M. M. Wintrobe, Leukokinetic
    studies. IV. The total blood, circulating and marginal
    granulocyte pools and the granulocyte turnover rate in normal
    subjects, J Clin Invest 40:989 (1961).
5. R. McMillan and J. L. Scott, Leukocyte labeling with Chromium-51.
    I. Technique and results in normals, Blood 32:738.
6. C. Dresch, Y. Najean, J. Bauchet, Kinetic studies of Cr-51- and
    DF^{32}P-labeled granulocytes, Br J Haematol 29:67 (1975).
7. E. R. Clark, E. L. Clark, Observations on changes in blood
    vascular endothelium in the living animal, Am J Anat 57:385
    (1935).
8. G. Vejlens, The distribution of leukocytes in the vascular
    system, Acta Pathol Microbiol Scand (Suppl 33) (1938).
9. J. Winkelman, C. J. Collica, S. G. Sandler, The delineation of
    abscesses by scintiphotography using Cr-51-labeled leukocytes.
    Am J Roentgenol Radium Ther Nucl Med 103:881 (1968).
10. M. Deysine, R. G. Robinson, J. R. Wilder, Abscess detection by
    radioactive chromium-labeled autologous white cells, Surg
    Gynecol Obstet 131:216 (1970).
11. W. C. Harvey and J. Silva, Detection of abscesses with Cr-51-
    labeled leukocytes, J Nucl Med 15:375 (1974).
12. R. L. Burleson, M. C. Johnson, H. Head, Scintigraphic
    demonstration of experimental abscesses with intravenous Ga-67
    citrate and Ga-67-labeled blood leukocytes, Ann Surg 178:446
    (1973).
13. R. L. Burleson, B. L. Holman, D. E. Tow, Scintigraphic
    demonstration of abscesses with radioactive gallium-labeled
    leukocytes, Surg Gynecol Obstet 141:179 (1975).
14. T. Uchida, S. Kariyone, Organ distribution of the Tc-99m-labeled
    white cells, Acta Haematol Jap 36:78 (1973).
15. D. English and B. R. Andersen, Organ distribution of canine
    leukocytes labeled with Tc-99m-sulfur colloid, J Nucl Med
    18:289 (1977).
16. H. J. Schroth, K. P. Muller, R. Berberich, E. Oberhausen, H.
    Wilhelm, The application of radioactive labeled leukocytes for
    the proof of inflammations, Eur J Nucl Med 4:359 (1979).

17. N. Colas-Linhart, M. Barbu, M. A. Gougerot, B. Bok, Five
     leukocyte labeling techniques: a comparative in vitro study,
     Br J Haematol 53:31 (1983).
18. J. G. McAfee and M. L. Thakur, Survey of radioactive agents for
     in vitro labeling of phagocytic leukocytes. I. Soluble
     agents, J Nucl Med 17:480 (1976).
19. M. L. Thakur, J. P. Lavender, R. N. Arnot, D. J. Silvester, A. W.
     Segal, Indium-111-labeled autologous leukocytes in man. J
     Nucl Med 18:1014 (1977).
20. M. L. Thakur, R. E. Coleman, M. J. Welch: Indium-111-labeled
     leukocytes for the localization of abscesses: preparation,
     analysis, tissue distribution and comparison with Gallium-67-
     citrate in dogs, J Lab Clin Med 89:217 (1977).
21. A. M. Peters, S. Karimjee, S.H. Saverymuttu, J. P. Lavender, A
     comparison of Indium-111-oxine and Indium-111-acetylacetone-
     labeled leukocytes in the diagnosis of inflammatory disease,
     Br J Radiol 55:827 (1982).
22. H. J. Danpure, S. Osman, F. Brady, The labeling of blood cells in
     plasma with In-111-tropolonate, Br J Radiol 55:247 (1982).
23. S. H. Saverymuttu, A. M. Peters, H. J. Danpure, H. J. Reavy,
     S. Osman, J. P. Lavender, Lung transit of In-111-labeled
     granulocytes: relationship to labeling techniques, Scand J
     Haematol 30:151 (1983).
24. A. M. Peters, S. H. Saverymuttu, H. J. Reavy, H. J. Danpure, S.
     Osman, J. P. Lavender, Imaging inflammation with Indium-111-
     tropolonate-labeled leukocytes, J Nucl Med 24:39 (1983).
25. S. O. Raab, J. W. Athens, O. P. Haab, D. R. Boggs, H.
     Ashenbrucker, G. E. Cartwright, M. M. Wintrobe, Granulo-
     kinetics in normal dogs, Am J Physiol 206:83 (1964).
26. P. C. Vincent, Granulocyte kinetics in health and disease, Clin
     Haematol 6:695 (1977).
27. J. T. Dancey, K. A. Deubelbeiss, L. A. Harker, C. A. Finch,
     Neutrophil kinetics in man, J Clin Invest 58:705 (1976).
28. A. S. Deinard and A. R. Page, An improved method for performing
     neutrophil survival studies, Blood 36:98 (1970).
29. P. C. Vincent, A. D. Chanana, E. P. Cronkite, D. D. Joel, The
     intravascular survival of neutrophils labeled in-vivo, Blood
     43:371 (1974).
30. B. J. Weiblen, L. Forstrom, J. McCullough, Studies of the
     kinetics of Indium-111-labeled granulocytes, J Lab Clin Med
     94:246 (1979).
31. J. G. McAfee, G. M. Gagne, G. Subramanian, Z. D. Grossman, F. D.
     Thomas, M. L. Roskopf, M. L. Fernandes, B. J. Lyons,
     Distribution of leukocytes labeled with In-111-oxine in dogs
     with acute inflammatory lesions, J Nucl Med 21:1059 (1980).
32. H. R. Bierman, K. H. Kelly, F. L. Cordes, The release of
     leukocytes and platelets from the pulmonary circulation by
     epinephrine, Blood 7:683 (1952).
33. H. R. Bierman, R. L. Byron, K. H. Kelly, N. L. Petraki, The lung
     removal mechanism for leukocytes, Blood 6:770 (1951).

34. B. A. Martin, J. L. Wright, H. Thomasen, J. C. Hogg, Effect of pulmonary blood flow on the exchange between the circulating and marginating pools of polymorphonuclear leukocytes in the dog lungs, J Clin Invest 69:1277 (1982).

35. A. S. Weisberger, R. W. Heinle, J. P. Storaasli, R. Hannah, Transfusion of leukocytes labeled with radioactive phosphorus, J Clin Invest 29:336 (1950).

36. S. H. Saverymuttu, A. M. Peters, H. J. Reavy, J. P. Lavender, Measurement of the migration and accumulation of granulocytes in inflammation in man, Clin Exp Immunol 52:607 (1983).

37. A. C. Issekutz and H. Z. Movat, The in vivo quantitation and kinetics of rabbit neutrophil leukocyte accumulation in the skin in response to chemotactic agents and Escherichia Coli, Lab Invest 42:310 (1980).

38. T. A. Lane, P. W. Bergum, J. P. Lichter, R. G. Spragg, The labeling of rabbit neutrophils with In-111-oxine, J Immunol Methods 51:293 (1982).

39. S. L. Poston, R. B. Jones, P. J. Hilton, Sodium transport in polymorphonuclear leukocytes: effect of isolation by the Ficoll/Triosil method, Clin Sci 62:563 (1982).

40. R. Uchida and S. Kariyone, Intravascular granulocyte kinetics and spleen size in patients with neutropenia and chronic splenomegaly, J Lab Clin Med 82:9 (1973).

41. C. R. Bishop, G. Rothstein, H. E. Ashenbrucker, J. W. Athens, Leukokinetic studies. XIV. Blood neutrophil kinetics in chronic steady state neutropenia, J Clin Invest 50:1678.

42. L. H. Brubaker and C. A. Johnson, Correlation of splenomegaly and abnormal neutrophil pooling (margination), J Lab Clin Med 92:508 (1978).

43. H. R. Bierman, R. L. Byron, K. H. Kelly, The role of the spleen in the leukocytosis following the intra-arterial administration of epinephrine, Blood 8:153.

44. J. L. Spivak and S. Perry, Evaluation of Cr-51 as a leukocyte label, Br J Haematol 25:321 (1973).

45. D. L. Williams, Improvement in quantitative data analysis by numerical deconvolution techniques, J Nucl Med 20:568 (1979).

46. A. M. Peters and J. P. Lavender, Factors controlling the intrasplenic transit of platelets, Eur J Clin Invest 12:191 (1982).

47. S. H. Saverymuttu, A. M. Peters, J. P. Lavender, H. J. Hodgson, V. S. Chadwick, Indium-111-autologous leukocytes in inflammatory bowel disease, Gut 24:293 (1983).

48. S. H. Saverymuttu, A. M. Peters, H. J. Hodgson, V. S. Chadwick, J. P. Lavender, Indium-111 leukocyte scanning in small-bowel Crohn's disease, Gastrointest Radiol 8:157 (1983).

49. P. Harris and D. Heath, The measurement of blood volume in the lungs, in: "The Human Pulmonary Circulation," Churchill Livingstone, (1977).

50. A. Atherton and G. V. R. Born, Quantitative investigations of the adhesiveness of circulating polymorphonuclear leukocytes to blood vessels walls, J Physiol 222:447 (1972).

51. A. Atherton and G. V. R. Born, Relationship between the velocity
     of rolling granulocytes and that of the blood flow in venules,
     J Physiol 233:157 (1973).
52. P. Gorog, B. Kovacs, G. V. R. Born, Suppression of the
     intravascular adherence of granulocytes by N-acetyl neuraminic
     (sialic), acid Br J Exp Pathol 61:490 (1980).
53. H. N. Mayrovitz, M. P. Weideman, R. F. Tuma, Factors influencing
     leukocyte adherence in microvessels. Thromb Haemost 38:823
     (1977).
54. L. Grant, The sticking and emigration of white blood cells in
     inflammation, in: "The Inflammatory Process," B.W. Zweifach,
     L. Grant, R. T. McCluskey, eds., Academic Press, New York
     (1973).
55. R. R. MacGregor, P. J. Spagnuolo, A. L. Lentnek, Inhibition of
     granulocyte adherence by ethanol, prednisone and aspirin
     measured with an assay system, N Eng J Med 291:642 (1974).
56. J. E. Garvin, Factors affecting the adhesiveness of human
     leukocytes and platelets in vitro, J Exp Med 114:51 (1961).
57. K. J. Wozniak and E. M. Silverman, Granulocyte adherence in
     chronic alcoholism, Am J Clin Pathol 71:269 (1979).
58. R. R. MacGregor, E. J. Macarak, N. A. Kefalides, Comparative
     adherence of granulocytes to endothelial monolayers, J Clin
     Invest 61:697 (1978).
59. J. D. Pearson, J. S. Carleton, J. E. Beesley, A. Hutchings, J. L.
     Gordon: Granulocyte adhesion to endothelium in culture, J Cell
     Sci 38:225 (1979).
60. H. S. Jacob, P. R. Craddock, D. E. Hammerschmidt, C. F. Moldow,
     Complement-induced granulocyte aggregation: an unsuspected
     mechanism of disease, N Eng J Med 302:789 (1980).
61. J. C. Marsh, D. R. Boggs, G. E. Cartwright, M. M. Wintrobe,
     Neutrophil kinetics in acute infection, J Clin Invest 46:1943
     (1967).
62. J. McCullough, B. J. Weiblen, M. E. Clay, L. Forstrom, Effect of
     leukocyte antibodies on the fate in vivo of Indium-111-labeled
     granulocytes, Blood 58:164 (1981).
63. P. Murphy, "The Neutrophil," Plenum Medical Book Co., London and
     New York (1976).

IMAGING INFECTION WITH INDIUM-111-LABELED LEUKOCYTES

Richard C. Reba and Paul L. Chandeysson*

Division of Nuclear Medicine
George Washington University Medical Center
Washington, D.C., USA 20037

The use of radiolabeled leukocytes for the detection and the localization of infections in patients depends on the inherent physiologic function of these cells to migrate to sites of inflammation. Techniques for radiolabeling leukocytes without destroying their biologic properties have been under development for almost three decades. This development has resulted in the Indium-111 leukocyte scan which has now been widely examined as a promising means for imaging focal infection.

The study reported by Grob et-al in 1947 suggested that diisopropylfluorophosphate (DFP-32) could be used to label erythrocytes (1). In 1956, Leeksma and Cohen (2) demonstrated that leukocytes contained significant uptake of radioactivity 6 hours after injection of phosphorus-32-labeled DFP, but the detection method was insensitive and they were unable to continue counting after 24 hours. The first successful method describing DFP-32 for leukokinetic studies was reported by Athens et al (3). Diisopropylfluorophosphate-32-labeled leukocytes are not suited to localization of infection by external scanning because phosphorus-32 has no gamma ray or photon emission. In 1955, McCall et al (4) described the labeling of human leukocytes with the gamma emitter Chromium-51 (Cr-51). In 1968, Winkelman et al (5) reported the use of Cr-51-labeled leukocytes for localization of abscesses induced in rabbits.

Because of its long half-life (27.3 days), relatively high gamma ray energy (320 keV), and low probability of emission (7%), Cr-51 is not a good tracer for external scanning. In 1973, Burleson and associates (6) reported the use of Ga-67-labeled leukocytes in patients with suspected abscesses. The labeling efficiency was low, with only 26 percent of the gallium activity bound to the cells. Gallium was visualized in liver, bone, and normal colon contents, a biodistribution pattern of radioactivity similar to that observed with Ga-67-citrate.

In 1975, Anderson et al (7) described the use of leukocytes labeled with technetium-99m (Tc-99m) sulfur colloid to image infections induced in dogs. The labeling efficiency was low, not exceeding 40 percent, and significant nonspecific uptake in the reticuloendothelial system occurred. The maximum abscess-to-blood ratio reported was 4.4.

In 1976, McAfee and Thakur (8) surveyed the compounds available for labeling leukocytes and selected Indium-111-8-hydroxyquinoline (oxine) as potentially the best compound for labeling leukocytes to image infection. The gamma emissions from Indium-111 (In-111) (171 keV, 89% and 245 keV, 94%) are favorable for detection with a gamma camera, and the half-life of 67 hours is sufficient to allow the leukocytes to accumulate at the site of infection and short enough to minimize the radiation dose to the patient. When chelated with oxine, In-111 diffuses through cell membranes and apparently becomes firmly bound to components of the cytoplasm and nucleus without disrupting the function of the leukocytes.

In 1977, Thakur et al reported the comparison of In-111-oxine-labeled leukocytes with Ga-67-citrate for abscess imaging in dogs (9). The range of WBC labeling efficiency was found to be 75 percent to 95 percent, and abscess-to-blood ratios as high as 117 (range 35 to 117) were obtained 24 hours after injection, while Ga-67 ratios were 1.2 to 8. Since then, several clinical and animal studies using In-111-oxine-labeled leukocytes have appeared in the literature.

Many variations of Thakur's technique have been used with apparent success, but all the methods that use In-oxine label the leukocytes in saline because of the higher binding affinity of Indium for plasma transferrin than for oxine. Typically, 40 milliliters of heparinized blood is allowed to settle for an hour, using a high molecular weight settling agent such as methyl cellulose. The leukocyte-rich plasma is centrifuged to separate the leukocytes, which are then washed and suspended in saline.

The In-111-oxine is prepared from In-111-chloride by complexing with oxine. Then it is extracted with chloroform, evaporated to dryness, and dissolved in an ethanol solution. The In-111-oxine is now available from several commercial sources. This solution is added slowly to the leukocyte suspension, and the cells are incubated at room temperature for about 20 minutes. The cells then are separated from the In-111-oxine by centrifugation, washed, resuspended in saline, and injected into the patient intravenously. The cell-labeling process requires approximately 2 hours.

Patients are typically scanned at 16 to 24 hours after injection of the leukocytes. The normal blood clearance half-time for leukocyte-bound radioactivity is approximately 5 to 7 hours. Therefore, blood pool activity on the later images suggests labeling

of erythrocytes, platelets, or plasma proteins because these blood
elements have longer clearance times than leukocytes.

At the time that clinical scans are recorded, leukocyte radioactivity is normally seen in the liver, spleen, and bone marrow. The
mechanism for the bone marrow uptake is unknown. It is possibly
related to uptake of normal or damaged leukocytes by the bone
marrow or possibly by the localization of Indium that has become bound
to transferrin. Lung activity is seen immediately after injection,
but this normally clears by 4 hours. Markedly increased lung activity
occurs if the cells are damaged, e.g., excess heat or agitation, and
then lung activity persists for 24 hours. In patients with suspected
occult or pulmonary infections, early scanning at 1 to 2 hours after
injection will display the physiologic pulmonary distribution and a
repeat scan after 4 hours should indicate those areas in which leukocytes will translocate in a normal fashion. Pulmonary uptake has been
reported to correlate in a number of patients known to have adult
respiratory distress syndrome (ARDS), pneumonitis, empyema, and lung
abscess (10-14). Cook and coworkers (15) found 48 examples of
pulmonary uptake at 24 hours in a retrospective review of 306 In-111
leukocyte studies in 232 patients with suspected occult infection.
The pattern of distribution, i.e., focal or diffuse lung uptake, was
not reliable as a predictor of either an infectious or noninfectious
process, and there was a significant incidence of false-positive
pulmonary localization, 7/48, in which no cause could be determined.

Abnormal leukocyte localization can be seen on a scan obtained
several hours after injection, but it is more likely to be seen on the
later images because the abscess to blood background ratio progres-
ively increases during the 24 hours after injection as the blood
radioactivity decreases. Abnormal leukocyte localization has been
consistently detected in abscesses as early as 30 minutes after the
injection of In-111-labeled donor leukocytes into leukopenic leukemia
patients (16). Although early visualization of abscesses in nonleukopenic patients using autologous leukocytes has been reported, "small"
collections generally are detected only on images obtained later,
i.e., 16 to 24 hours (18).

Several clinical studies of the effectiveness of In-111 leukocyte
scanning in the detection and localization of abscesses have now been
published (16-25). The results of these studies are summarized in
Table 1. In general, reports that contained less than 20 studies were
not included except for those of Deutcher and coworkers (16) and Alavi
et al (17) describing the use of homologous donor granulocytes in
patients with severe neutropenia. The type of abscess, such as wound
abscess or hepatic abscess, has been tabulated separately when the
study specified the type of abscess. All patients studied for
possible abscess have been included in the "All" column, regardless of
whether the location was specified or not. An attempt has been made
to avoid counting the same result more than once when the study was

Table 1. Summary of Indium-111 WBC Scan for Suspected Abscess

Study (Ref.)	Scan	Abdomen and Pelvis	Wound Infection	Head and Neck	Thorax	Other or Unspecified	All
Deutcher '81 (16)	Scan + (T.P. / F.P.)	5	1	11		8	25
Alavi '80 (17)	Scan − (F.N. / T.N.)	1			2	1	4
Coleman '80 (18)	Scan +	22				7	29
	Scan −	91				28	119
Forstrom '80 (19)	Scan +	63	6	2	30	11	112
	Scan −	19				182	201
Goodwin '80 (20)	Scan +	19	2	3	1	116	141
	Scan −	9	5	2	24	118	158
Knochel '80 (21)	Scan +	30				5	30
	Scan −	96					96
Carroll '81 (22)	Scan +	27				6	27
	Scan −	125				5	125
Rovekamp '81 (23)	Scan +	14				8	22
	Scan −	14				8	22
Peters '82 (24)	Scan +	18		1	3	27	49
	Scan −	16			3	3	22
Seabold '84 (25)	Scan +	28		3		23	51
	Scan −	63				5	63
Totals	Scan +	226		20		46	486
	Scan −	434		23		56	810

reported more than once in the literature. Nevertheless, the totals
given are only approximate. For example, the data published in 1979
by Ascher et al (10) and by McCullough et al (26) were presumed to
have been included in the more comprehensive report by Forstrom and
coworkers (19); that by McDougall et al (12) and Bicknell et al (27)
included by Goodwin et al (20); and the report by Coleman et al (18)
included data reported about the same time elsewhere (11). However,
the data of Carroll et al (22) were considered to be significantly
different from those reported by the same group in another publication, so they are listed separately.

The overall sensitivity for detecting abscess by In-111 leukocyte
scanning in the studies listed in Table 1 is 90 percent. The sensitivity and specificity for detection of an abscess in the abdomen or
pelvis are calculated separately. These data agree with the report by
Forstrom et al who summarized the collaborative work of the
Minneapolis group. These workers now have records of 1178 patients
which have been retrospectively analyzed. In 144 patients with
abdominal or pelvic abscess and in 65 patients with lung abscess the
sensitivity has been 88 percent. Specificity was 88 percent in the
patients with abdominal abscess. It is of interest that in 92
patients in this review with "an ill-defined infectious process," the
sensitivity was only 50 percent but specificity remained high -- 96
percent. Although we know little about the prevalence of abscess in
the patient populations reported, we would expect that as patients
are studied with vague signs and symptoms of abscess and with
low probability of having an abscess, despite the high sensitivity and
high specificity of the In-WBC scan, the posterior probability, or
predictive value, of a positive WBC test will not be high (28). In
the report by Seabold et al (25), 128 patients with suspected abscess
in the abdomen or pelvis were scanned, and the sensitivity (88%) and
specificity (90%) were similar to that reported by others. Retrospective analysis revealed that 16 percent of the total group had
unexpected sites of infection outside of the abdomen. Furthermore, 20
of the 58 true positive scans (34%) produced a significant change in
patient management as a result of the In-111-WBC study.

There are several causes of false-negative In-111 leukocyte
scans. One cause is infections which do not evoke a neutrophil

Intrinsic Power of In-111 WBC Scan for Infection

	DISEASE	NO DISEASE
Scan +	486	46
Scan -	56	810

Sensitivity = 486 ÷ 542 = .90
Specificity = 810 ÷ 856 = .95

Sensitivity and Specificity for Abscess Detection in the Abdomen and
Pelvis

	ABSCESS	NO ABSCESS
Scan +	226	23
Scan −	20	434

Sensitivity = .92
Specificity = .95

response, such as a tuberculous abscess, because it is primarily the
polymorphonuclear neutrophil that concentrates in an acute septic
abscess. The use of In-111-oxine-labeled eosinophils has been found
more effective than labeled neutrophils in mice for the localization
of abscesses caused by parasites (29). If techniques of harvesting
eosinophils can be developed, it may be possible to improve the sensitivity of In-111-leukocyte scanning in parasitic disease.

Abscesses in or around the liver or spleen are more difficult to
detect because of the high background activity in these organs. The
use of a Tc-99m-sulfur colloid scan and computer subtraction techniques in conjunction with the In-111 leukocyte scan is believed to
improve the detection of abscesses in these areas (30, 31).

The specific problem of WBC antibody interference has been
addressed. In the reports by Deutcher et al (16), In-111-WBC scans
using homologous random donor cells were positive for a focal
infection in all 20 leukopenic patients who had no evidence of
alloimmunization; however, only 3 of 12 scans were positive in those
patients with an abscess who were alloimmunized. McCullough et al
(26) believe that the presence of granulocytotoxic and lymphotoxic
antibodies did not interfere significantly with in vivo WBC functions
but that granulocyte agglutinating (GA) antibodies are a problem.
Several patients in their report with GA antibodies had decreased WBC
recovery, decreased survival, increased localization in the liver, and
false-negative In-111-WBC scans for abscess localization. The
significance of the presence of antibodies to antibiotics producing
neutropenia and altered leukocyte function (32) is unknown with regard
to the observation that false-negative scans occur in patients
receiving antibiotics (21).

Causes of altered leukocyte migration ability, the apparent
mechanism whereby white cells accumulate at sites of infection, and,
therefore, possible circumstances in which false-negative In-111-WBC
scans may occur have been discussed in detail (33, 34). Chronic
infection other than osteomyelitis has been listed frequently as a
cause for false-negative WBC scans (10,35). Poor blood supply and a
thick membrane are mentioned as responsible mechanisms, but in the
many conditions associated with false-negative WBC scan reports --
e.g., chronic infection, hyperalimentation, steroid administration,

hemodialysis, and immunosuppression -- a fundamental and common denominator may simply be malnutrition or lack of some critical subcellular nutrient.

The overall specificity of In-111 leukocyte scanning for abscess of 95 percent compares favorably with that of computed tomography (CT) and ultrasound (US) (21, 22, 25, 35). It is superior to that of gallium scanning because of the lack of nonspecific uptake in normal bowel contents. Indium leukocyte bowel activity is not normally seen, but it can occur with inflammatory bowel or ischemic disease, GI bleeding, multiple enemas, or other causes of leukocytes in the bowel such as sinusitis or inflammatory lesions of the mouth or esophagus draining leukocytes into the bowel (37). Such conditions may result in a false-positive interpretation. However, these circumstances produce diffuse, rather than focal, bowel activity, and clinical signs and symptoms associated with an abdominal infection are absent or minimal. Therefore, Indium activity in the bowel should not be confused with abscess localization. Indeed, the diffuse In-111-WBC accumulation by inflamed mucosa in patients with inflammatory bowel disease is so reliable (39), several groups have proposed In-111 quantitation as a dependable noninvasive index of disease activity and response to treatment (40, 41). An inflammatory reaction about an area of infarction -- that is, necrosis without infection, such as focal accumulation in an area of bowel infarction (38) or pancreatitis (12, 23) -- is another matter. Several publications have reported that an accessory spleen has been mistaken for an abscess in the left upper quadrant of the abdomen.

Uptake in noninfectious inflammatory processes such as rheumatoid arthritis (10), myocardial infarction (42), and malignant tumor (12, 23, 25, 31, 43) have been reported. However, In-111-WBC uptake in primary or metastatic malignancies is uncommon, unlike that seen with Ga-67-citrate scanning.

The In-111 leukocyte scan has not been particularly helpful in detecting osteomyelitis and joint infection. Coleman et al (18) have reported the results of 45 In-111 leukocyte scans in patients with suspected bone or joint infections. Of 14 patients with proven infection, 7 had positive scans, a sensitivity of only 50 percent. In five patients with bone or joint infections who had In-111 leukocyte and Ga-67 citrate scans, only two had abnormal In-111 leukocyte scans while all had positive Gallium scans. McDougall et al (12) reported four cases of acute bone or joint infection detected by In-111 leukocyte scans, but two patients with chronic osteomyelitis had normal scans.

Recently, Raplopoulos et al (44) have reported that In-111 leukocyte scans become positive more quickly than Tc-99m MDP scans in acute osteomyelitis induced in rabbits by injecting bacteria. If confirmed clinically, this may enable detection of osteomyelitis early enough to allow treatment to be started before bone destruction has occurred.

Other applications of In-111 leukocyte scanning have been reported. McKeown et al (45) reported nine patients studied for possible infected aortofemoral grafts. There were three true-negative studies and five true positive studies. The ninth patient, who had an abnormal scan, was found at surgery to have an abscess in a colon diverticulum rather than an infected graft. McDougall et al (12) reported 13 patients studied for fever of unknown origin with no strong clinical evidence of abscess. There were nine true-negative scans. The scans were positive in two patients with pneumonia, one with Crohn's disease, and one with a probable postsurgical abscess. Ascher et al (10) reported true-positive scans in peritonitis, pneumonia, and jugular vein phlebitis and a false-negative scan in an infected aortofemoral graft. The technique is also believed useful to diagnose peritoneal dialysis catheter tunnel infections (46). It has been suggested that the WBC scan may be helpful to differentiate cerebral abscess from other intracranial lesions even when the CT diagnosis is uncertain. Peters et al (47) report a patient with a positive WBC study for cerebral abscess in whom there was peripheral ring enhancement on the CT following IV contrast.

The overall sensitivity and specificity of In-111-WBC scanning for abscess localization appears to be similar to that of US and CT. In some publications, CT is reported to be superior to the WBC scan or US, and others indicate the WBC scan is better. For example, Knochel et al (21) report a CT sensitivity of 98%, but US, 82%, and WBC scan, 86%, were not quite as good. Yet Seabold and coworkers' (25) publication records a sensitivity of 68% for US, 79% for CT, and 88% for the WBC scan. In most reports, specificity is high -- greater than 90%.

Since the sensitivity and specificity of the three imaging techniques are approximately the same, the clinical circumstances best dictate the sequence in which the tests should be considered. A strategy for the selection of the most appropriate imaging test for patients suspected of having an abscess has been proposed by Knochel et al (21). These workers suggest that in patients who are not critically ill and have no focal signs or symptoms, In-111 leukocyte scanning should be performed first. A negative study is sufficient basis to observe the patient without further imaging. A positive study should be followed with US or CT to confirm the presence of an abscess. In patients who are more ill but have no localizing findings, In-111 leukocyte scanning is also recommended first; however, a negative study should then be followed by US or CT because the clinical suspicion for abscess is high. US or CT may also be recommended after a positive scan to identify the three-dimensional extent of the abscess which is useful in the surgical management of the abscess. In critically ill patients, CT or US is recommended first because each is accomplished more rapidly than the In-111

leukocyte scan. If the CT or US is positive and typical of abscess,
the scan is not required; however, if the study is negative or
equivocal, the In-111 leukocyte scan is recommended. We have found
this scheme useful.

Several generalizations are apparent from the results of studies
that have attempted to compare the efficiency of CT, gray-scale US,
and radionuclide imaging (RN) for the detection of malignant tumor or
abscess. (A) Radionuclide imaging is the least operator dependent and
is, therefore, the simplest procedure to perform effectively, whereas
US is the most difficult. (B) The percentage of patients who cannot
be examined because of appliances, i.e., drains, surgical dressings,
internal gas, obesity, or motion, is highest for US, ranging from 5%
to 30% for hospitalized patients. The number of patients not able to
be examined by RN is virtually zero and by fast (2 to 4 sec) CT is
only a few percent. (C) Optimal results from CT of the liver are
obtained using unenhanced and enhanced techniques in each patient.
Without contrast medium some of the more subtle lesions will be
missed. (D) The spatial resolution of RN is poorest, especially for
deep hepatic lesions. In areas accessible to the US transmission, the
best US devices are almost competitive with CT; however, the fast (2
sec), fine matrix (512 x 512) CT scanners remain superior to US in
spatial resolution. (E) RN has improved over the past five years but
has made the least dramatic technological advances. The promise of
single emission tomography is great but as yet is unproven. The limit
of technological advance possible in CT and US hardware is near. (F)
Because diagnostic US cannot penetrate bone or air-filled lungs, the
posterior-superior aspect of the liver constitutes a sonographic
"blind spot" in patients who cannot (or will not) inspire deeply or
who have an abnormally elevated right hemidiaphragm. No similar
"blind spots" exist for RN or CT. (G) The cost of CT is generally
highest, and the cost of US is generally lowest. RN is closer to US
in this regard. (H) In virtually all institutions, the waiting period
for US and RN is minimal compared with the waiting period for CT,
which generally varies from 3 days to 3 weeks. (I) RN is sensitive
but usually not specific, in characterizing a lesion once identified;
US is far more effective for lesion characterization, and CT is
usually even more effective than US.

The liver, spleen, and bone marrow are the critical organs for
radiation dose in In-111 leukocyte scans. Thakur et al (48) report
that the radiation dose was 6 to 18 rads to the spleen and 1 to 5 rads
to the liver per millicurie In-111 administered. Goodwin et al (20)
estimated 5 rads to the spleen and 1.2 rads to the liver per mCi.
Although the recommended administered dose of 300 to 500 microcuries
reduces these stated body burdens to those comparable with clinical
doses received from Ga-67, repeated In-111-WBC studies in children
should be considered with caution or could be carried out with not
greater than 100 μCi administered dose.

The usual technique of preparing leukocytes for labeling results in a great concentration of the polymorphonuclear leukocytes. However, some lymphocytes are present, and since the In-111-oxine technique is nonspecific, all cells present in the mixture, including the lymphocytes, will be radiolabeled. Since lymphocytes are more sensitive to the effects of radiation, and T-lymphocytes have a long biological life with the ability to reproduce, the theoretical possibility for malignant transformation exists. Indeed, ten Berge et al (49) have reported extensive chromosome damage of lymphocytes following In-111-oxine labeling. The damage was equivalent to that following 200 rads from x-rays.

It has been recommended for some time to expose all blood components to 5,000 rads before administration to immunosupressed patients, i.e., those with severe combined immunodeficiency disease (50) or to infants who have previously received intrauterine transfusions (51). The purpose of this irradiation is to eliminate the ability of contaminating lymphocytes to divide and, thereby, to reduce the likelihood of inducing graft versus host disease. Such a maneuver should probably be performed before injection of In-111-labeled WBC.

The acceptance of In-111 leukocyte scanning as a routine diagnostic tool in patients suspected of having an abscess will not depend on its effectiveness in detecting abscess. Its efficacy has been demonstrated in several clinical studies. The future of In-111 leukocyte scanning will depend on the acceptance of a radiopharmaceutical preparation procedure which is far more complicted than those currently used in clinical nuclear medicine. Acceptance may also depend on additional experience to provide confidence that the technique is free of the long-term complication of induced neoplasia.

---

*Present Address: Department of Nuclear Medicine, 110 Irving Street, N.W., Washington Hospital Center, Washington, D. C.   20010

REFERENCES

1. D. Grob, J. L. Lilienthal Jr., A. M. Harvey, B. F. Jones, The administration of diisopropylfluorophosphate (DFP) to man. I. Effect on plasma and erythrocyte cholinesterase; general systemic effects; use in study of hepatic function and erythropoiesis; and some properties of plasma cholinesterase, Johns Hopkins Med J 81:217, (1947).
2. C. H. W. Leeksma, J. A. Cohen, Determination of the life span of human blood platelets using labeled diisopropylfluorophosphate, J Clin Invest 35:964, (1956).
3. J. W. Athens, A. M. Mauer, H. Ashenbeuker, G. E. Cartwright, M. M. Wintrobe, Leukokinetic studies. I. A method for labeling leukocytes with diisopropylfluorophosphate (DFP-^{32}), Blood 14:303, (1959).

4. M. S. McCall, D. A. Sutherland, A. M. Eisentraut, H. Lanz, The
    tagging of leukemic leukocytes with radioactive chromium and
    measurement of the in vivo cell survival,   J Lab Clin Med
    45:717 (1955).
5. J. Winkelman, C. J. Collica, S. G. Sandler, The delineation of
    abscessess by scintiphotography using Cr-51-labeled
    leukocytes, Am J Roentgenol 103:881 (1968).
6. R. L. Burleson, B. L. Holman, D. E. Tow, Scintigraphic
    demonstration of abscesses with radioactive Gallium-labeled
    leukocytes, Surg Gynecol Obstet 141;379 (1975).
7. B. R. Anderson, D. English, H. E. Akalin, Inflammatory lesions
    localized with Technetium- (Tc-99m) labeled leukocytes, Arch
    Int Med 135:1067 (1975).
8. J. G. McAfee, M. L. Thakur, Survey of radioactive agents for in
    vitro labeling of phagocytic leukocytes. I. Soluble agents, J
    Nucl Med 17:480 (1976).
9. M. L. Thakur, R. E. Coleman, M. J. Welch, Indium-111-labeled
    leukocytes for the localization of abscesses: Preparation,
    analysis, tissue distribution and comparison with Gallium-67
    citrate in dogs, J Lab Clin Med 89:217 (1977).
10. N. L. Ascher, D. H. Ahrenholz, R. L. Simmons, B. Weiblen, L.
    Gomez, L. A. Forstrom, M. P. Frick, C. Henke, J. McCullough,
    Indium-111 autologous tagged leukocytes in the diagnosis of
    intraperitoneal sepsis, Arch Surg 114:386 (1979).
11. R. E. Coleman, R. E. Black, D. M. Welch, J. G. Maxwell,
    In-111-labeled leukocytes in the evaluation of suspected
    abdominal abscess, Am J Surg 139:99 (1980).
12. I. R. McDougall, J. E. Baumert, R. L. Lantieri, Evaluation of
    In-111 leukocyte whole body scanning, Am J Roentgenol 133:849
    (1979).
13. A. W. Segal, M. L. Thakur, R. N. Arnot, J. P. Lavender,
    In-111-labeled leukocytes for localization of abscesses,
    Lancet 2:1056 (1976).
14. J. E. Powe, S. Alastair, W. J. Sibbald, A. A. Driedger, Pulmonary
    accumulation of polymorphonuclear leukocytes in the adult
    respiratory distress syndrome, Crit Care Med 10:712 (1982).
15. P. S. Cook, F. L. Datz, M. A. Disbro, N. P. Alazraki, A. T.
    Taylor, Pulmonary uptake in Indium-111 leukocyte imaging:
    Clinical significance in patients with suspected occult
    infections, Radiology 150:557 (1984).
16. J. P. Deutcher, C. A. Schiffer, G. S. Johnston, Rapid migration
    of 111-Indium-labeled granulocytes to sites of infection, New
    Eng J Med 304:586 (1981) and Letter-to-the-Editor, New Eng J
    Med 304:1547 (1981).
17. J. B. Alavi, M. M. Staum, A. Alavi, Indium-111 granulocyte
    labeling in neutropenic patients, in: "Indium-111-Labeled
    Neutrophils, Platelets, and Lymphocytes," M. L. Thakur and A.
    Gottschalk, eds., Trivirum Publishing, New York (1980).
18. R. E. Coleman, D. M. Welch, W. J. Baker, R. W. Beightol, Clinical
    experience using Indium-111-labeled leukocytes, in:

"Indium-111-Labeled Neutrophila, Platelets, and Lymphocytes,"
M.L. Thakur and A. Gottschalk, eds., Trivirum Publishing, New
York (1980).

19. L. A. Forstrom, B. J. Weiblen, L. Gomez, N. L. Ascher, D. R.
     Hoogland, M. K. Loken, J. McCullough, Indium-111-oxine-labeled
     leukocytes in the diagnosis of occult inflammatory disease,
     in: "Indium-111-Labeled Neutrophila, Platelets, and
     Lymphocytes," M.L. Thakur and A. Gottschalk, eds., Trivirum
     Publishing, New York (1980).

20. D. A. Goodwin, P. W. Doherty, I. R. McDougall, Clinical use of
     Indium-111-labeled white cells: An analysis of 312 cases, in:
     "Indium-111-Labeled Neutrophila, Platelets, and Lymphocytes,"
     M.L. Thakur and A. Gottschalk, eds., Trivirum Publishing, New
     York (1980).

21. J. Q. Knochel, P. R. Koehler, T. G. Lee, D. M. Welch, Diagnosis
     of abdominal abscesses with computed tomography, ultrasound,
     and In-111 leukocyte scans, Radiology 137:425 (1980).

22. B. Carroll, P. M. Silverman, D. A. Goodwin, I. R. McDougall,
     Ultrasonography and Indium-111 white blood cell scanning for
     the detection of intraabdominal abscesses, Radiology 140:155
     (1981).

23. M. H. Rovekamp, M. R. Hardeman, J. B. van der Schoot, J. Belfer,
     Indium-111-labeled leukocyte scintigraphy in the diagnosis of
     inflammatory disease -- first results, Br J Surg 68:150
     (1981).

24. A. M. Peters, S. Karimjee, S. H. Saverymuttu, J. P. Lavender, A
     comparison of Indium-111-oxine and Indium-111-acetyl-
     acetone labeled leucocytes in the diagnosis of inflammatory
     disease, Brit J Rad 55:827 (1982).

25. J. E. Seabold, D. G. Wilson, L. M. Lieberman, C. M. Boyd,
     Unsuspected extraabdominal sites of infection: Scintigraphic
     detection with Indium-111-labeled leukocytes, Radiology
     151:213 (1984).

26. J. McCullough, B. J. Weiblen, M. E. Clay, L. Forstrom, Effect of
     leukocyte antibodies on the fate in vivo of Indium-111-labeled
     granulocytes, Blood 58:164 (1981).

27. T. A. Bicknell, S. Kohatsu, D. A. Goodwin, Use of Indium-
     111-labeled autologous leukocytes in differentiating
     pancreatic abscess from pseudocyst, Am J Surg 142:312 (1981).

28. R. C. Reba, J. C. Kleinman, The sensitivity, specificity and
     predictive value of diagnostic tests, in: "Quality Control in
     Nuclear Medicine," B. A. Rhodes, ed., C. V. Mosby, St. Louis
     (1977).

29. V. M. Runge, T. H. Rand, J. A. Clanton, J. P. Jones, D. G.
     Colley, C. L. Partain, A. E. James Jr., Work in progress:
     Radionuclide imaging of Indium-111-labeled eosinophils in
     mice, Radiology 147:563 (1983).

30. M. H. Rovekamp, E. A. van Royen, Reinders Folmer SCC, J. B. van
     Schoot, Diagnosis of upper-abdominal infections by
     In-111-labeled leukocytes with Tc-99m colloid subtraction
     technique, J Nucl Med 24:212 (1983).

31. H. D. Fawcett, R. L. Lantieri, A. Frankel, I. R. McDougall, Differentiating hepatic abscess from tumor: Combined In-111 white blood cell and Tc-99m liver scans, Am J Roentgenol 135:53 (1980).

32. M. F. Murphy, R. M. Minchinton, P. Metcalf, P. Grint, A. H. Waters, Neutropenia due to β-lactamine antibodies, Brit Med J 288:795 (1984) (letter-to-the-editor).

33. P. A. Ward, Leukotaxis and leukotactic disorders, Am J Path 77:520 (1974).

34. R. Synderman, E. J. Goetzl, Molecular and cellular mechanisms of leukocyte chemotaxis, Science 213:830 (1981).

35. M. Korobkin, P. W. Callen, R. A. Filly, P. B. Hoffer, R. R. Shimshak, H. Y. Kressel, Comparison of computed tomography, ultrasonography, and Gallium-67 scanning in the evaluation of suspected abdominal abscess, Radiology 129:89 (1978).

36. G. N. Sfakianakis, W. Al-Sheikh, A. Heal, G. Rodman, R. Zeppa, A. Serafini, Comparison of scinitigraphy with In-111 leukocytes and Ga-67 in the diagnosis of occult sepsis, J Nucl Med 23:618 (1982).

37. R. E. Coleman, D. Welch, Possible pitfalls with clinical imaging of Indium-111 leukocytes: Concise communication, J Nucl Med 21:122 (1980).

38. H. W. Gray, I. Cuthbert, J. R. Richards: Clinical imaging with Indium-111 leukocytes: Uptake in bowel infarction. J Nucl Med 22:701 (1981).

39. A. W. Segal, J. Ensell, J. M. Munro, M. Sarner, Indium-111-tagged leukocytes in the diagnosis of inflammatory bowel disease, Lancet 2:230 (1981).

40. D. T. Stein, G. M. Gary, P. B. Gregory, M. Anderson, D. A. Goodwin, I. M. McDougall, Location and activity of ulcerative and Crohn's colitis by Indium-111 leukocyte scan. A prospective comparison study, Gastroenterology 84:388 (1983).

41. S. H. Saverymuttu, A. M. Peters, H. J. Hodgson, V. S. Chadwick, J. P. Lavender, Indium-111 autologous leukocyte scanning: Comparison with radiology for imaging the colon in inflammatory bowel disease, Brit Med J 285:255 (1982).

42. R. A. Davies, M. L. Thakur, H. J. Berger, F. J. Th. Wackers, A. Gottschalk, B. Zaret, Imaging the inflammatory response to acute myocardial infarction in man using In-111-labeled autologous leukocytes, Circulation 63:826 (1980).

43. G. N. Sfakianakis, W. Mnaymneh, L. Ghandur-Mnaymneh, W. Al-Sheikh, M. Hourani, A. Heal, Positive Indium-111 leukocyte scintigraphy in a skeletal metastasis, Am J Roentgenol 139:601 (1982).

44. V. Raptopoulos, P. W. Doherty, T. P. Gross, M. A. King, K. Johnson, N. M. Gantz, Acute osteomyelitis: Advantage of white cell scans in early detection, Am J Roentgenol 139:1077 (1982).

45. P. P. McKeown, D. C. Miller, S. W. Jamieson, R. S. Mitchell, B. A. Reitz, C. Olcott IV, J. T. Mehigan, R. J. Silberstein, I. R. McDougall, Diagnosis of arterial prosthetic graft infection

by Indium-111-oxine white blood cell scans, _Circulation_ 66 (Suppl I): I130 (1982).

46. S. L. Kipper, R. W. Steiner, K. F. Witztum, R. M. Basarab, M. S. Kipper, S. E. Halpern, W. L. Ashburn, In-111-leukocyte scintigraphy for detection of infection associated with peritoneal dialysis catheters, _Radiology_ 151:491 (1984).

47. A. M. Peters, J. P. Lavender, J. Macdermott, Diagnosing cerebral abscess with Indium-111 labeled leukocytes, _Lancet_ 2:309 (1980).

48. M. L. Thakur, J. P. Lavender, R. N. Arnot, D. J. Silvester, A. W. Segal, Indium-111-labeled autologous leukocytes in man, _J Nucl Med_ 18:1014 (1977).

49. R. J. M. ten Berge, A. T. Natarajan, M. R. Hardeman, E. A. van Royen, Schellekens PThA, Labeling with Indium-111 has detrimental effects on human lymphocytes: Concise communication, _J Nucl Med_ 24:615 (1983).

50. L. N. Button, W. C. DeWolf, P. E. Newberger, M. S. Jacobson, S. V. Kevy, The effects of irradiation on blood components, _Transfusion_ 21:419 (1981).

51. R. Parkman, D. Mosier, I. Umansky, W. Cochran, C. B. Carpenter, F. S. Rosen, Graft-versus-host disease after intrauterine and exchange transfusions for hemolytic disease of the newborn, _New Eng J Med_ 290:359 (1974).

LYMPHOCYTE MIGRATION STUDIES IN MAN

John Wagstaff

Cancer Research Campaign Department
Medical Oncology
Christie Hospital and Holt Radium Institute
Wilmslow Road
Manchester, M20 9BX, England

INTRODUCTION

Over 200 years ago William Hewson (1) shrewdly observed that,

> "Vast numbers of central particles (Lymphocytes)
> made by the thymus and lymphatic glands, are poured
> into the blood vessels through the thoracic duct
> and if we examine the blood attentively we can see
> them floating in it. Nature surely would not make
> so infinitely many particles to answer no purpose!
> What then becomes of these particles after they are
> mixed with the circulating blood are they
> immediately destroyed? No. They are we believe,
> carried with the blood to the spleen, not that the
> spleen has any elective attraction over them but
> that being equally and uniformly diffused through
> the general mass of blood, a due proportion is
> received by the spleen with its arterial blood, and
> that when arrived there, the spleen has the power
> of separating them from the other parts of the
> blood...".

Later, in 1885, Walter Flemming (2) noted that the efferent
lymph vessel draining a lymph node was always much richer in
lymphocytes than the lymph arriving at the node. He proposed two
explanations for this. Firstly, large numbers of cells were added to
the lymph because of rapid division within the node. Secondly,
lymphocytes could be added to lymph directly by diapedesis into the
node from the blood. He concluded that there was no evidence to

support either hypothesis, and it was not until the 1960's, when
radioisotopic cell labels became available, that this problem was
resolved -- at least in animals. Even 200 years after William Hewson
expounded his ideas, we still have not answered many basic questions
about the migration and recirculation of the lymphocyte in man either
in health or disease. In this paper I will outline what we know of
the migration of lymphocytes in animals and man and attempt to
demonstrate how a better grasp of these phenomena may help us
understand the pathophysiology of a number of human diseases.

THE INCREDIBLE JOURNEY OF THE LYMPHOCYTE

Until the recent past most immunological research in man has
concentrated on lymphocytes obtained from the blood. Since the
lymphocyte is a ubiquitous cell, of which those found in the blood are
only a small proportion, this state of affairs is clearly
unsatisfactory. However, the blood seems a good place to begin to
trace the journey of the lymphocyte through the organism. The
experimental evidence for lymphocyte migration has been extensively
reviewed (3-6). Labeled thoracic duct lymphocytes migrate initially
to the spleen (55%), bone marrow (15%), lymph nodes (20%) and the
nonlymphoid tissues (10%) (Fig. 1). The rate of entry of both T and B
cells into these organs is similar, but the transit times through them
vary considerably. T cells have a mean transit time through the
spleen of 4 to 6 hours, whereas B lymphocytes take about eight times
longer. The mean transit time across lymph nodes from blood to
efferent lymph is shorter for T cells (18 hours) than for B cells.
The data available for B cells are less precise than for T cells but
indicate a mean transit time of 36-48 hours across lymph nodes. The
transit time through bone marrow for thoracic duct lymphocytes (80-90%
T cells) is approximately 2-3 hours, but the time for recirculation of
B cells through this compartment is not known. It is remarkable that
the distribution and tempo of lymphocyte migration seem similar for
widely varying species, for example, the plaice (7), the duck (8), and
the sheep (9). Human peripheral blood lymphocytes consist mainly of T
cells (60-90%) with 10-30% B cells and a small percentage of "Null"
cells.

Small lymphocytes leave the bloodstream and enter the lymph node
tissue by crossing between specialized high-walled endothelial cells
(HEV) found in the post-capillary venules. Although T and B cells
appear to cross these structures at similar rates, once within the
lymph node parenchyma they have diverse migratory pathways. Some
tissues, through which lymphocytes readily migrate, do not possess HEV
(e.g., spleen, bone marrow, and liver). Again T and B cells seem to
enter these tissues at similar rates and by the same route but have
different pathways of migration once within the tissue.

Evidence is accumulating that subsets of lymphocytes possess
different migratory properties. For example, lymphocytes obtained

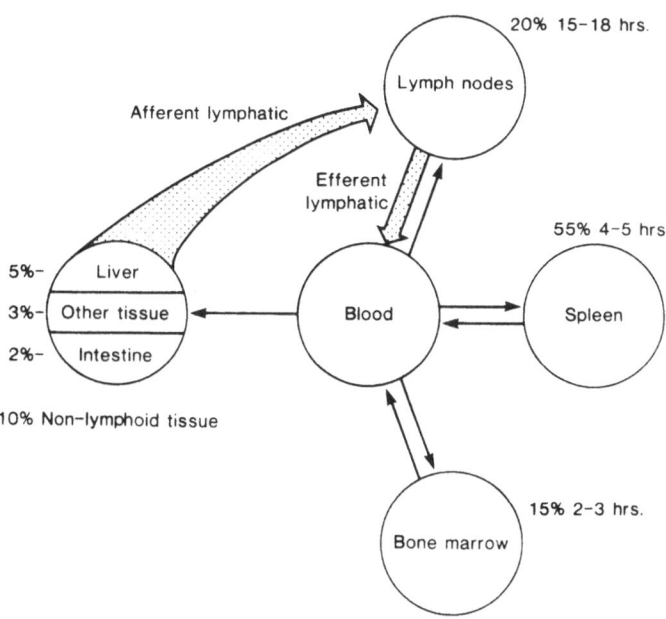

Fig. 1   The distribution and tempo of recirculation of intravenously
         injected syngeneic thoracic duct lymphocytes in the rat.

from gut associated lymphoid tissue (GALT) will "home" preferentially
back to GALT rather than to peripheral lymphoid tissue and vice versa
(10). This phenomenon seems to be dependent neither on whether the
cells are T or B, nor entirely on the presence of antigen within the
gut (11).These observations may have important implications if they
are born out by studies performed in humans.

Factors responsible for the specific migration of lymphocytes are
beginning to be better understood. It seems that cell surface
glycoprotein structures are important with this regard (12-15).
Malignant lymphocytes are known to possess altered cell surface sugars
(16), and this may account for some of the clinical features of these
interesting diseases. This will be discussed later. The control of
the progress of lymphocytes throught the tissues is a complex matter
involving lymphocyte-lymphocyte interaction, the production of
"modulation factors" (17), and the interaction between lymphocytes and
the so-called accessory cells within the tissues, such as the
interdigitating reticulum cells and the dendritic reticulum cells. As
will be demonstrated later, all these aspects of lymphocyte migration
have importance in interpreting the data obtained from studying
lymphocyte kinetics in man.

RADIOISOTOPIC CELL LABELS: WHICH RADIOISOTOPE?

The radioactive labeling of cells has been chiefly responsible for the advances in our understanding of the migration of lymphocytes in experimental animals. In man the direct study of this phenomenon has been hampered by the lack of appropriate techniques. The methods used in animals are invasive, usually require the sacrifice of the animal, and are clearly not applicable to humans. For clinical application the ideal characteristics for a radioactive tracer would be:

1. Lack of toxicity in man either acutely or in terms of oncogenic potential.

2. The procedure of cell labeling would not impair the normal migratory properties of the lymphocytes.

3. Capable of external imaging with conventional radionuclide scanning equipment.

4. A half-life that is long enough to allow studies to be performed over several days but not so long that the subject would receive unnecessary irradiation after the end of the experiment.

5. Radioactivity should not be lost from the cells in such a way that a misleading impression of cellular localization would be obtained.

The most widely used radionuclide in cell labeling experiments has been Chromium-51 (Cr-51). This isotope does not fulfill the above criteria. It has low gamma photon emissions which cannot usefully be detected with a gamma camera, the labeling efficiency is low, its half-life of 28 days is excessively long, and a significant proportion dissociates from the cells within the first 24 hours after labeling. Tritiated thymidine- and uridine-labeled lymphocytes have been used in man, but these agents will only label lymphocytes which are cycling (S phase), and the majority of peripheral blood lymphocytes are in GO (resting). Furthermore, they are pure beta emitters, which makes external detection impossible. Technetium-99m (Tc-99m) is already in widespread use in nuclear medicine and most commercially available imaging systems are designed around this isotope. It has been used to label lymphocytes in mice, but getting the Technetium into the cells has proved difficult (18).

Indium-111-oxine has been used in laboratory animals to study the migration and recirculation of lymphocytes. It has been used, with success, in man to study the kinetics of granulocytes and platelets. The physical characteristics of this radioisotope are almost ideal for gamma camera imaging. It has a high labeling efficiency and low spontaneous release once the cells have been labeled. Its half-life of 2.8 days is convenient for most clinical purposes. Other chelates of this isotope, such as Indium-tropolone may allow labeling of cells in plasma, which would be a further advantage. It seemed, therefore, that In-111-oxine was a suitable substance to use to study the migration and recirculation of human lymphocytes.

DOSE RECOMMENDATIONS FOR STUDIES IN MAN

A number of studies have compared In-111 with Cr-51 in animal systems. Chisholm et al (19) demonstrated that in the rat the dose of In-111 per 10^8 lymphocytes were critical. At doses of In-111 greater than 20 μci per 10^8 cells the migration and recirculation were impaired. Issekutz et al (20) demonstrated that at a dose of 10 μCi per 10^8 cells In-111-labeled cells behaved identically to those labeled with Cr-51. Sparshott et al (21) confirmed the results of Chisholm and concluded that at doses greater than 20 μCi per 10^8 cells, recirculation is impaired before 24 hours after reinjection of the cells. They also examined the sites of deposition of the radioactivity using autoradiography. They showed that after the lymphocyte had died, the radioactivity it contained was transferred to tissue macrophages and histiocytes, thus remaining in the tissue where cell death had occurred. It therefore seems clear that for human studies doses greater than 20 μci per 10^8 lymphocytes should be used with caution since interpretation of the physiological data obtained would be difficult due to artefacts resulting from cell death. However, primary localization of the lymphocytes occurs within the first few hours after injection, and the radiation damage which immobilizes the cells is cumulative over a similar time scale. It is, therefore, reasonable that higher radiation doses per cell could be entertained if external imaging is all that is required, rather than an additional concern that the lymphocytes should behave in a nearly perfect physiological manner.

Frost et al (22) were able to obtain excellent images of lymphoid tissues in sheep by labeling thoracic duct lymphocytes with In-111-oxine. The first study using In-111 for studying human lymphocytes was performed by Lavender et al (23) who were able to obtain images of lymph nodes in normal subjects and patients with Hodgkin's disease (HD). They were, however, unaware at that time of the critical doses as described above, and they used high doses, thus making interpretation of their kinetic data difficult. They were unable to demonstrate any differences in localization between lymph nodes involved with HD and those that were not.

DOSIMETRY AND THE RISK OF MALIGNANCY

Although it has been calculated that the total body dose to humans as a result of lymphocyte labeling is in the range produced by other diagnostic nuclear medicine investigations, the dose to the individual lymphocyte may be very high. This is because In-111 produces Auger electrons, which have a very short path length in the tissues -- much less than one cell diameter. Since lymphocytes are radiosensitive cells, it was originally assumed that the cumulative radiation dose to the lymphocyte would be such that the cells would not survive. If this were the case, then risk of radiation-induced malignancy from cell labeling would be zero. However, a recent report from ten Berge (24) has shown that although the proliferative capacity of labeled cells does decline with increasing doses of In-111, some of the lymphocytes do retain the ability to divide. Furthermore, they examined the chromosomes of labeled cells and discovered that there was dose-dependent chromosomal damage. These data are clearly of concern to investigators who would like to study the lymphocytes of patients who have diseases with a long survival. However, the magnitude of the risk has not as yet been calculated. In collaboration with Dr. Scott of the Patterson Laboratories in Manchester we have determined a biological estimate of the dose to the genetic material of the cell by measuring the amount of chromosome damage (dicentric aberrations) induced by In-111 in vitro (Fig. 2) and expressing this in terms of "X-ray equivalents" using a calibration curve of dicentric yield against dose of acute X-rays (Fig. 3). Figure 4 shows the combined results of our studies and those of the Dutch group. They confirm the dose-dependent nature of the damage.

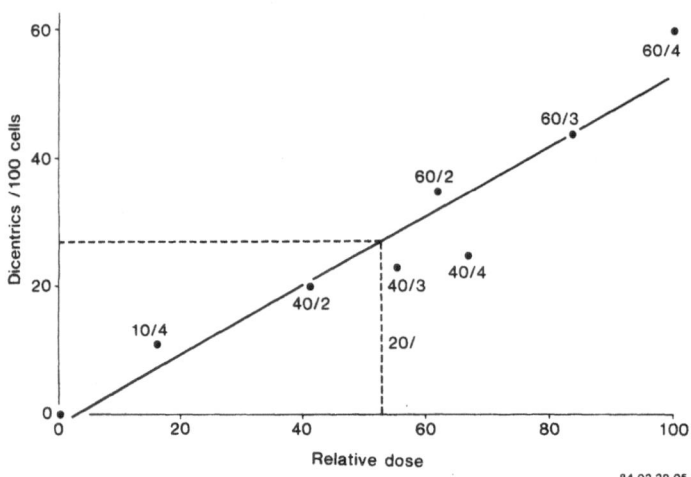

84.02.29.05

Fig. 2   The number of dicentric chromosome aberrations related to radiation dose for human peripheral blood lymphocytes labeled with In-111-oxine. The numbers at each point on the graph represent the dose of In-111 (μCi) per 10^8 lymphocytes and the number of days of in vitro storage before stimulation with PHA. The dashed line represents the relative dose for 20 μCi per 10^8 cells to total decay of the radionuclide and the number of aberrations that it would be expected to produce.

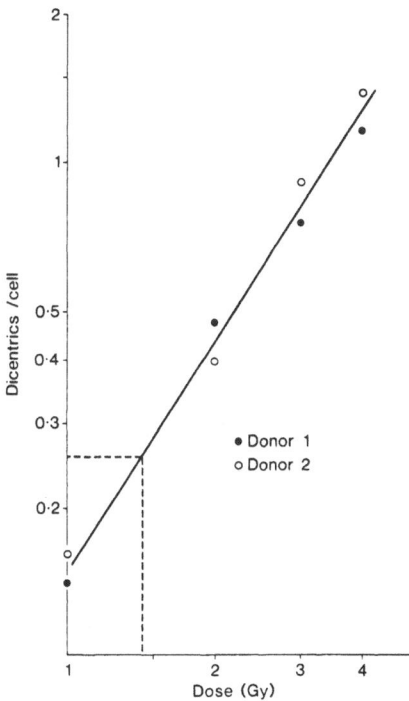

Fig. 3. The calibration curve of dicentric chromosome aberrations in
human lymphocytes irradiated with 300 KV X-rays. The hatched
line shows the X-ray dose that would have caused the number
of aberrations observed when 10^8 lymphocytes were labeled
with 20 µCi of In-111-oxine (see Fig. 2).

Lymphocytes were labeled with In-111-oxine at different activities
and cultured for 0, 1, or 2 days before PHA stimulation for an
additional 2 days, the dose rate decreasing exponentially during the
entire culture period. Relative doses were calculated after
designating as 100 units the highest activity for the longest
incubation time (60 µCi per 10^8 lymphocytes for 4 days). On this
scale the damage produced by 20 µCi per 10^8 cells to total decay was
53 units which induced 27 dicentrics per cell (see Fig. 2). This is
the level of damage induced by 1.45 Gy of X-rays (see Fig. 3). If
lymphocyte monitoring is confined to the 24-hour period after
injection, the dose received will be equivalent to 0.32 Gy. This is
less than one third the dose calculated from the physical data.

Using estimates of risk of lymphoid malignancy from whole-body
exposure to ionizing radiation (International Conference on Radiation
Protection. Publications 26 and 27, Pergamon Press [1977]) and
allowing for the fact that only about 0.1% of the patient's
recirculating lymphocytes are labeled, we calculate the excess risk to
be 2.8×10^{-6} from the Auger electrons. The TOTAL cancer risk from

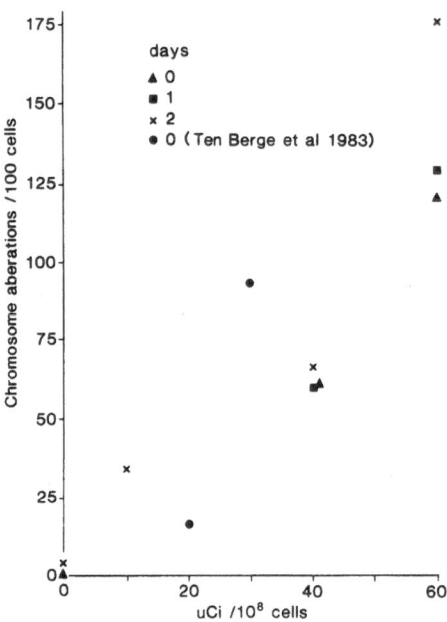

Fig. 4.   Combined data from Manchester and Holland for chromosome
         aberrations in In-111-labeled human lymphocytes. These data
         demonstrate a linear dose/damage curve but also show that
         storage in vitro prior to PHA stimulation increases the
         amount of damage. Cells stimulated with PHA immediately
         after labeling are receiving irradiation during cell
         division, but those in storage are being irradiated in G0.
         The latter is analogous to the in vivo situation.

the gamma emission is calculated to be 1.5×10^{-5}. In other words, if
one million patients were studied using this technique, then we might
expect that 2.8 of them will develop a lymphoid malignancy as a
result. This risk is very small and certainly of the same order of
magnitude as the risk from other diagnostic imaging procedures.

METHODOLOGY

     The method we have used to obtain, purify, and label human
lymphocytes for reinjection has been published in detail previously
(25). I do not intend to describe the methodology in detail but will
confine myself to describing some modifications in the technique and
commenting on some important aspects of the procedure.

     It is recommended that only 20 µCi/10^8 cells or less of In-111 is
used, based upon the animal experiments, showing that doses greater
than this perturb the migration of the lymphocytes. Because of this
restriction it is necessary to obtain large numbers of lymphocytes for
labeling in order that enough total radioactivity is reinjected to

allow adequate gamma camera imaging. Consequently, it becomes
impractical to obtain cells by small volume venesection. We have,
therefore, used a cell separator to obtain a large number of
lymphocytes in a relatively small and easily manageable volume. By
collecting the buffy coat, a cell suspension is obtained which will
contain in excess of 10^9 lymphocytes but which will be heavily
contaminated with platelets (Fig. 5). Relatively few granulocytes
will be present in this suspension. This procedure is well tolerated
by the patients and generally takes about 1 hour to perform. We have
used both a Haemonetics Model 30 separator and an IBM machine with
equal success.

In our original method we depleted the buffy bag of platelets by
low-velocity centrifugation. In order to reduce the number of
centrifugation steps and to keep the system closed for longer we have
subsequently used a glass bead column to deplete platelets (Fig. 6).
This system has the added advantage that monocytes will adhere to the
glass beads, thus removing them. However, some subpopulations of
lymphocytes are also known to adhere to glass beads, but in practice
we found this to be a minor problem. The acid citrate dextrose
anticoagulation used in the cell separator was reversed with calcium
gluconate since glass bead adherence is calcium dependent.
Anticoagulation was continued with Heparin. The glass bead column was
prepared by inserting a bung into the bottom of a glass tube (15 cm
long and 1 cm diameter) with a glass tube passing through the bung. A
small plug of glass wool was placed in the bottom. Glass beads
(Ballotini grade A) of 0.8 mm diameter were then poured into the
column to the 10 cm mark. A small plug of glass wool was placed over
the beads in order to stop them moving during autoclaving. A similar
bung, with protruding tube, was inserted into the top of the tube.
The tube was then autoclaved to ensure sterility.

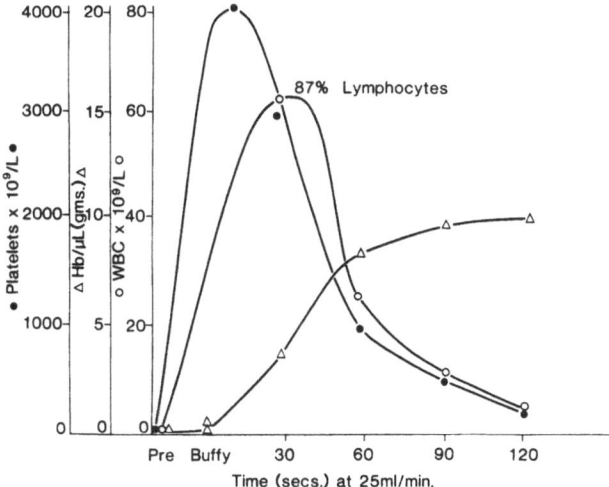

Fig. 5. The breakdown of the buffy bag contents obtained from the
Haemonetics cell separator.

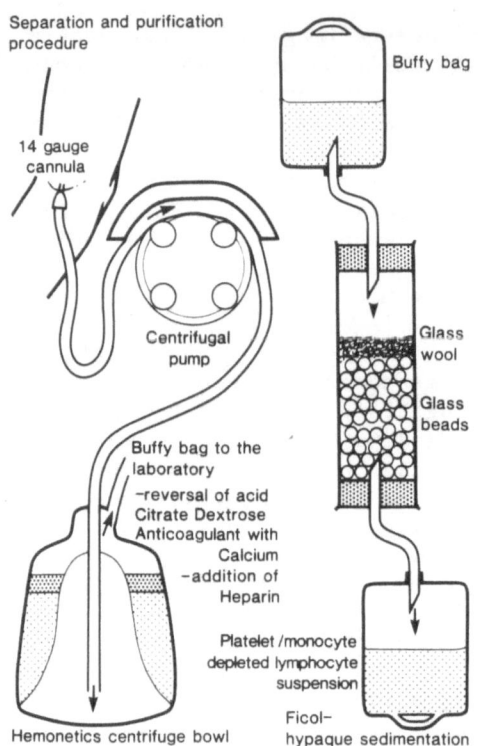

Fig. 6. The modified procedure for obtaining human lymphocytes for
    migration studies.

The buffy bag was connected to the top tube by way of a platelet
administration set (Travenol Ltd., England) and a collection bag
connected to the bottom by the same method. The buffy coat suspension
was then allowed to run slowly through the column at room temperature.
This usually took about 15 minutes. The rest of the procedure
involving Ficol-Hypague sedimentation was carried out as previously
described. All our studies are now performed using In-111-oxine
obtained from Amersham Radiochemicals. This product gives high
labeling efficiencies of 80-90%.

NORMAL SUBJECTS

The blood clearance curve for a normal subject is shown in
Fig. 7. The lymphocytes leave the blood in an exponential manner
initially, reaching a nadir of between 15 and 25% at 1 to 4 hours
after injection. There then follows a rise in the number of cells in
the blood which reaches its peak between 12 and 24 hours. Gamma
camera imaging shows that the lymphocytes are distributed initially to
the liver and spleen (26). The activity in the spleen reaches a

maximum at about 4 to 6 hours after injection and over the next 24 to 48 hours falls progressively (Fig. 7). This suggests that the lymphocytes at first enter and then traverse the splenic white pulp only to reenter the blood via the splenic vein with a mean transit time of 4-24 hours. As previously mentioned, human peripheral blood lymphocytes are predominantly T cells, and in animals the transit time of T cells through the spleen is of the order of 4-6 hours. Thus it appears that human lymphocytes are behaving similarly to those of other animal species. By 24 hours after injection it becomes possible to image normal lymph nodes (Figs. 8 and 9) in the neck and inguinal regions. Imaging of axillary and mediastinal nodes has been less successful due to technical difficulties with positioning of the gamma camera. In our studies there is little change in the radioactivity in the lymph nodes between 24 and 48 hours and we have, therefore, not been able to demonstrate definitively that the cells are recirculating through lymphoid tissues.

The blood clearance curves in normal subjects reported by Hersey (27) are rather different from those reported by us. His curves fell to between 2 and 3% of the injected cells by 2 hours after injection, whereas our normals had 15-25% remaining in the blood. Subsequently, his curves continued to show a gradual decline whilst ours demonstrated a secondary rise. This secondary rise was also shown in the work of Scott et al (28), but their results were not expressed as

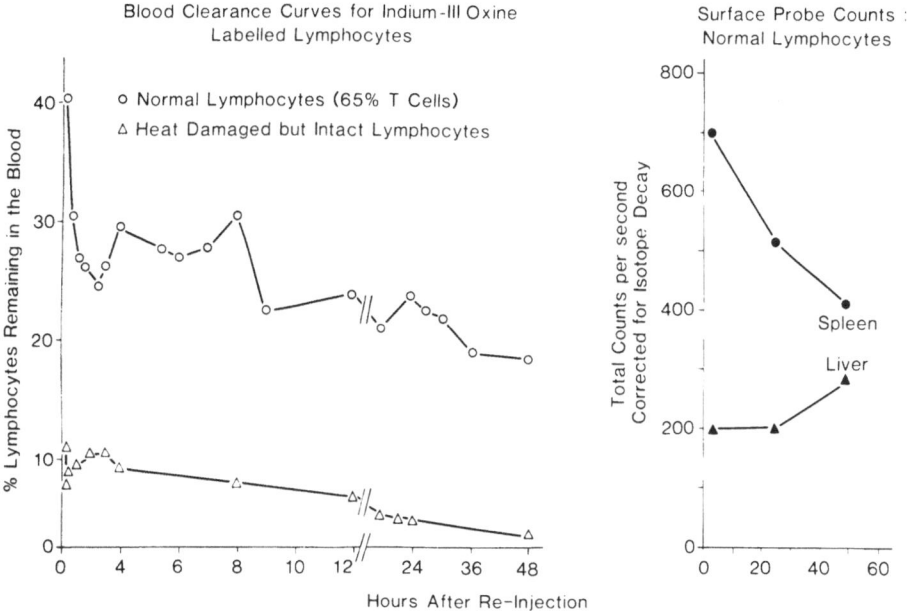

Fig. 7. The blood clearance curves and surface probe count profiles of IV reinjected normal peripheral blood lymphocytes labeled with 20 μCi per 10^8 cells of In-111-oxine in a normal volunteer.

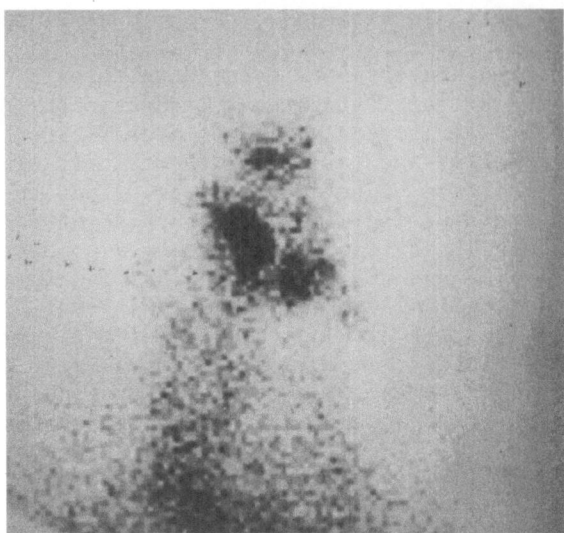

Fig. 8. A right oblique view of the neck of a normal subject showing imaging of cervical lymph nodes 24 hours following the reinjection of In-111-labeled peripheral blood lymphocytes.

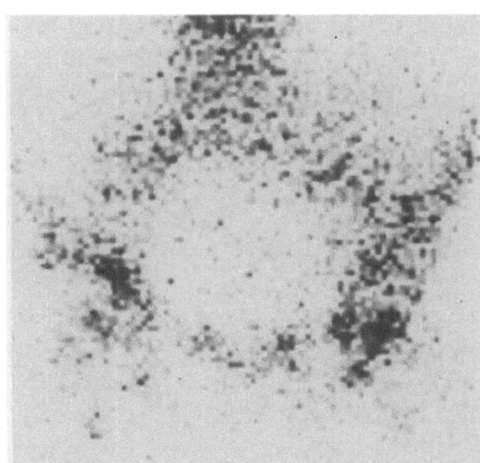

Fig. 9. An antero-posterior view of the pelvis of a normal subject showing lymph node (inguinal) and bone marrow imaging 24 hours after the reinjection of In-111 labeled peripheral blood lymphocytes.

a percentage of the cells injected and so are not directly comparable with ours in absolute terms. We believe that the reappearance of the lymphocytes in the blood of our normal subjects has a physiological explanation which accounts for the patterns of blood clearance seen and which is consistent with the data from other species. Hersey's blood clearance data in normal subjects suggest that the lymphocytes are not returning to the blood in detectable numbers after primary migration. Indeed, the splenic surface probe count results in his paper do not show the same amount of fall as either our results or the data obtained from a number of other species.

CHRONIC LYMPHOCYTIC LEUKEMIA (CLL)

Previous studies of lymphocyte kinetics in CLL, using various isotopic labels, have suggested that CLL lymphocytes leave the blood less readily than normal cells (29). More recently, Bazerbashi et al (30) have pointed out that although the clearance of individual cells from the peripheral blood in CLL was slower than normal, the total number leaving each day was increased. This latter observation is consistent with the electron microscopic findings of Manaster et al (31). Bazerbashi also suggests that "the transit time of CLL lymphocytes through the tissues must also be increased." We have performed the first study of CLL lymphocytes using In-111-oxine-labeled cells (32).

In CLL patients, mainly B lymphocytes are labeled. These cells leave the peripheral blood in an exponential manner, and there is no reappearance "hump", as seen in normal subjects. The splenic activity is high initially, as in normals, but does not show the same fall over the subsequent 40 hours. This suggests that CLL B cells either take longer than 48 hours to traverse the spleen or remain sequestered therein. The initial rate of clearance of lymphocytes from the blood of our patients with CLL showed some degree of variation. In some instances the rate was slower and in others similar to normal. Scott et al (33) found that at 2 hours the numbers of lymphocytes remaining in the blood was higher than normal, but by 24 and 48 hours the levels were at the lower end of the normal range or below it.

Bone marrow imaging has been seen both in normal subjects and in those with CLL. This suggests that, as in animals (34), there is significant migration of lymphocytes to this compartment. Because of the diffuse distribution of the marrow it is difficult to quantify the extent of this migration in man, but the images obtained in CLL patients were more intense than in normals. This suggests that the proportion of the injected cells migrating to, or the transit time through, the marrow of CLL B cells is increased. Although it was easy to visualize the enlarged lymph nodes of CLL patients, the intensity of the radioactivity in the nodes was less than in normal nodes. This might be explained by the greater propensity of T cells to migrate to the lymph nodes than B cells, due to the greater sequestration of the latter by the spleen and possibly the bone marrow.

It does, however, remain uncertain whether the differences in lymphocyte kinetics observed between normal subjects and those with

CLL are due to "B" cell nature of the CLL lymphocyte or to its
malignant characteristics. Experiments will be required where the
kinetics of normal "B" lymphocytes are determined in order to resolve
the point.

THE NON-HODGKIN'S LYMPHOMAS (NHL)

The non-Hodgkin's lymphomas are a heterogeneous group of lymphoid
malignancies. The phenotype and biology of the malignant cells vary
widely and the histological classification is complicated and fraught
with difficulties. Perhaps not surprisingly, therefore, the clinical
patterns of disease that we see in these patients are quite diverse.
The so-called low-grade lymphomas tend to disseminate early and widely
throughout the body. Some patients who present with disease of
Waldeyers ring will relapse only in the gastrointestinal tract, and
patients with gastrointestinal NHL often remain localized to the
abdomen for long periods, despite having very large tumors (16). We
know from animal experiments that lymphocytes obtained from gut tend
to "home" preferentially back to gut rather than to peripheral lymph
nodes (10-11). It may be that lymphoma cells from a particular site
retain the migratory characteristics of their normal counterpart, thus
accounting for the clinical patterns of disease that we observe.

We have studied a number of patients with NHL. We have not as
yet been able to demonstrate preferential localization of lymphoma
cells to one site rather than to another. It is clear, however, that
both normal and malignant lymphocytes are in a state of flux with the
tissues involved with the disease (Fig. 10). Because of these
observations it occurred to us that it might be possible to use the
"homing" properties of lymphocytes to deliver cytotoxic drugs or
irradiation to sites of disease. In the department of cellular
immunology at the University of Manchester, Professor Ford and his
team have carried out some experiments in rats in order to determine
the feasibility of this idea. They have labeled thoracic duct
lymphocytes with a strongly beta emitting radioisotope, namely,
Indium-114m chelated with oxine. After reinjection of the labeled
cells, the rats were rendered lymphopenic without a significant effect
on the other bone marrow elements. The rats continued to thrive
similarly to the control animals (Fig. 11). Other experiments were
performed in which labeled thoracic duct lymphocytes were injected
simultaneously with Roser leukemia cells. The injection of the
In-114m labeled cells delayed the development of leukemia in these
animals in a dose-dependent fashion (Fig. 12). We intend to use this
system in patients with refractory NHL, using In-114m-labeled
autologous human lymphocytes as a novel form of radiotherapy.

HODGKIN'S DISEASE

Despite over a century of medical research, HD remains an enigma.
The histology of this tumor shows a spectrum of change from lymphocyte

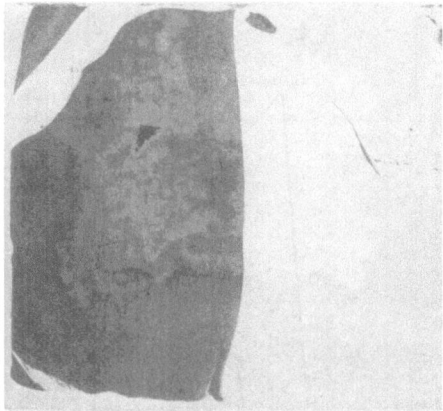

Fig. 10. An antero-posterior view of the pelvis of a patient with
         bulky inguinal lymphadenopathy and bone marrow infiltration
         from a low-grade non-Hodgkin's lymphoma. The scan shows that
         the labeled normal peripheral blood lymphocytes have migrated
         readily to the sites of lymphomatous involvement.

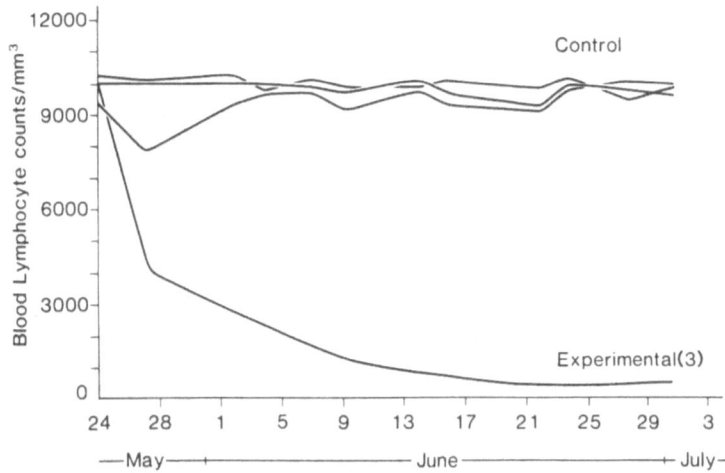

Fig. 11. Changes in the blood lymphocyte counts of rats intravenously
         injected with syngeneic thoracic duct lymphocytes either
         labeled with In-114m or not. The rats injected with labeled
         cells became lymphopenic, showing that their recirculating
         lymphocyte pool had become depleted by the irradiation.

predominant to lymphocyte depleted. The tumor cells (Reed-Sternberg
or Hodgkin cells) are usually in the minority, and the tissues are
replaced by a variety of cell types of nonmalignant phenotype. At the
same time there is often peripheral blood lymphopenia even in the

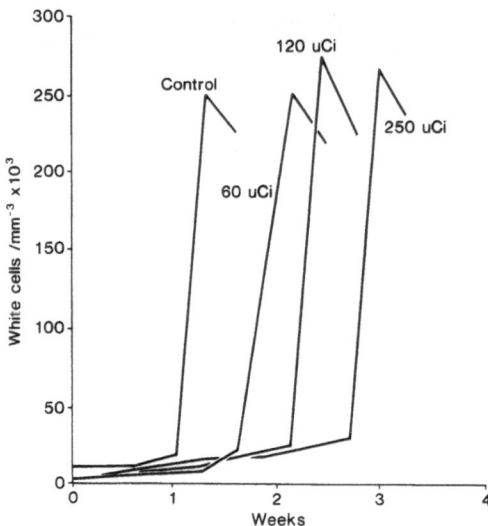

Fig. 12. The effect on the development of leukemia of injecting
In 114m labeled (vailous doses) tnoracic duct lymphocytes
into rats simultaneously inoculated with Roser leukemia
cells. There is a dose-dependent delay in the development of
leukemia.

early stages of the disease. A number of workers have now
demonstrated that there is an alteration in the distribution of T cell
subsets in this disease. T helper cells are proportionately reduced
in the peripheral blood, and T suppressor cells are increased. The
reverse is true for the spleen (35-37). In accord with this, mitogen
responses of peripheral blood lymphocytes are diminished, but those of
splenic lymphocytes are normal. Anergy to cutaneously administered
delayed hypersensitivity recall antigens is often present, and
"reactive" lymphocytes are not infrequently observed in the blood.
Maria De Sousa (38) put all this information together and hypothesized
that "Ecotaxopathy" may be occurring in HD. In other words, there may
be a redistribution of lymphocytes from one compartment to another.
We set out to investigate this by studying lymphocyte migration HD
patients. The blood clearance curves are shown in Fig. 13. It can be
seen that some of them appear normal, but others fall to levels below
the controls (see Fig. 7). Scanning of the involved lymph node areas
demonstrated quite marked preferential accumulation of the injected
cells at the sites of disease (Figs. 14 and 15). We have, therefore,
confirmed that "ecotaxopathy" is occurring in this disease. The
mechanism of this phenomenon remains to be elucidated, but the use of
lymphocyte labeling is proving a useful tool in the investigation of
this intriguing disease.

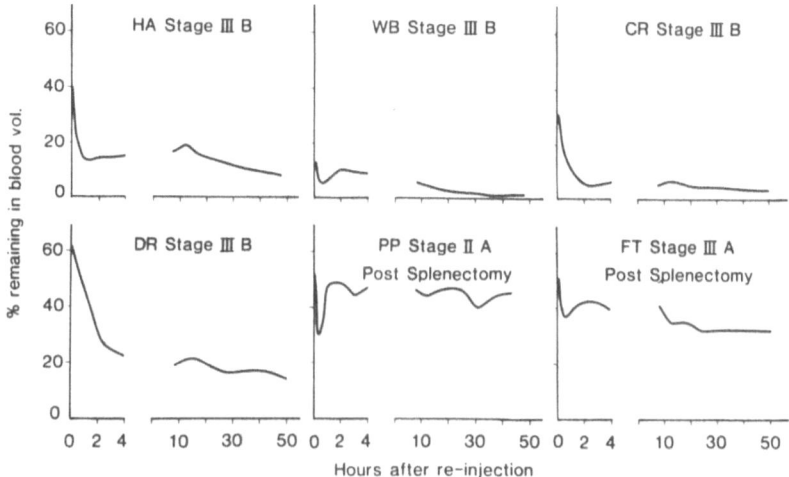

Fig. 13. The blood clearance curves of six untreated patients with
Hodgkin's disease intravenously injected with In-111-labeled
lymphocytes.

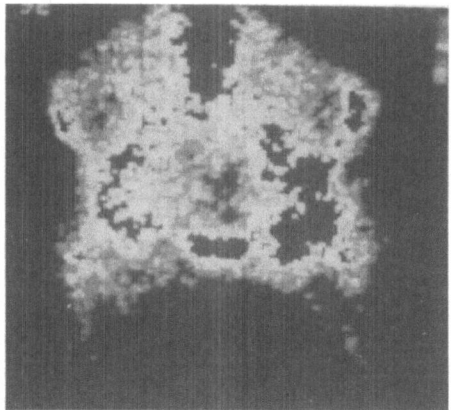

Fig. 14. Intense localization of labeled lymphocytes in the involved
left cervical lymph nodes of a patient with mixed cellularity
Hodgkin's disease.

OTHER DISEASES WHERE LYMPHOCYTE LABELING MIGHT HELP

Solid tumors are frequently infiltrated by large numbers of
lymphocytes and macrophages. Indeed, the prognosis has in some
instances been shown to be related to the degree of the lymphocytosis
in the tumor (39). The specialized structure through which
lymphocytes enter lymph nodes and some other tissues is seen in solid

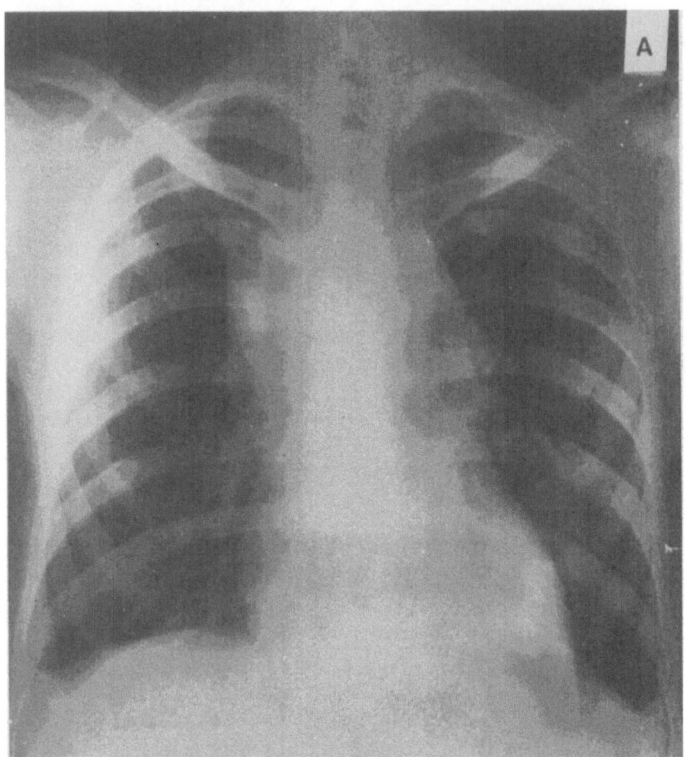

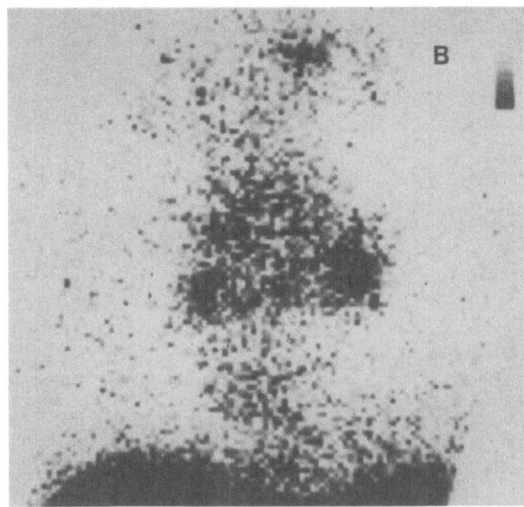

Fig. 15. The preferential accumulation of In-111-labeled peripheral
blood lymphocytes in the involved mediastinal and left
supraclavicular lymph nodes of a patient with nodular
sclerosing Hodgkin's disease.

tumors. Zatz et al (40) demonstrated that there was significantly greater localication of lymphocytes to the lymph nodes draining Maloney Sarcoma virus induced tumors than in the contralateral nodes. We have been able to demonstrate lymphocyte trapping in one patient with a sarcoma (41) but have failed to show preferential localization in several other patients with other solid tumors. It seems possible that lymphocytic infiltration in solid tumors may have some importance with regard to survival of these patients. A better understanding of the mechanisms involved in determining whether lymphocyte infiltration will occur might allow beneficial pharmacological intervention.

Alterations in T and B lymphocytes in a number of connective tissue diseases are well documented. In rheumatoid arthritis (RA) the proportion of T cells is increased in the synovial fluid as compared with the blood (42-47). Many of these T cells are activated and have been used to image the joints of patients with RA using In-111-labeled lymphocytes (48-49). As van de Putte (47) pointed out, "the meaning of the relative increase of T cells and the relative decrease of B cells in synovial fluid remains speculative, one possibility is that the predominance of T cells in the extravascular fluid is an expression of their recirculation pattern." In systemic lupus erythematosis there is frequently a peripheral blood lymphopenia associated with exacerbations of the disease (50). One explanation of this is that these cells are being sequestered in another compartment and that it is these cells which are responsible for the tissue damage in this disease. A similar situation pertains to systemic sclerosis in which lymphocyte infiltration into the involved tissues may be quite florid (51-53).

Abnormalities of peripheral blood and peripheral immune function similar to those seen in Hodgkin's disease may also be found in patients with Crohn's diseae (54-58). Strickland et al (59) addressed the possibility that this was due to a maldistribution of lymphocytes from one compartment to another by studying the proportions of T and B lymphocytes in the blood and bowel of these patients. He found that within two to six months after resection of the affected bowel the T lymphocyte lymphopenia in the peripheral blood had disappeared. Examination of the affected bowel revealed that lymphocyte infiltration in all bowel layers was increased and that in the deeper layers of the bowel the infiltrate was predominantly of T cells. Douglas et al (60) found substantially higher levels of radioactivity in the stools of patients with both Crohn's disease and gluten-sensitive enteropathy following intravenous injection of Cr-51 labeled lymphocytes. In the latter group this increase in the radioactivity excreted returned to normal when the patient went on a gluten-free diet. These studies show that "ecotaxopathy" is probably occurring in this disease and that an alteration in the migratory properties of lymphocytes may be involved in their pathogenesis.

CONCLUSION

The advent of reliable radiosotopic cell labels means that
the migration of cells may be studied in man for the first time.
Lymphocytes are cells of vital importance providing the corner
stone of the immune response. I have attempted to highlight
areas, not only in my own discipline of oncology, but also other
medical s-ecialities where alterations in lymphocyte kinetics may
be important in the pathophysiology of disease. A more complete
understanding of these phenomena might allow pharmacological
intervention to alter the course of the disease to the benefit of
the patient. One example of this is our proposal to use a Beta
emitting radioisotope to produce selective irradiation of
lymphomatous tissues. It may also be feasible to produce clones
of autologous lymphocytes which have been sensitised in vitro to
antigens in order to modify the immune response. Such studies
have been initiated on a preliminary basis in patients with cancer
and in animals with viral infections. The introduction of new
will hopefully be the case with the technology discussed at this
meeting.

REFERENCES

1. W. Hewson, in: The works of William Hewson", Gulliver, FRS,
    p 275, London, 1846. Printed for the Sydenham Society.
2. W. Fleming, "Studien uber regeneration der gewebe. I. Die
    Zellrermehrung in den Lymphadrusen und verwandten Organen, und
    ihr Einfluss auf Deren Bau," Max Cohen und Sohn, Bonn 1885.
3. M. De Sousa, "Lymphocyte circulation: Experimental and clinical
    aspects," John Wiley & Sons, Chichester, England 1981.
4. W. L. Ford, Lymphocyte migration and immune response, Prog
    Allergy 19:1, 1981.
5. J. L. Gowans, The recirculation of lymphocytes from blood to
    lymph in the rat, J Physiol 146:54, 1959.
6. J. Sprent, Recirculating lymphocytes. in: "The Lymphocyte:
    Structure and Function," Marcel Dekker, New York, 1977.
7. A. E. Ellis and M. De Sousa, Phylogeny of the lymphoid system.
    I. A study of the fate of circulating lymphocytes in plaice,
    Eur J Immunol 4:338, 1974.
8. R. G. Bell and J. J. Lafferty, The flow and cellular
    characteristics of cervical lymph from anaesthetized ducks.
    Aust J Exp Biol Med Sci 50:611,1972.
9. H. Frost, R. N. P. Cahill, Z. Trnka, The migration of
    recirculating autologous and allogenic lymphocytes through
    single lymph nodes, Eur J Immunol 5:839,1975

19. P. M. Chisholm, H. J. Danpure, G. Healy, S. Osman, Cell damage resulting from the labeling of rat lymphocytes and Hela S3 cells with In-111-oxine, J Nucl Med 20:1308, 1979.

20. T. Issekutz, W. Chin, J. B. Hay, Measurement of lymphocyte traffic using In-111, Clin Exp Immunol 39:216, 1980.

21. S. M. Sparshott, H. Sharma, J. D. Kelly, W. 1. Ford, Factors influencing the fate of In-111-labeled lymphocytes after transfer to syngenic rats, J Immunol methods 41:303, 1981.

22. H. Frost, P. Frost, C. Wilcos, Lymph node scanning in sheep with In-111-labeled lymphocytes Int J Nucl Med Biol 6:60, 1979.

23. J. P. Lavender, J. M. Goldman, R. N. Arnot, M. L. Thakur, Kinetics of In-111-labeled lymphocytes in normal subjects and patients with Hodgkin's disease, Br Med J 2:797, 1977.

24. R. J. M. ten-Berrge, A. T. Natarajan, M. R. Hardeman, E. A. Van Royen, P. TH. A. Schellekens, Labeling with In-111 has detrimental effects on human lymphocytes: concise communication, J Nucl Med 24:615, 1983.

25. J. Wagstaff, C. Gibson, N. Thatcher, W. L. Ford, H. Sharma, W. Benson, D. Crowther, A method for following human lymphocyte traffic using Indium oxine labeling, Clin Exp Immunol 43:435, 1981.

10. J. G. Hall, An essay on lymphocyte circulation and the gut, Monogr Allergy 16:100, 1979.

11. D. Guy-Grand, C. Griscelli, P. Vassalli, The mouse gut T lymphocyte, a novel type of T cell; nature, origin, and traffic in mice in normal and GVH conditions, J Exp Med 148:1661, 1978.

12. J. J. Woodruff and B. M. Gesner, The effect of neuraminidase on the fate of transfused lymphocytes, J Exp Med 129:361, 1969.

13. W. L. Ford, Lymphocyte migration and immune response, Prog Allergy 19:1, 1975.

14. R. W. Gillette, G. O. McKenzie, and M. M. Swanson, Effect of concanavalin A on the homing of labeled T lymphocytes, J Immunol III:1902, 1973.

15. A. A. Freitas and M. de Sousa, Controlled mechanisms of lymphocyte traffic. Modification of the traffic of Cr-52-labeled mouse lymph node cells by treatment with plant lectins in the intact and splenectomized host, Eur J Immunol 5:831, 1975.

16. G. R. Blackledge, Flow cytometric investigations of DNA and the cell surface in normal and malignant cells, PHD Thesis, University of Manchester, Manchester, England, 1981.

17. A. S. G. Curtis, Clues concepts and possible answers from other systems in lymphocyte circulation, Auth de Sousa, John Wiley and Sons, Chichester, England, 1981.

18. R. F. Barth and O. Singla, Organ distribution of Tc-99m and Cr-51-labeled lymphocytes, J Nucl Med 16:633, 1975.

26. J. Wagstff, C. Gibson, N. Thatcher, D. Crowther, A method for
     studying the dyanamics of primary migration of human
     lymphocytes using In-lll-oxine cell labeling, in: :In vivo
     Immunology," Nieuwenhuis, van den Broek and Hanna, eds.,
     Plenum, p 153, 1982.

27. P. Hersey, The separation of Cr-51 labeling of human lymphocytes
     with in vivo studies of survival and migration, Blood 38:360,
     1971.

28. J. L. Scott, J. G. Davidson, J. V. Marino, R. MacMillian,
     Leukocyte labeling with Cr-51. III. The kinetics of normal
     lymphocytes, Blood 40:276, 1972.

29. P. A. Strykmans, L. Debusscher, E. collard, Cell kinetics in CLL,
     Clin Haematol 6:159, 1977.

30. M. B. Bazerbashi, J. Reeve, I. Chanarin, Studies in CLL. The
     kinetics of Cr-51-labeled lymphocytes, Scand J Haematol 20:37,
     1978.

31. J. Manaster, J. Fribling, P. Strylsmans, Kinetics of lymphocytes
     in chronic lymphocytic leukaemia. I. Equilibrium between
     blood and a "readily accessible pool," Blood 41:425, 1973.

32. J. Wagstaff, C. Gibson, N. Thatcher, D. Crowther, The migratory
     properties of In-lll-oxine-labeled lymphocytes in patients
     with CLL, Br J Haematol 49:283, 1981.

33. J. L. Scott, R. MacMillan, J. V. Manno, J. G. Davidson, Leukocyte
     labeling with Chromium-51. IV. The kinetics of CLL
     lymphocytes, Blood 41:55, 1973.

34. G. H. Rannie and K. J. Donald, Estimation of the migration of
     thoracic duct lymphocytes to non-lymphoid tissues, Cell Tissue
     Kinet 10:523, 1977.

35. L. Moretta, M. Ferrarini, M. C. Mingari, A. Moretta, S. R. Webb,
     Subpopulations of human T cells identified by receptors for
     immunoglobulins and mitogen responsiveness, J Immunol
     117:2171, 1976.

36. S. Romagni, E. Maggi, Biaggiotti, M. Guidizi, A. Anadori, M.
     Ricci, Altered proportion of Tu and T cell subpopulations in
     patients with Hodgkin's disease, Scand J Immunol 7:511, 1978.

37. S. Gupta and C. T. C. Tan, Subpopulations of human T lymphocytes.
     XIV. Abnormality of T cell locomotion and of distribution of
     subpopulations of T and B lymphocytes in peripheral blood and
     spleens from children with untreated Hodgkin's disease, Clin
     Immunol Immunopathol 15:133, 1980.

38. M. De Sousa, M. Yamg, E. Lopes-Corrales, C. Tan, J. A. Hansen, B.
     Dupont, R. A. Good Ecotaxis: The principle and its
     application to the study of Hodgkin's disease, Clin Exp
     Immunol 27:143, 1977.

39. M. M. Black, S. R. Opler, F. D. Speer, Survival in breast cancer
     cases in relation to the structure of the primary tumor and
     regional lymph nodes, Surg Gyn Obst 100:543, 1955.

40. M. M. Zatz, A. White, A. L. Goldstein, Alterations in lymphocyte
     populations in tumorigenesis, J Immunol 3:706, 1974.

emitting radionucliedes for in vivo study, British Institute
Radiology, 1980.

50. R. Fleischmajer, V. Damiano, A. Nedwich, Alteration of
     subcutaneous tissue in systemic scleroderma, <u>Arch Dermatol</u>
     105:59, 1972.

51. A. J. Barnett, "Scleroderma," Charles C. Thomas, Springfield,
     Il., p 14, 1974.

52. M. A. Scheinberg, E. S. Cathcut, B and T cell lymphopenia in
     systemic lupus erythematosus, <u>Cell Immunol</u> 12:309, 1974.

53. G. P. Rodnan and T. A. Medoger, Musculoskeletal involvement in
     progressive systemic sclerosis (scleroderma), <u>Bull Rheum dis</u>
     17:419, 1966.

54. W. J. Williams, A study of Crohn's syndrome using tissue extracts
     and the kveim and mantoux tests, <u>Gut</u> 6:503, 1965.

55. J. V. Jones, J. Housley, P. M. Ashurst, C. F. Hawkins,
     Development of delayed hypersensitivity to
     dinitrochlorobenzene in patients with Crohn's disease, <u>Gut</u>
     10:52, 1969.

41. J. Wagstaff, C. Gibson, N. Thatcher, W. L. Ford, H. Sharma, D.
     Crowther (1981) Human lymphocyte traffic assessed by In-111-
     oxine labeling: clinical observations, <u>Clin Exp Immunology</u>
     43:443, 1981.

42. S. S. Froland, J. B. Natvig, G. Husby, Immunological
     characterization of lymphocytes in synovial fluid from
     patients with rheumatoid arthritis, <u>Scand J Immunol</u> 2:67,
     1973.

43. R. J. Winchester, J. B. Winfield, F. Siegel, P. Wernet, Z.
     Bentwich, H. G. Kinkel, Analysis of lymphocytes from patients
     with rheumatoid arthritis and systemic lupis erythematous, <u>J
     Clin Invest</u> 54:1082, 1974.

44. P. J. Sheldon, M. Pappamichail, E. J. Holborow, Studies on
     synovial fluid lymphocytes in rheumatoid arthritis, <u>Ann Rheum
     Dis</u> 33:509, 1974.

45. B. Vernon-Roberts, H. L. F. Currey, J. Penin, T and B cells in
     the blood and synovial fluid of rheumatoid arthritis patients,
     <u>Ann Rheum Dis</u> 33:430, 1974.

46. A. I. Brenner, M. A. Scheinberg, E. S. Cathcart, Surface
     characteristics of synovial fluid and peripheral blood
     lymphocytes in inflammatory arthritis, <u>Arthritis Rheum</u> 18:297,
     1975.

47. L. B. A. Van de Putte, C. J. L. M. Meijer, G. J. M. Lafebar, R.
     Kleinjan, A. Cats, Lymphocytes in rheumatoid arthritis and
     non-rheumatoid synovial fluids, <u>Ann Rheum Dis</u> 35:451, 1976.

48. U. Galili, L. Rosentahl, N. Galilia, E. Klein, Activated T cells
     in synovial fluid of arthritic patients: Characterization and
     comparison with in vitro activated human and murine T cells in
     cooperation with monocytes in cytotoxicity, <u>J Immunol</u> 122:878,
     1979.

49. D. A. Goodwin, Indium-labeled leukocytes in inflammatory
     diseases, Symposium on cell and protein labeling with gamma-

56. K. Parent, J. Barrett, I. D. Wilson, Investigation of the
     pathogenic mechanisms in regional enteritis with in vitro
     lymphocyte cultures, Gastroenterology 61:431, 1971.
57. P. J. Guillou, T. G. Brennan, G. R. Giles, Lymphocyte
     transformation in the mesenteric lymphnodes of patients with
     Crohn's disease, Gut 14:20, 1973.
58. D. B. Sachar, R. N. Taub, S. M. Brown, D. H. Present, B.
     Kovelitz, H. D. Janowitz, Impaired lymphocyte responsiveness
     in inflammatory bowel disease, Gastroenterology 64:203, 1973.
59. R. G. Strickland, G. Husby, W. C. Black, R. C. Williams,
     Peripheral blood and intestinal lymphocyte populations in
     Crohn's disease, Gut 16:847, 1975.
60. A. P. Douglas, A. P. Weetman, J. W. Haggith, The distribution and
     enteric loss of Cr-51-labeled lymphocytes in normal subjects
     and in patients with coeliac disease and other disorders of
     the small intestine, Digestion 14:29, 1976.

INDIUM-111 LABELED CELLS: NEW APPROACHES AND RADIATION DOSIMETRY

D. A. Goodwin and C. F. Meares*

Veterans Administration Medical Center
Palo Alto, California 94304,
Stanford University School of Medicine
Stanford, California 94305, and
*Department of Chemistry
University of California
Davis, California 95616

INDIUM-111 LABELED MONOCLONAL ANTIBODIES FOR CELL LABELING

Introduction

A large number of mouse monoclonal antibodies (MoAbs) are now available* to human leukocyte cell surface antigens. An incomplete list includes MoAbs specific for T lymphocytes; B cell associated antigen; a 24,000 dalton surface structure present on bone marrow lymphohemopoietic precursors and lymphoid leukemic cells; antigens primarily expressed on cells of T lymphocyte and monocyte/macrophase lineage; and "pan-T" reagent detecting a 65,000 dalton antigen (T65) common to all normal and leukemic T lymphocytes. This list is being increased monthly by several innovative companies* concerned with the development of new monoclonal antibodies. Using bifunctional chelates it is now possible to label MoAbs with In-111 for in vivo imaging(1). The advantages and disadvantages of using MoAbs are listed in Table 1.

---

*Hybritech, Inc., San Diego, California; Cetus Immune Corporation, 3400 W. Bayshore, Palo Alto, CA 94303; Monoclonal Antibodies, Inc., 2319 Charleston Road, Mountain View, CA 94040: See Linscott's Directory of Immunological and Biological Reagents, Second Edition, 40 Glen Drive, Mill Valley, CA 94941

Table 1.

Advantages

    1. Highly specific; e.g., T & B lymphocytes, lymphoma cells
    2. Label cells in plasma, or inject directly IV, or
        subcutaneously
    3. No cell separation needed
    4. Less nuclear radiation, therefore better lymphocyte viability
        than In-111-oxine
    5. Stable and irreversible in vivo

Disadvantages

    1. Labeling yield in vivo 50% (cf: 95% with oxine)
    2. Allergenicity of animal proteins
    3. Expensive

The Problem of In-111 Toxicity to In-111 Oxine Labeled lymphocytes

Although In-111 is a desirable marker for following cells in a variety of experimental settings, it has become increasingly clear that the conventional In-111 oxine labeling is toxic to the labeled population. This severel limits its usefulness for more refined long-term clinical applications.

Rannie et al(2) were the first to point out that $2\mu Ci/10^6$ ($1\mu Ci/10^6$ cells = 1 pecocurie/cell) rat lymphocytes almost completely abolished their capacity to mount a graft versus host (GVH) reaction. Frost(3) showed $0.1\mu Ci/10^6$ human tumor cells was toxic and also reported that doses of In-111 oxine in excess of $1\mu Ci/10^6$ mouse lymphocytes resulted in a normal initial organ distribution (useful in scanning) but diminished recirculation of the cells(4). This showed that homing was a less sensitive parameter of function than recirculation. Chisholm(5) reported abolition of colony formation for HeLa cells by $0.25\mu Ci/10^6$ cells and Kraal(6) reported impairment of colony formation when bone marrow cells were labeled with as little as $0.01\mu Ci/10^6$ cells.

Kinetics identical to cells labeled with Cr-51 have been obtained with sheep lymphocytes labeled with In-111 oxine $0.1\mu Ci/10^6$ cells(7). These authors also showed In-111 oxine at this level did not impair the proliferative response observed in the normal lymphocyte transfer reaction compared to control unlabeled cells. However, autoradiographic findings on tissue taken 24 hours after injection of In-111-oxine-labeled rat lymphocytes indicated $0.2\mu Ci$ In-111/10^6 lymphocytes was the upper limit compatible with survival of the labeled cells(8).

Also, the migration of guinea pig T cells to a skin inflammatory site was markedly reduced, compared with Cr-51-labeled cells, with even low doses of In-111 oxine ($0.01\mu Ci - 0.01\mu Ci/10^6$ cells) which did not impair lymph node localization(9).

These various experiments show that radiosensitivity from In-111-oxine labeling varies considerably, depending on both the cell type and the function measured. Curiously, the neutrophil appears uniquely resistant among nucleated cells.

The "upper limits: of In-111 labeling of lymphocytes discussed above (approximately 0.1 to 0.2 pCi/cell) are suprisingly close to the levels reported(10) to produce radiotoxicity with the nuclear labels bromine-77 bromodeoxyuridine (0.13 pCi/cell), I-125-5-iododeoxyuridine (0.1 pCi/cell), and H3 thymidine (1.64 pCi/cell) in Chinese hamster ovary cells in culture. The close agreement obtained from these experiments using specific nuclear DNA labels containing nuclides emitting low energy beta emissions and Auger electrons in the same energy range and concentration as In-111 (2.4 keV LMM Auger electrons, mean range $\simeq 0.23\mu$) gives support to the hypothesis that the radiotoxicity seen with In-111-oxine labeling of lymphocytes is at least partly due to concentration of the label in the radiosensitive nucleus of the cell.

We and others have recently shown(11,12) that the subcellular location of In-111 may be resolved to a distance of approximately 0.2μ using the short tracks produced by the low energy Auger electrons seen on high resolution electron microscope autoradiographs (EMAR). Using this method, we have shown that a large fraction ($\simeq 80\%$) of the In-111 in human lymphocytes (as well as neutrophils and macrophages) labeled with In-111-oxine is localized in the nucleus. The "self" irradiation dose to In-111-labeled cells that results mostly from these Auger electrons is an extremely large value; approximately 0.138 rads/decay deposited in a very small volume(13,14).

Improvements in the chemistry of In-111-oxine cell labeling techniques have mainly involved the development of related water-soluble chelates. Presumably, the mechanism of cell labeling is identical for all these derivatives and would therefore produce the same nuclear labeling as oxine. The monoclonal antibody approach, however, offers some very promising ways of circumventing In-111 radiotoxicity to lymphocytes. These include labeling the membrane rather than the nucleus and the direct injection of In-111 MoAbs IV or subcutaneously.

Historical Review of Protein Labeling with Bifunctional Chelates

An outline of the chemical and biological studies leading to In-111-labeled MoAbs is given in Table 2.

Table 2.

Date	Chemistry	Biological Studies
1969	Physical/chemical studies using In-111 - PAC as a macromolecular probe (15).	In vivo, animal and clinical tracer studies with In-111 (transferrin and simple chelates) (16). Develop dual channel gamma camera for increased In-111 counting efficiency(17).
1972	PAC studies in vivo in mice showed In-111 EDTA stable(18).	Study of In-111 citrate and EDTA. In vivo in mice using PAC(18).
1973	Synthesis of the first bifunctional chelate and its diazonium coupling to albumin and fibrinogen(19,20).	Studies of In-111 chelate conjugates of albumin as prototype molecule; comparison with data available on I-131 albumin (21).
1974	Improved methods for coupling, purifying, and attaching chelate conjugates to albumin and fibrinogen.	Biological properties of In-111 protein conjugates establish in vivo stability of the label and lack of toxicity (21).
1975	Quantitative chemical studies suggest conformational changes likely in Azo-Albumin(22).	Half-life plasma disappearance in humans InAzo-Alb 7 days (vs I-131 Alb 19 days(22).
1976	Chelate complexes of metals shown to be kinetically inert in vitro(23).	Metal chelate-protein conjugates extremely stable at tracer concentrations in living organisms(22).
1977	Alkylabumin synthesis using a new bifunctional chelate BAPE (mild conditions) and characterization by metal ion titration using both PAC and fluorescence(25).	Half-life plasma disappearance 18 days; identical to I-125 albumin: stable tracer 30 days in vivo. Reacts with specific antibody in vitro with similar affinity as I-125 albumin (24).

Table 2. Continued

1979	New synthetic route to bifunctional chelates from any alpha alpha amino acid: new p-carboxyl coupling agent(26).	
1980	Synthesis using a new benzyl derivative (BABE)(25).	
1982	Development of an isothicyanate reagent. Labeling of IgG and transferrin with isothiocyanate reagent. Labeling of IgG and transferrin with isothiocyanate and BABE reagents.	Biological studies of chelate labeled transferrin and IgG: high tumor uptake in mouse model. In vivo stability $> I-125$ label(27,28).
1983		In-111 MoAb to BALB/k B lymphocyte shows high specificity and lymphocyte uptake in vivo (1).

Results With Labeled MoAbs

Most of the information on radiolabeled MoAbs has been obtained with radioiodine; either I-125 or I-131 and more recently also I-123. All of these radionuclides of iodine have physical limitations for imaging, but more importantly the labeled MoAbs are unstable in vivo due to rapid deiodination by the enzyme deiodinase in the liver (29,30). Bifunctional chelation techniques present an alternative way of radiolabeling these proteins(1,31). Table 3 compares chelate conjugation and conventional halogenation for labeling MoAbs.

Excellent lymph node uptake has been reported for I-125-labeled mouse monoclonal anti-Thy 1.1 of both the intact IgG molecule and the Fab fragment(32). These workers found at optimal times (20 hours post injection), that the lymph nodes of AKR/j mice (Thy 1.1) contained nine fold more antibody than the lymph nodes of AKR/cu mice (Thy 1.2). Even more striking was the localization of the $F(ab')_2$ fragment of the monoclonal antibody showing 144-fold increase in the lymph nodes of mice with the Thy 1.1 antigen. Lymph nodes contained 4.2% of the total injected dose per gram of the intact antibody. Weinstein et al (33) have obtained similar results with subcutaneous injection of I-125 labeled MoAbs in mice, with greater concentrations in regional draining nodes.

We have obtained excellent lymph node uptake and images of mouse regional lymph nodes following both IV and subcutaneous injection of

Table 3. Comparison of MoAb Labeling Techniques

Characteristics of Labeled MoAbs	Halogenated (Directly Labeled)	Radiometal (Chelate Conjugates)
Stable bond: No composition in vivo	No: Rapid dehalogenation (Bond stability $I < Br < F$)	Yes
Choice of "site selective" reagents	No	Yes: Away from Ab binding site
Mild labeling conditions	ICL and enzymes No: Chloramine T	Yes: BABE, CITC No: Diazonium
High radiolabeling yield	ICL $\approx$ 80% Chlormaine T $\approx$ 90%	Chelation – 100%
High specific activity	Yes: < 100Ci/mM (1–10μCi/μg)	Yes: < 100Ci/mM (1–10μci/μg)
"Instant Kits" available	No	Yes
Optimum gamma energy and low patient radiation (Imaging)	Yes: I-123	Yes: In-111
Optimum physical characteristics for possible therapy	Yes: At-211	Choice: Bi-210, Y-90, etc.
Positron emitters available	Yes: F-18, Br-76	Yes: Ga-68
Labeled molecule biologically active	Yes	Yes
Decomposition products rapidly excreted	Yes: (if thyroid blocked) through kidney	Yes: through kidney, No thyroid, salivary gland, or stomach uptake

*Obtained from Professor H. McDevitt, Department of Immunology, Stanford University School of Medicine, Stanford, CA 94305

In-111 MoAb(1). We labeled mouse monoclonal anti-1a^k* (BALB/k)
lymphocyte antibody (MoAb) with bromacetamidobenzyl-EDTA (BABE)
In-111. The labeled MoAb contained approximately 1 chelating group
per molecule and had a specific activity of approximately 300Ci/m
mole. The organ and lymph node uptakes were measured 24 hours after
i.v. and subcutaneous (both hind footfpads) injection in BALB/k and
control BALB/c mice. The regional node uptake of subcutaneous In-111
MoAb was compared with In-111 colloid. After i.v. injection the
BALB/k spleen (SP), peripheral (PN) and mesenteric nodes (MN) all had
significant higher In-111 MoAb concentration than BALB/c ($p < .01$); SP
116%/gm (10.5% of dose), PN 18%/gm (0.4% of dose) and MN 17.4%/gm
(0.9% of dose). Following subcutaneous injections, the regional
draining nodes contained 244%/gm (1.6% of dose), MN 17%/gm, cervical
nodes 20%/gm, and SP 95%/gm (8.3% of total dose) ($p < .01$). The
BALB/k to BALB/c ratios were: regional nodes 11:1, MN 5.3:1, cervical
nodes 4.4:1. After subcutaneous In-111 colloid, there was no significant difference in organ and lymph node uptake between BALB/c and
BALB/k. Absorption of subcutaneous In-111 MoAb was better than In-111
colloid ($p < .001$): approximately 10% remaining vs approximately 30%
colloid.

This study demonstrated the high degree of specificity of In-111
labeled MoAb in vivo and showed the feasibility of using IV and
subcutaneously injected In-111 MoAbs to T and B lymphocytes and
lymphoid leukemia cells as well as other tumor cell markers for
imaging in humans.

Discussion

Antibodies on the surface of lymphocytes will probably not
prevent in vivo migration per se. It has been shown that it is
possible to transfer B memory of T cell function by immunoglobulin
(Ig) coated cells after fluorescence activated cell sorter (FACS)
separation. It has also been shown that there is only a slight
influence of anti-Ig antibodies(34) or of anti-T 200 antibodies on
lymphocyte homing in the mouse. The technique has several potential
advantages when compared with the analogous antitumor antibody
technique for tumor imaging, such as the use of radiolabeled anti-
carcinoembryonic antigen (anti-CEA)(36). The leukocyte determinants do
not elute from the cells, and therefore there will not be a high
circulating plasma background such as may occur with CEA.

In principle, using the appropriate conditions and bifunctional
reagents, it should be possible to direct the chelate label away from
the Fab antigen binding site by attaching it to another part of the
molecule such as the Fc fragment, thereby preserving antibody binding
affinity. We have already used this principle in attaching a chelate
to the terminal amine group of bleomycin (which is variable in nature)
and in this way have successfully preserved its biological ability to
concentrate in tumor cells(37).

Halpern et al(29) have used the bifunctional chelate technique to label monoclonal anti-CEA antibody and compared it with conventional I-125-labeled antibody in a nude mouse human colon tumor model. They found the chelate-labeled antibody not only retained its binding properties with high tumor uptake but was also considerably more stable in vivo than I-125 antibody. Other workers have used murine erythroleukemia as a model system of the study of targeting of monoclonal antibody labeled by a variety of chelate conjugation techniques(30). They also reported little loss of binding affinity with in vitro tests using tumor cells and in vivo imaging studies of leukemic mice. They also found chelate-labeled antibody more stable in vivo than I-125 antibody(38).

When labeled MoAbs are used directly intravenously or subcutaneously to label leukocytes, high blood background at the time of imaging will be a problem just as it is for radioimmunodetection of tumors. We have shown injection of a second antibody (specific for the first) will clear the blood activity in approximately 5 minutes when given just prior to imaging(39). The immune complex formed in the circulation is rapidly removed by the liver. Of course, if cells are labeled in vitro, the unbound activity can be removed prior to reinjection of the labeled cells.

Summary

The use of In-111 monoclonal antibodies has the potential for providing a highly specific cell label that will make separation and purification of cells unnecessary and allow labeling in plasma. Monoclonal antibodies are highly specific not only for cell type but also for subtypes, such as T and B lymphocytes; they may be less toxic than nuclear labels (In-111-oxine), and, as mentioned, the indium chelate is more stable than radioiodine in vivo. In addition to labeling lymphocytes in vitro (followed by injection of labeled cells), these monoclonal antibodies may also be injected directly intravenously or subcutaneously to localize lymphocytes and lymph nodes (as well as other normal or leukemic cells), thus avoiding toxicity from cell manipulations and radiation incurred during cell labeling and migration.

THE DISTRIBUTION AND DOSIMETRY OF IN-111 LABELED LEUKOCYTES AND PLATELETS IN HUMANS

Introduction

The indium oxine method introduced in 1976 by McAfee and Thakur (40) has proven useful for labeling a wide variety of living cells for both in vitro and in vivo use. The most important clinical application is the identification and localization of abscesses by scintillation scanning after injection of In-111-labeled granulocytes ("WBC Scan"). Because of the high sensitivity and specificity of the

method, its use is becoming more common in large medical centers
(41,42,43,44). Indium-111-labeled platelets and lymphocytes are also
beginning to be used more commonly in clinical investigation and
research, and their use will probably continue to increase as these
studies have many promising clinical applications(45,46,47,11).

It is important, therefore, to obtain accurate estimates of the
radiation dose received by human subjects undergoing these procedures.
Data on radiopharmaceutical distribution in animals may, in certain
cases with suitable conversions (usually some power function of weight
and surface area)(48,49), be extrapolated to obtain likely human
distributions for the purpose of radiation dose calculation. Marked
interspecies differences in peripheral blood cell counts (e.g.,
rabbit; 60% lymphocyte count) and function e.g., dog; hyperactive
coagulation - fibrinolytic system, a large contractile spleen) make
this method inaccurate when extrapolating autologous blood cell
distributions from animal to man. Direct human observations are
therefore necessary. In this paper, we report the distribution and
dosimetry of In-111-labeled autologous leukocytes and platelets
measured in human patients.

Methods

The subjects were patients referred to the Nuclear Medicine
department for abscess localization or entered in investigative
protocols to study carotid atherosclerotic ulcers with In-111
platelets or chronic inflammatory diseases with In-111 lymphocytes.
An average of 1.0mCi In-111 WBC's (mixed leukocytes, predominantly
neutrophils ≈ 80%), and 0.5 mCi of In-111 lymphocytes and platelets
were used for each study. The resultant calculated radiation doses
were normalized to 0.5mCi administered In-111 activity. Several
patients were studied immediately or shortly (1-4 hours) following
injection, and all were studied 18-24 hours following injection.

Whole body scanning was performed with an Ohio Nuclear large
field of view gamma camera equipped with a parallel hole, medium
energy (250 keV) collimator. With this system in the whole body mode
the detector is electronically masked to give a sensitive rectangle
20cm x 32cm. The detector makes two sweeps: one down and one up the
body with one index of the bed to include all parts of the body. To
perform the posterior view, the camera head is repositioned underneath
the bed and the procedure repeated without any movement of the patient
being necessary. All scans were done at the same speed (20cm/min).
The entire procedure including anterior and posterior views takes 20
minutes per view, or 40 minutes total. The final image produced on
the oscilliscope (and film) by the camera positioning circuitry (which
corrects for the scanning motion of the detector) is a minified
rectangle about 1/20th actual size, containing whole body.

The X and Y coordinate signals producing the imaging on the
oscilliscope were simultaneously digitized and stored in list mode on

the disk of an Informatek computer. The list of X and Y coordinates
were than framed in a 64 x 64 matrix which was displayed in a color
coded format on a TV monitor. Regions of interest were manually
chosen over the liver, spleen, body background region, and, where
applicable, lungs, abscess, and injection site. The net counts from
each region were calculated assuming that the background, obtained
from a region near the spleen but containing no underlying bone
marrow, was uniform throughout the whole body, and then subtracting
this value for the appropriate area (number of pixels) underlying the
chosen ROI. The geometric mean of the net counts from the anterior
and posterior view ROIs was calculated. This value was expressed as a
percent of the geometric mean of the total counts in the anterior and
posterior views (minus activity in the injection site if any).

Dosimetry calculations were performed by taking the activity in
the liver, spleen, and, in applicable cases, lung, abscess, etc., as a
percentage of the injected dose. The remaining activity, concentrated
mainly in the bone marrow, or in the blood in the case of platelets,
is termed "rest of body." The cumulated activity, $\tilde{A}$, was calculated
for the ROI assuming the effective half-life was equal to the physical
half-life for WBC's and lymphocytes. The effective half-life for
platelets was taken as 1.6 days, accounting for the linear disappearance of the cells from the blood to the liver and subsequent decay.
The doses were calculated using the "S values" as tabulated in MIRD 11
(50). The WBC's, the "S values" for the red marrow, considered as
both a source and a target organ, were used for "rest of body." For
lymphocytes, the "S values" for red marrow were modified to include
the weight of the lymphatic tisue and were then used for the "rest of
body.F"

Results

In-111 WBC's (Predominantly Neutrophils: 80%). The highest
concentration is seen in the spleen, followed by the liver and bone
marrow. The organ distribution and dosimetry for 6 patients with
normal scans and 4 patients with abnormal scans is shown in Table 4.
There is no significant difference in the organ distribution in these
two groups with approximately 20% in spleen, 20% in liver, and 60%
predominantly in bone marrow with a small amount of this fraction
distributed through the body tissues.

Occasionally ($\approx$ 5% of the time) lung activity is seen especially
early after injection, probably secondary to clumping of the cells
produced by trauma during the labeling procedure. In 2 of our 10
cases shown in Table 1, this amounted to an average of 15 $\pm$ 1% of the
cells producing a dose of $\approx$ 1.25 rads to the lungs. In the case with
a liver abscess, the background ROI was taken over an adjacent area of
normal liver and subtracted from the abscess ROI. The abscess
contained $\approx$ 6% of the injected WBC's.

Table 4. Six Patients with Normal Scans (24 Hrs)

	Liver	Spleen	Lung (Bad Prep.)	Rest of Body*
% Distribution	19 ± 6 (6)	19 ± 6 (6)	0	62 (4)
Mean ± SD # of Pts = ( )			15 (2)	47 (2)
Rad/500μCi	1.32 (4) 1.36 (2)	8.51 (4) 8.54 (2)	0 1.24 (2)	2.30 (4) 1.83 (2)

4 Patients with Abnormal Scans (24 Hrs)

	Liver	Spleen		Rest of Body*
% Distribution	18 ± 5 (4) 6 (1 liver abscess)	21 ± 10 (4)		61 (3) 55 (1)
Rad/500μCi	1.26 (3) 1.39 (1)	9.38 (3) 9.38 (1)		2.30 (3) 2.09 (1)

*Mainly bone marrow (85%), so dose is that to bone marrow.

(Reprinted with permission from the Third International Radiopharmaceutical Symposium, 1980, FDA 81-8166, pp. 88-101.)

The distribution of WBC's in 2 patients who had undergone a
splenectomy is shown in Table 5. After background subtraction, the
net activity in the splenic area was very nearly zero (≈ 0.7%). The
normal splenic uptake (≈ 20%) was shifted almost equally to liver (↑
liver 0.4 rad, ↑ bone marrow 0.4 rad). Some idea of the reproducibility of the method is given by the results of three separate studies
on the same patient. This patient had previously undergone surgical
drainage of a large abscess in the splenic bed diagnosed on a previous
WBC scan, so some resolving activity was seen on the first postoperative scan of 6/19/80: Table 6. Also shown in Table 6 is the 4 hour
and 24 hour distribution in another splenectomized patient showing
little change in distribution during this time.

Indium-111 Lymphocytes. Due to the high probability of a variable
amount of radiation damage to the highly radio-sensitive lymphocytes,
(2,5) the results presented in Table 7 are considered only preliminary
and subject to revision. The similarity to WBC distribution is seen,
with the additional distribution in lymphatic tissue of a small amount
we were unable to quantitate in the present study due partly to the
low resolution of the whole body image and partly to overlying bone
marrow structures. The additional uptake in lymphatic tissue may

partly explain the suggested slightly lower uptake in liver and spleen
compared to WBC's.

Table 5.   Distribution of In-111 WBC's at 24 hours: 2 Patients with
           Splenectomy, Otherwise Normal Scans

	Liver	Spleen	Lung (Bad Prep.)	Rest of Body*
% Distribution	26 ± 2 (2)	0.7 ± .5 (2)	0	73.3 (1)
Mean ± SD # of Pts = ( )			11 (1)	62.3 (1)
Rad/500μCi	1.74 (1) 1.77 (1)	0 0	0 0.96 (1)	2.70 (1) 2.35 (1)

*Mainly bone marrow (85%), so dose is that to bone marrow.

(Reprinted with permission from the Third International Radiopharmaceutical Symposium, 1980, FDA 81-8166, pp. 88-101.)

Table 6.   Reproductability of the Method: In-111 WBC Distribution
           (Both Patients Splenectomized) (Patient No. 1 Had as
           Abscess in the Splenic Bed)

Patient	Time of Scan	Date of Scan	Liver	Spleen	Rest of Body	Lung
1	24 hours	06/19/80	29%	4% (abscess)	67%	0
	24 hours	07/03/80	43%	1% (abscess resolved)	56%	0
	24 hours	07/30/80	29%	0.4%	61%	0
2	4 hours		17%	0.4%	63%	20%
	24 hours		20%	1%	63%	17%

*Mainly bone marrow (85%)

(Reprinted with permission from the Third International Radiopharmaceutical Symposium, 1980, FDA 81-8166, pp. 88-101.)

Table 7. Distribution of In-111 Lymphocytes at 24 Hours in Two
          Patients With Normal Scans

	Liver	Spleen	Rest of Body*
5 Distribution	11 ± 2	15 ± 3	74 ± 12
Rad/500µCi	0.76	6.70	1.40

*Mainly bone marrow (35%) plus lymphatic tissue, so dose is that to
bone marrow.

Table 8. Distribution of In-111 Platelets. Time = 24 Hours

	Liver	Spleen	Rest of Body*
2 normal patients % Distribution	8.5 ± 0.05	23.5 ± 4.5	68 ± 4
Rad/500µCi	3.15	8.60	0.32
1 Patient with spleen 3 x normal size			
% Distribution	4	33	63
Rad/500µCi	2.64	5.04	0.26

*Blood; dose is that total body from circulating blood.

(Reprinted with permission from the Third International
Radiopharmaceutical Symposium, 1980, FDA 81-8166, pp. 88-101.)

In-111 Platelets. The most striking difference in the distribution from that of leukocytes is the high blood concentration and
abscence of bone marrow uptake at 24 hours. Blood disappearance
curves in normal patients show that between 40 and 70% of the platelets remain circulating in the intravascular compartment initially,
and that they disappear in a linear arithmetic fashion with approximately a 9-day life-span. The distribution in two normal patients and
one with a spleen enlarged three times normal size is seen in Table 8.
The amount in the whole body minus liver and spleen at 24 hours (68%)
correlates well with the independently determined blood levels (40-
70%) at the same time. In the kinetic model used to determine the
integral radiation dose, the liver was the main site of destruction
and sequestration of platelets leaving the blood as shown by others

(51). In our experimental subjects no lung uptake was seen, although
in our clinical experience it occurs with about the same frequency ($\approx$
5%) as with labeled leukocytes with approximately the same resultant
radiation exposure shown in Table 6 WBC's.

Discussion

Normally no radioactivity is excreted following injection of
In-111-labeled cells. Indium-111 does not elute from the living
cells. Following cell death and release of the In-111, it is probably
either firmly bound by proteins such as transferrin or is hydrolyzed
to the extremely insoluble hydroxide $In(OH)_3$, and sequestered in the
reticuloendothelial system(52). Patients studied immediately or
shortly after i.v. injection, and again at 18-24 hours, had decay
corrected total body counts at 24 hours equal to the 1-4 hour total
body counts $\pm$ 10%.

The accuracy of our whole body counting method was tested in
other pharmacokinetic experiments using In-111 BLEDTA in patients in
whom urine excretion was carefully measured over 24 hours(53). Since
tne urine is the major route of excretion of In-111 BLEDTA, containing
> 90% of the radioactivity lost to the body, we were able to correlate
percentage dose excreted by urine counting and percentage dose
retained by whole body counting. These values were within 10%, the
larger excretion value (50%) being obtained by whole body counting.
This may be explained partly by some excretion of In-111 BLEDTA in the
feces (not collected) in addition to the urine. Absolute calibration
with a whole body phantom was not performed since our system was very
sensitive ($\pm$ 10%) to relative changes of activity in the human
subject. The techniques we employed partly correct for variations in
organ location and depth within the body by using two views 180°
opposed and a mathematical combination (in our case the geometric
mean), relatively independent of source depth and volume. The
validity of the geometric mean technique has been confirmed by Heyns
et al and Scheffel et al by comparing this estimate of organ
In-111-oxine-labeled platelet concentration in baboons and rabbits
with the activity measured in the excized organs(54). Errors reported
for this method are on the order of $\pm$ 10%, similar to that which we
observed here(55,56,57). A great advantage for clinical use is the
simplicity and availability of the equipment to many Nuclear Medicine
laboratories. Other commercially available gamma cameras with whole
body scanning capability and computers could be employed as well.

More elaborate methods have been published for absolute measurement of activity using a gamma camera and computer combination(58,59).
These methods require corrections for the varying attenuation of gamma
photons in body tissue and usually require a measurement of transmission of gamma photons of the same energy through the body, or in
tissue equivalent material to obtain an absorption coefficient.

Errors with these methods have also been reported to be on the order of ± 10%.

Another source of error in our method is the determination of organ boundaries by the manual technique(60). The scaled color display with background subtraction helped by enhancing the edges, but inconsistancies in manual drawing of the ROI are unavoidable. A method that might improve consistancy employs a computer algorithm to detect organ boundaries based on a preselected percentage of maximum levels, so called "thresholding"(61).

The usual method of predicting the radiopharmaceutical distribution of a new agent prior to human use is to extrapolate from animal data, preferably from two species. Various mathematical transformations, such as power functions of body weight ratios between animal and "standard" man or simple transforms of the time variable are used to correct for species differences(49). These methods work fairly well for water soluble compounds distributed primarily in extracellular fluid and excreted mainly by glomerular filtration. The method is less successful with intracellular or actively transported agents. Marked species differences have been observed in the half-life in circulating blood of labeled autologous blood cells(62). Because of this and also notable species differences in types and function of circulating cells already mentioned for rabbit lymphocytes and dog platelets, extrapolation from animal data is likely to give erroneous results for estimating human cell distributions.

Williams et al(63) have reported results in six normal patients for the distribution of In-111-labeled WBC's at 4 hours and 24 hours. They employed a single posterior view and a liver phantom correction for liver uptake. Their results (rad/0.5mCi) were liver, 1.6 (14% higher than our result); spleen, 11.7 (38% higher than our result); and bone marrow, 0.325 (86% lower than our result). The posterior view may have contributed to their high spleen value due to its proximity to the back. Our visual estimate of 85% concentration in bone marrow may be overestimating the true value and contribute to our higher bone marrow dose.

Van Reenen et al(64) have recently reported the radiation dose from human platelets labeled with In-111: spleen, 13.7 rads/0.5mCi; liver, 1.8 rads/0.5mCi; whole body, 0.45 rads/0.5mCi. These values are 55% higher for spleen and 45% lower for liver than our results. As these authors demonstrated by conjugate counting, their one-view surface counting method overestimates the spleen and underestimates the liver concentration by about 20% each. Additional errors in their method arise from the use of diverging collimator and lack of background subtraction. Heyns et al(54), using the geometric mean, report equilibrium percentage distribution values in six humans much closer to those we observed in our smaller sample (spleen, 31.1 ± 6.1; liver, 9.6 ± 1.2) at 24 hours.

In addition to estimating radiation dose, the whole body scanning method should provide a clinical approach to quantitating cell migration and kinetics. For example, the approximate number of granulocytes entering an abscess or platelets incorporated in a clot could be calculated from the blood cell count and the fractional uptake in 24 hours. This type of quantitation promises to provide new information on lymphocyte kinetics and migration as well.

Acknowledgments

We thank Dr. H. McDevitt for the generous gift of anti Ia^k MoAb. This work was supported by a Veterans Administration Research Grant and PHS Grant Number 5 ROI CA 28343 (D Goodwin) and CA 16861, RCDA CA 00462 (C Meares).

REFERENCES

1. D. A. Goodwin, C. F. Meares, M. J. McCall, H. O. McDevitt, M. McTigue, C. I. Diamanti, Lymphoscintigraphy with In-111 labeled monoclonal antibodies, Clin Nucl Med 8P26m (Abstr) (1983).
2. G. H. Rannie, M. L. Thakur, W. L. Ford, An experimental comparison of radioactive labels with potential application to lymphocyte migration studies in patients, Clin Exp Immunol 29:209 (1977).
3. P. Frost, R. Wiltrout, Z. Maciarowski, N. R. Rose, An isotope release cytotoxicity assay applicable to human tumors, Oncology 34:102 (1977).
4. P. Frost, J. Smith, H. Frost, The radiolabeling of lymphocytes and tumor cells with Indium-111, Proc Soc Exp Biol Med 157:61 (1978).
5. P. M. Chisholm, J. Danpure, G. Healey, Cell damage resulting from the labeling of rat lymphocytes and HeLa S3 cells with In-111 oxine, J Nucl Med 20:1308 (1979).
6. G. Kraal, A. A. Geldof, Radiotoxicity of Indium-111, J Immunol Methods 31:193 (1979).
7. T. Tssekutz, W. Chin, J. B. Hay, Measurement of lymphocyte traffic with Indium-111, Clin Exp Immunol 39:215 (1980).
8. S. M. Sparshott, H. Sharma, J. D. Kelley, W. L. Ford, Factors influencing the fate of Indium-111-labeled lymphocytes after transfer to syngeneic rats, J Immunol Methods 41:303 (1981).
9. A. C. H. M. Van Dinther-Janssen, R. J. Scheper, Restriction to the use of Indium-111 oxine as a radiolabel in lymphocyte migration studies, J Immunol Methods 46:353 (1981).
10. A. I. Kassis, S. J. Adelstein, C. Haydock, Uptake and radiotoxicity of Rb-77 bromodeoxyuridine in mammalian cells, J Nucl Med 22:P44 (Abstr) (1981).
11. D. A. Goodwin, J. R. Heckman, L. F. Fajardo, A. Calin, S. J. Propst, C. I. Diamanti, Kinetics and migration of Indium-111 labeled human lymphocytes, in: "Proceedings of the

International Symposium on Medical Radionuclide Imaging, International Atomic Energy Agency," SM-247/95, Vienna, pp. 437-497 (1981).

12. H. H. Davis II, R. M. Senior, G. L. Griffin, C. Kuhn III; Indium-111-labeled human alveolar macrophages and monocytes: function and ultrastructure, J Immunol Methods 36:99 (1980).

13. D. J. Silvester; Consequence of Indium-111 decay in vivo: calculated absorbed radiation dose to cells labeled by indium-111 oxine; J Labeled Compds and Radiopharma 16:193 (1979).

14. D. A. Bassano, J. G. McAfee; Cellular radiation doses of labeled neutrophils and platelets, J Nucl Med 20:255 (1979).

15. C. F. Meares, M. W. Sundberg, J. B. Baldeschwieler, Perturbed angular correlation study of a haptenic molecule, in: "Proceedings of the National Academy of Sciences USA," 69:3718 (1972).

16. D. A. Goodwin, R. Goode, L. Brown, C. J. Imbornone, In-111 labeled transferrin for the detection of tumors, Radiology 100:175 (1971).

17. D. A. Goodwin, D. Menzimer, R. DelCastilho, A dual-spectrometer system for high efficiency imaging of multi gamma emitting nuclides with the auger gamma camera, J Nucl Med 11:221 (1970).

18. D. A. Goodwin, C. H. Song, C. F. Meares, The study of Indium-111 labeled compounds in mice using perturbed angular correlations of gamma radiation, Radiology 105:699 (1972).

19. M. W. Sundberg, C. F. Mears, D. A. Goodwin, C. I. Diamanti, Selective binding of metal ions to macromolecules using bifunctional analogs of EDTA, J Med Chem 17:1304 (1974).

20. M. W. Sundberg, C. F. Meares, D. A. Goodwin, C. I. Diamanti, Chelating agents for the binding of metal ions to macromolecules, Nature 250:587 (1974).

21. D. A. Goodwin, C. F. Meares, C. I. Diamanti, M. W. Sundberg, Bifunctional chelates for radiopharmaceutical labeling, Nuclear Medizin XIV: 365 (1975).

22. C. S-H Leung, C. F. Meares, D. A. Goodwin, The Attachment of metal-chelating groups to proteins: Tagging of albumin by diazonium coupling and use of the products as radiopharmaceuticals, Intl J Appl Radiat Isot 29:687 (1978).

23. S. M. Yeh, C. F. Meares, D. A. Goodwin, Decomposition rates of radiopharmaceutical indium chelates in serum, J Radioanalytical Chem 53:327 (1979).

24. D. A. Goodwin and C. F. Meares, Bifunctional chelates for radiopharmaceutical labeling, in; "Radiopharmaceuticals: Structure-Activity Relationships," R. P. Spencer, ed., Grune & Stratton, New York (1981).

25. L. DeRiemer, C. F. Meares, D. A. Goodwin, C. I. Diamanti, BLEDTA: Tumor localization by a bleomycin analog containing a meta-chelating group J Med Chem 22:1019 (1979).

26. S. M. Yeh, D. G. Sherman, C. F. Meares, A new route to "bifunctional" chelating agents: conversion of amino acids to

analogs of ethylenedinitrilotetraacetic acid, Anal Biochem 100:152 (1979).

27. D. A. Goodwin, C. F. Meares, C. I. Diamanti, J. J. McCall, C. D. Lai, F. M. Torti, B. C. Martin, Use of specific antibody for rapid clearance of circulating blood background from radiolabeled tumor imaging proteins, J Nucl Med 24:P31 (abstr) (1983).

28. D. A. Goodwin, C. F. Mears, C. I. Diamanti, M. J. McCall, H. H. Sussman, C. D. Lai, C. H. Song, Indium-111 chelate conjugates of transferrin for tumor imaging, J Nucl Med 24:P32 (abstr) (1983).

29. S. E. Halpern, P. L. Stern, P. L. Hagen, A. Chen, S. G. David, W. J. Desmond, T. H. Adams, R. M. Bartholomew, J. M. Frincke, C. E. Brautigam, Radiolabeling of monoclonal antitumor antibodies. Comparison of I-125 and In-111 anti CEA with Ga-67 in a nude mouse-human tumor model, Clin Nucl Med 6:453 (abstr) (1981).

30. D. A. Scheinberg and M. Strand, Leukemic cell targeting and therapy by monoclonal antibody in a mouse model system, Cancer Res 42:44 (1982).

31. D. J. Hnatswich, W. W. Layne, R. L. Childe, D. Lanteigne, M. ^ Davis, T. W. Griffin, P. W. Doherty, Radioactive labeling of antibody: a simple and efficient method, Science 220:613 (1983).

32. L. L. Houston, R. C. Nowinski, I. D. Berstein, Specific in vivo localization of monoclonal antibodies directed against the Thy 1.1 antigen, J Immunol 125:837 (1980).

33. J. N. Weinstein, R. J. Parker, A. M. Keenan, S. K. Dower, H. C. Morse, S. M. Sieber, Monoclonal antibodies in the lymphatics: toward the diagnosis and therapy of tumor metastases, Science 218:1334 (1982).

34. Y-H. Chin, G. D. Carey, J. J. Woodruff, Lymphocyte recognition of lymph node high endothelium IV. Cell surface structures mediating entry into lymph nodes, J Immunol 129:1911 (1982).

35. W. M. Gallatin, I. L. Weissman, E. C. Butcher, A cell surface molecule involved in organ specific homing of lymphocytes Nature 303:30-34 (1983).

36. F. H. DeLand, E. E. Kim, G. Simons, D. M. Goldenberg, Imaging approach in radioimmunodetection, Cancer Res 40:3046 (1980).

37. D. A. Goodwin, C. F. Meares, L. H. DeRiemer, C. I. Diamanti, G. L. Goode, J. E. Gaumert, D. J. Sartoris, R. L. Lantieri, H. D. Fawcett, Clinical studies with indium-111 BLEDTA, a tumor imaging conjugate of bleomycin with a bifunctiona chelating agent, J Nucl Med 22:787 (1981).

38. D. A. Scheinberg, M. Strand, O. A. Gansow, Tumor imaging with radioactive metal chelates conjugated to monoclonal antibodies, Science 215:511 (1982).

39. D. A. Goodwin, C. F. Meares, C. I. Diamanti, M. McCall, C. D. Lai, F. M. Torti, M. McTigue, B. C. Martin, Use of specific antibody for rapid clearance of circulating bold background

from radiolabeled tumor imaging protein, <u>Eur J Nucl Med</u> 9:209-215 (1984).

40. J. G. McAfee and M. L. Thakur, Survey of radioactive agents for in vivo labeling of phagocytic leukocytes. I. Soluble agents, <u>J Nucl Med</u> 17:480 (1976).

41. M. L. Thakur, T. P. Lavender, R. M. Arnot, Indium-111 labeled autologous leukocytes in man, <u>J Nucl Med</u> 18:1012 (1977).

42. D. A. Goodwin, P. W. Doherty, I. R. McDougall, Clinical use of indium-111 labeled white cells - an analysis of 312 cases, in: "Indium-111 labeled Neutrophils, Platelets, and Lymphocytes, Proceedings of the Yale Symposium: Radiolabeled Cellular Blood Elements," M. L. Thakur and A. Gottschalk eds., Trivirum Publishing Company, New York (1981).

43. I. R. McDougall, J. E. Baumert, R. L. Lantieri, Evaluation of In-111 leukocyte whole body scanning, <u>Am J Radiology</u> 133:849, (1979).

44. L. Forstrom, D. R. Hoagland, L. Gomez, Indium-111 oxine labeled leukocytes in the diagnosis of occult inflammation or abscess, <u>J Nucl Med</u> 20:659 (1979).

45. D. A. Goodwin, J. T. Bushberg, P. W. Doherty, In-111 labeled autologous platelets for localization of vascular thrombi in humans, <u>J Nucl Med</u> 19:626 (1978).

46. J. P. Lavender, J. M. Goldman, R. N. Arnot, M. L. Thakur, Kinetics of indium-111 labeled lymphocytes in normal subjects and patients with Hodgkin's disease, <u>Br Med J</u> 2:797 (1977).

47. D. A. Goodwin, J. R. Heckman, L. F. Fajardo, A. Calin, S. L. Propst, C. I. Diamanti, Kinetics and migration of indium-111 labeled human lymphocytes, <u>Br J Radiology</u> 53:930 (1980).

48. J. M. Thomas and L. L. Eberhardt, Can results from animal studies be used to estimate dose or lower dose effects in humans? in: "Proceedings of the Third International Radiopharmaceutical Dosimetry Symposium," Oak Ridge, Tennessee, October 1980, pp. 259-282.

49. J. G. McAfee and G. Subramanian, Interpretation of interspecies differences in the biodistribution of radioactive agents, in: "Proceedings of the Third International Radiopharmaceutical Dosimetry Symposium, Oak Ridge, Tennessee, October 1980, pp. 292-306.

50. W. S. Snyder, M. R. Ford, G. G. Warner, S. B. Watson, "S" absorbed dose per unit cumulated activity for selected radionuclides and organs, in: "MIRD pamphlet #11, "New York Society of Nuclear Medicine, (1975).

51. P. A. du Heyns, M. G. Lotter, P. N. Badenhorst, Kinetics, distribution and sites of destruction of In-111 indium-labeled human platelets, <u>Br J Haematol</u> 44:269 (1980).

52. D. A. Goodwin, M. W. Sundberg, C. I. Diamanti, C. F. Meares, In-111 labeled radiopharmaceuticals and their clinical use, in: "Radiopharmaceuticals," Subramanian, Rhodes, Cooper, and Sodd, eds., Society of Nuclear Medicine, New York (1975).

53. D. Sartoris, D. A. Goodwin, C. F. Meares, L. H. DeRiemer, Pharmacodynamics of In-111 BLEDTA in man, <u>Invest Radiol</u> 19:221-227 (1984).

54. P. A. du Heyns, M. G. Lötter, H. F. Kotzê, H. Pieters, P.
    Wessels, U. Scheffel, R. Hill-Zobel, M. F. Tsan,
    Quantification of in vivo distribution of platelets labeled
    with indium-111 oxine, J Nucl Med 23:943 (1982).
55. R. E. Johnston, Quantitative measurement of radioactivity in vivo
    in: "The Physics of Clinical Nuclear Medicine," AAPM Annual
    Summer School, Lexington (1977).
56. E. D. Williams, H. I. Glass, R. N. Arnot, A. C. DeGarete, A dual
    detector scanner for quantitative uptake and organ volume
    studies, Medical Radioisotope Scintigraphy 1:665, IAEA, Vienna
    (1969).
57. N. Arimizu and A. C. Morris, Quantitative measurement of
    radioactivity in internal organs by area scanning, J Nucl Med
    10:265 (1969).
58. J. S. Fleming, A technique for the absolute measurement of
    activity using a gamma camera and computer, Phys Med Biol
    24:176 (1979).
59. L. S. Graham and R. Neil, In vivo quantitation of radioactivity
    using the Auger camera, Radiology 112:441 (1974).
60. E. D. Williams, H. I. Glass, A. W. G. Golden, S. Satyavanich,
    Comparison of two methods or measuring the thyroidal uptake of
    Tc-99m, J Nucl Med 12:159 (1972).
61. B. R. Line, A. E. Jones, R. G. Crystal, G. S. Johnston, J. J.
    Bailey, An algorithm for the selection of lung margins in
    scintigraphic ventilation-perfusion studies in: "Proceedings
    of the Sixth Symposium on Sharing of Computer Programs and
    Technology in Nuclear Medicine," Atlanta GA, 1976, Society of
    Nuclear Medicine, New York.
62. O. W. Schalm, N. C. Jain, E. J. Carroll, "Veterinary Hematology,"
    Chapters 8 and 10, Lea Febiger Philadelphia (1975).
63. L. E. Williams, L. A. Forstrum, B. J. Weiblen, "Proceedings of
    Mayo Symposium on In-111 Labeled Platelets and Leukocytes,"
    173-188 (1981).
64. O. R. van Reenen, M. G. Lötter, P. C. Minnaar, et al, Radiation
    dose from human platelets labeled with indium-111, Br J Radiol
    53:790 (1980).

RADIOLABELED RED BLOOD CELLS: STATUS, PROBLEMS,

AND PROSPECTS

Suresh C. Srivastava

Medical Department
Brookhaven National Laboratory
Upton, New York 11973

INTRODUCTION

Of the various cellular blood elements, red cells (RBC) are 1) most abundant; 2) easy to separate and handle; 3) less susceptible to damage from physical or chemical manipulations; 4) not as dependent on energy and nutritional requirements in vitro, and 5) more amenable to labeling with radionuclides due to the availability of a variety of cellular transport mechanisms and of hemoglobin, which is rich in active metal-binding sites. Consequently, red cells have served as simple, convenient, and useful models for the study of, among other things, cellular transport phenomena and membrane structure and function(1, 2).

Red cells dispersed in plasma are not truly living cells. They are composed of water (65%), hemoglobin (32%), and other protein and lipid stroma (3%) and possess properties (circular, non-nucleated biconcave discs; marked elasticity, etc.) that are ideally suited for their functions to rapidly absorb oxygen in the lungs, to pass through smallest capillaries without damage, and to give up oxygen rapidly to the tissues. The average RBC count is about 5×10^9 per ml of blood and their surface area approximately 3×10^3 sq.m. (or 1500 times the body surface area). Normal red cells have a life-span of 110-120 days; the usual rate of replacement is 0.8-1% per day. Hemoglobin, the most important component of the red cells, remains stable and does not undergo degradation or resynthesis during the life of the cell. The hemoglobin molecule contains four heme and globin molecules and has an average molecular weight of 68000. The ferrous iron in heme is bound covalently to the four porphyrin nitrogen atoms. The fifth bond is to the imidazole nitrogen of the histidine of globin, and the sixth is in a reversible binding to oxygen.

Radioactive labels that have been utilized for red cells mostly bind to hemoglobin but can also bind to other intracellular components as well as to surface proteins on the membrane. The desirable properties of an ideal radionuclide label, especially for diagnostic nuclear medicine applications, are as follows: 1) the radionuclide can be incorporated without altering the physical or biochemical properties of the cells and their in vivo function; 2) the radionuclide should have a gamma energy emission in high abundance suitable for imaging, a minimum of cell-damaging low-energy Auger and conversion electrons, and a physical half-life matched to the time frame of the study being performed; 3) the label should be reasonably stable in vitro as well as in vivo, and once incorporated into the cell should not elute during the study or get reutilized after cell destruction; and 4) the radionuclide should have little or no particulate emission in order to minimize patient radiation dose.

Radionuclidic labels for red cells can be divided into two main categories – cohort or pulse labels, and random labels. The cohort labels bind to marrow precursors but not to cells already in circulation   Labeled cells of uniform age appear in the circulation after a few days, and thus cohort labels are useful for the study of cell production rate and survival. The random labels are incorporated into circulating cells of all ages, and the labeling process is usually carried out in vitro on a small sample of venous blood. Except for some earlier studies on ferrokinetics and red cell production, etc., using iron radionuclides, most of the red cell labels developed so far and those in predominant use at the present time involve random labeling and employ Tc-99m, Cr-51, In-111, and Ga-68, roughly in that order. A listing of the various labels appears in Table 1. Also included is information on the compounds used and whether the labeling is carried out in vivo or in vitro. The various diagnostic applications of randomly used labels are shown in Table 2. The extent of the usefulness depends, of course, on the properties of the label such as the half-life, decay mode, and in vivo stability. Labeled cells can be used for red cell survival measurements when the half-life of the radionuclide is sufficiently long. The major portion of this article will deal with random labels; only a brief discussion of the cohort labels will be provided.

COHORT LABELS

As mentioned above, the use of radionuclidic labels that get incorporated into bone marrow red cell precursors over a limited time period produces a labeled cohort of cells of approximately identical age. In an ideal situation, approximately 95% of the label would appear in circulation within 4-5 days, and the amount of the label in blood would remain constant until removed from the circulation due to aging of the cells, in a sigmoid fashion(3).

Table 1. Red Cell Labels

Nuclide	Compound Used	In Vivo	In Vitro	Random	Cohort
Fe-55, Fe-59	Ferric chloride Ferrous citrate	x			x
N-15, H-3, C-14	Glycine	x		x	x
N-15, H-3, C-14	DFP	x		x	x
P-32	DFP	x		x	x
Cr-51	Chromate	x	x	x	
In-111	Oxine, Acetylacetone, Tropolone		x	x	
Ga-68	"		x	x	
Tc-99m	Pertechnetate	x	x	x	

Table 2. Diagnostic Applications of Radiolabeled Red Cells

Normal
  1.   Nuclear cardiology
  2.   Blood pool imaging
  3.   Detection of vascular malformations
  4.   Detection of G.I. bleeding
  5.   Detection of hemangiomas
  6.   Red cell mass determination
  7.   Red cell life-span measurement

Heat-damaged
  1.   Spleen imaging
  2.   Accessory spleen localization
  3.   Detection of G.I. bleeding

Such an ideal situation is not obtained with the use of either labeled glycine or radioiron. Nevertheless, these tracers have provided useful information on the normal red cell life-span and hemoglobin synthesis(4). The most important application of radioiron

has been in ferrokinetic measurements(5, 6). Its use permits a
complete functional analysis of the red cells at all stages; for
example, quantification of erythropoiesis and life-span,
identification and evaluation of the sites of red cell production and
destruction, and quantitative determination of blood loss(7). In
combination with Cr-51, radioiron can be used for measuring,
simultaneously, both red cell production and destruction(8). Although
it is not suitable for routine studies, the use of DFP-32 (diisopropyl
fluorophosphate) has been advocated in special investigations, both as
a cohort and as a random label(9, 10).

RANDOM LABELS

Random labeling of human erythrocytes is used more widely than
cohort labeling, mainly because it involves procedures that are easier
to perform in a clinical setting. In addition, and perhaps more
importantly, radionuclides that possess favorable chemical and
physical properties and that are available for research and clinical
use label red cells randomly.

Chromium-51

Chromium-51 labeling of red cells is the most popular technique at
the present time for the measurement of red cell life-span. Gray and
Sterling in 1950 showed that Cr-51 in the form of chromate could be
used for effectively tagging red cells of various species, including
man(11). Their observations led to the rapid development of Cr-51-RBC
for measuring red cell survival in human subjects(12, 13). Two
problems are, however, associated with this technique. The chromium
label leaks from the red cells, and this leakage, normally 1% per day,
is often variable. Secondly, there appears to be an uneven labeling
of cells, depending on the age of the RBC and also on the nature of
the hemoglobin itself. Appropriate corrections can be made to account
for these problems, and reliable results can be obtained.

The chromium in the hexavalent form penetrates the red cell membrane and attaches to the globin part of the hemoglobin molecule. The
predominant attachment is to the beta chain of globin(14). The
cationic trivalent form of chromium does not cross the membrane but
binds to proteins in the plasma. Chromium-51 red cell survival curves
can be erroneous in situations where cells contain abnormal hemoglobin
and especially when greater or lesser amounts of the beta globin chain
are present. The half-life of Cr-51 of 27.7 days is convenient for
clinical studies, although not ideal. Gamma rays of 0.32 MeV energy
are emitted in 9.8% abundance, and this allows for measurements using
a sodium iodide well counter. In addition, this gamma emission is
useful when double isotope studies are carried out employing the
higher energy (1.10 and 1.29 MeV) Fe-59, with appropriate crossover
corrections(8).

The binding of Cr-51 to red cells in vitro is rapid and is greater than 90% complete within 30 minutes at room temperature in ACD solution. The label is quite firm and resists repeated washing of the cells or dialysis. Factors that decrease efficiency of binding are 1) prior contact of Cr-51 with ACD for periods exceeding an hour before addition of RBC; 2) increasing the pH of ACD solution or autoclaving it; 3) presence of calcium ions; and 4) prolonged exposure of the Cr-51-ACD mixtures to strong light. Erythrocytes can be damaged by excess metallic chromium (50 µg/ml RBC), and this was a limitation in the early work because of the low specific activity of Cr-51 available at that time. Alterations in the RBC are noticed even at 5 µg chromium per ml; however, such changes, mostly enzymatic, do not produce a significant effect on red cell survival. High specific activity Cr-51, presently available, permits the use of less than 0.1 µg chromium per ml blood, thus allowing for the use of more activity with minimal damage to the cells.

The labeling procedure, briefly, is as follows. Approximately 16-20 ml of blood is withdrawn from the patient and added to a sterile multi-injection bottle containing 4-5 ml of ACD solution (commercially available, containing, per ml, 13.2 mg dextrose, 25 mg anhydrous sodium citrate, and 8 mg anhydrous citric acid). Fifty to 100 µCi of Cr-51 as sodium chromate is then added and the mixture allowed to incubate for 30 minutes at room temperature with intermittent gentle swirling. Fifty milligrams ascorbic acid (0.2 ml of a 250 mg/ml solution) is added next to stop the tagging. The unlabeled Cr-51 can be removed by centrifuging and discarding the supernatant solution and injecting the saline-resuspended RBC or the mixture injected as such and allowance made for non-cell-bound Cr-51 in the calculations. The labeling efficiency should be determined on an aliquot of the well-mixed suspension in the latter case. Appropriate standards, properly diluted, are necessary for counting and comparing purposes.

The addition of ascorbic acid was proposed by Read and coworkers (13) in order to reduce the unbound chromate to trivalent chromium and thus prevent further labeling of the cells before or after injection (Cr^{3+} does not cross the cell membrane). This obviates the need for washing the cells prior to injection. By performing external body measurements for Cr-51 activity using a scintillation detector, the clinical usefulness of red cell survival studies can be enhanced further. For example, the potential sites of red cell sequestration in hemolytic states can thus be determined.

Damaging the cells (chemically[15,16] or by heat treatment[17,18]) following (or during) the Cr-51-labeling operation provides an agent suitable for the scanning of the spleen or for the study of splenic function. In early work, anti-D antibodies were also used to induce splenic trapping of the cells(18,19). The procedure for heat damaging the cells was developed later(17) and is almost exclusively used at the present time. The degree of heating (49°C, 15 min), however, has

to be controlled very carefully since excessive damage to the cells
would result in hepatic as well as splenic sequestration of the
activity.

Indium-Labeled and Gallium-Labeled RBC

Certain radionuclides of Indium and Gallium (e.g., In-111, Ga-67,
Ga-68 possess favorable properties for application in diagnostic
nuclear medicine. Cell labeling with these nuclides is mainly carried
out using a procedure developed by Thakur et al(20,21) which employs
the 8-hydroxyquinoline (oxine) chelates of these elements. Indium-
oxine-labeled leukocytes and platelets constitute an important class
of radiopharmaceuticals, and their value in clinical nuclear medicine
is already well established(22, 23). Labeling of erythrocytes has
also been carried out using the oxine chelates of Indium and Gallium
(20, 21, 24-26). The use of an acetylacetone (acac) complex of Indium
for labeling RBC and other cells has also been advocated (27-29). It
is claimed that the method of preparation of the acac complex is
simple and that the labeling procedure may be less damaging to the
cells.

Indium-oxine-labeled red cells have been used in the detection of
intermittent G.I. bleeding in animal models as well as in man
(26, 30, 31). The typical labeling procedure is as follows.
Approximately 10 ml of venous blood is drawn into a syringe containing
about 60 units of heparin, transferred to a round-bottom tube, and
centrifuged for 5 minutes at 1000-1500 G. The supernatant plasma and
buffy coat are removed and the red cells washed twice with 5 ml of
normal saline. The cells are then resuspended into 2 volumes of
saline, incubated with 0.5-2 mCi of In-111-oxine complex for 15
minutes and washed twice with saline. The labeling efficiency
generally is in the order of $90 \pm 5\%$. The cells are mixed with
autologous plasma and injected intravenously. If the In-111-oxine
complex is not commercially available, it can be prepared in a typical
case by the addition of 50 µg oxine to 1 mCi carrier-free
In-111-chloride adjusted to pH 5.5 with 0.3 M acetate buffer,
extracting into an equal volume of chloroform, evaporating the
chloroform to dryness, and finally dissolving the residue in 50 µl
propylene glycol and diluting with 150 µl saline. It is often
difficult to locate the source of intermittent G.I. bleeding by the
usual procedures such as angiography, endoscopy, or scanning with
Tc-99m-labeled sulfur colloid or red cells. These methods show
hemorrhage only at the time of bleeding. A tracer that would remain
for a long enough period in the circulation to permit repeat imaging
is preferable, and In-111-RBC fulfills this requirement. The half-
life of 67 hours and its abundant 173 and 247 keV gamma emissions make
In-111 ideal for this purpose. A major disadvantage is the relatively
faster in vivo elution of the label (approximately 7% per day in one
reported case in man(30), and much higher in animals). In rabbits(26),
70% of the activity in blood had a $t^1/^2$ of 3 and 30% a $t^1/^2$ of 75 hr

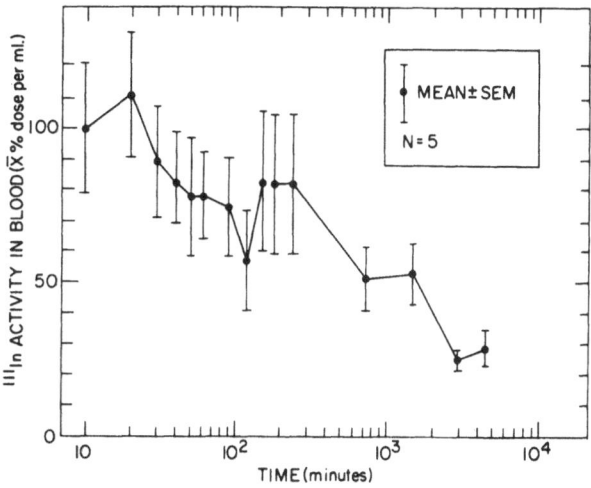

Fig. 1. Blood clearance of In-111-labeled red cells in rabbit. (Data from Winzelberg, et al(26), reproduced with permission.)

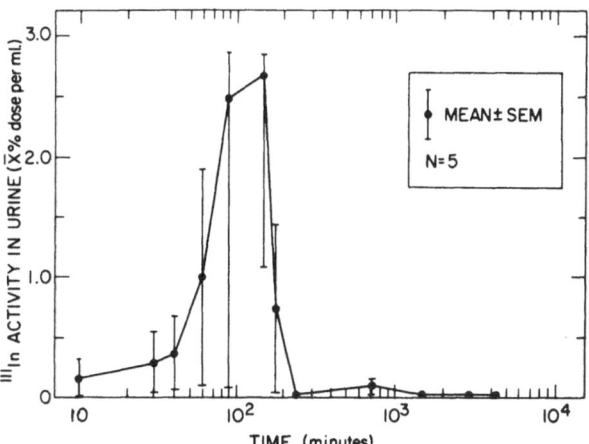

Fig. 2. Urine clearance of In-111-labeled red cells in rabbit. (Data from Winzelberg, et al(26), reproduced with permission.)

(Fig. 1). Urine excretion reached a maximum of 2.5% per ml at 60-150 minutes; it was very slow thereafter up to 72 hours (Fig. 2). It is claimed that despite the elution of the label, vascular imaging can be performed for as long as 120 hours and bleeding in the G.I. tract detected up to 48 hours after injection(30). A method to produce In-111-RBC with higher in vivo stability is highly desirable.

Labeling yields of greater than 90% have been reported in the Ga-68-oxine labeling of red cells(24). The oxine complex is prepared by a method very similar to that used for preparing In-111-oxine. If the Ge-68/Ga-68 generator provides 68-EDTA, the complex has to be dissociated following the usual procedures(32), before preparing the

oxine chelate. Gallium-68-RBC were shown to hold promise for imaging
blood pools using positron tomography; the results were found to
compare favorably with those obtained using carbon-11-monoxide as a
label(24).

A new approach was described recently for labeling the surface of
red cells with Ga-67(33). The washed cells are first treated with
tannic acid and then incubated with pH 7 Ga-67 chloride. The
incorporation of gallium was reported to be greater than 90% in 15
minutes and the in vitro stability of the label was claimed to be
good. Further work will be necessary to demonstrate the usefulness of
this approach for in vivo studies.

The use of In-111-tropolone complex for labeling platelets was
described in a recent report(34). The possibility of labeling other
cells with this complex was also suggested. Both oxine and tropolone
form neutral chelates with Indium and Gallium with comparable
stability constants, and the cell-labeling mechanisms of the two may
indeed by very similar.

Miscellaneous Labels

Red cells have been labeled both in vitro and in vivo (cohort
labeling) with various other compounds including C-11[CO](35),
K-42-[KCl](36, 37), P-32[Na$_2$HPO$_4$](37), I-125 or I-131-p-
iodobenzenesulfonamide (38), I-131-iodophenylhydroxylamine(39), Fe-
52(40), and Se-75 selenomethionine(41). Any significant use of most
of these has not resulted due to many obvious reasons.

Carbon monoxide, which competes with oxygen to form carboxyhemo-
globin, was used (by inhalation) in one of the earlier methods for
measuring blood volume with reasonable accuracy. Carbon-11, a
positron emitter with a 20-minute half-life, has been used recently
for blood volume measurements. The results are approximately 5-10%
higher than with Cr-51. Nevertheless, C-11[CO] could prove valuable
for conditions in which minimum radiation is desirable and repeat
blood volume studies are necessary. However, a cyclotron has to be
available on site, and this imposes severe restrictions on the
widespread use of C-11[CO].

Based on various biochemical and kinetic considerations, radioactive iron is the best nuclide for localizing erythropoiesis.
Imaging of Fe-59 in good resolution is extremely difficult because of
the high gamma energies of 1.10 and 1.29 MeV. Iron-52, which is not
commercially available, is a good candidate, although its short half-
life of 8.3 hours and annihilation radiation of 511 keV make it less
than ideal.

An interesting new method for labeling red cells using the enzyme-
inhibitor approach was described recently(38), which may have some

useful applications. In this study, radioiodinated p-iodobenzene-
sulfonamide, a lipophilic carbonic anhydrase inhibitor, was employed
to label red cells in whole blood rapidly and with high efficiency (95
$\pm$ 5%). In rats after I.V. administration, however, the activity was
found to elute with a $t^{1/2}$ of about 30 hours.

Technetium-99m-Labeled Red Cells

Rapid and convenient kit procedures are presently available for
labeling red cells with Tc-99m that provide essentially quantitative
labeling yields. The current methodology has resulted from a slow
progression of the various steps involved in the labeling process.
This has been due to our poor understanding of the chemistry of Tc-99m,
which in turn has been slow to evolve(42).

In vitro methods. Soon after Tc-99m was recognized as the "ideal"
radiotracer for use in nuclear medicine imaging, efforts were begun to
label red cells with this nuclide in vitro. Several workers attempted
to label RBC using the commonly available form of Tc-99m which is the
pertechnetate ion(43-45). Pertechnetate moves in and out of the RBC
rather freely and cannot be bound firmly to the cells in this chemical
form. It was recognized early on that a reduced Technetium species
would be necessary to bind irreversibly with hemoglobin or other red
cell components, and that reduction of pertechnetate within the cell
would be a more effective way of achieving this goal. The use of
stannous compounds to reduce pertechnetate for the purpose of labeling
red cells was reported by several workers in quick succession (46-51),
and stannous ion is still the most widely used reducing agent in the
currently available procedures(42, 52-54).

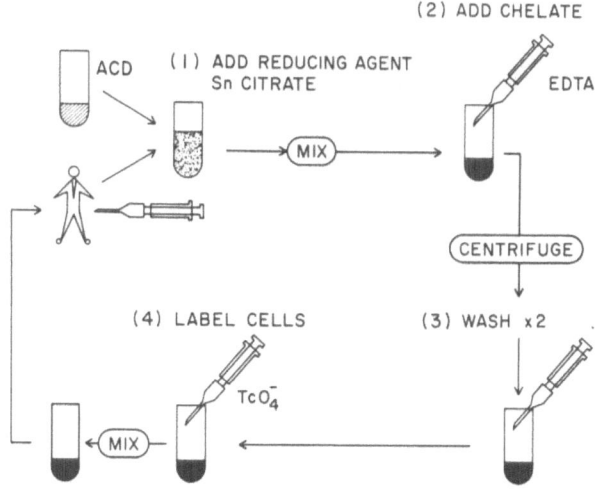

Fig. 3. An early scheme
of labeling RBC with
Tc-99m using the
"pretinning" method.
Data from Nouel and
Brunelle(47), reproduced
with permission.

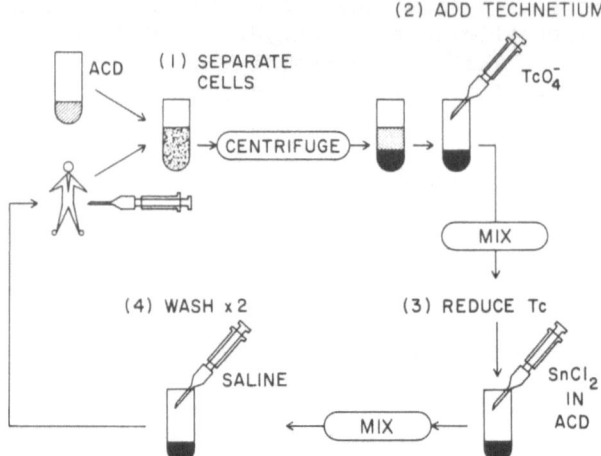

Fig. 4. An early BNL scheme of labeling RBC with Tc-99m using the "post-tinning" method. Data from Atkins, et al. (74), reproduced with permission

The labeling yields were initially limited to the 50-60% range, and this made it necessary to separate the unbound Tc-99m before injection. Several undesirable washing steps were required. The two earlier schemes involving "pretinning" and "post-tinning" of the cells are represented in Figures 3 and 4. These figures describe the various steps that were necessary to obtain good labeling yields and to remove unbound Tc-99m before injection, when necessary.

Most current kit or nonkit procedures now use tinning of the cells first, using a suitable compound, containing tin(II), such as pyrophosphate, glucoheptonate, DTPA, or citrate. The widely used BNL kit (approximately 20,000 kits are distributed annually to investigators worldwide) consists of the following: One Vacutainer reagent tube (100 x 15/16 mm–10 ml capacity), evacuated to draw up to 6 ml blood and containing a lyophilized preparation of 2.0 μg tin, 3.67 mg sodium citrate, 5.50 mg dextrose, and 0.11 mg sodium chloride (maximum).

The labeling procedure using this kit is as follows (use aseptic techniques):

1. Add 1-3 ml of saline Tc-99m pertechnetate to a sterile and pyrogen-free vial. Assay and store in lead shield.
2. Draw 4 ml of patient blood into a heparinized syringe and add to kit.
3. Mix immediately to dissolve the freeze-dried solids in the blood and gently rotate the tube for 5 minutes at room temperature.
4. Add 1 ml of a 4.4% EDTA solution. Draw an equal volume of air to avoid pressure build-up in the tube.

5. Mix by gently inverting about 5 times and centrifuge the tube
    upside down 5 minutes at approximately 1300 x G.
6. Maintain the tube in the inverted position to avoid disturbing
    the packed RBC's. Using a standard 20 g sterile needle and a 2-3
    ml sterile disposable syringe, withdraw 1.25 ml of RBC's and
    transfer to the premeasured Technetium prepared in 1.
7. Incubate the Technetium-RBC mxiture for 10 minutes at room
    temperature with gentle mixing.
8. Assay and dilute appropriately for injection. Cell separation
    and yield determination at this point consistently give 98%
    yields.
9. The described procedure yields an excellent agent for blood pool
    imaging and red cell mas studies. Substitution of the following
    for Step 7 produces an ideal splenic agent: incubate the
    Technetium-RBC mixture 15 minutes at 49°C with gentle mixing.

To prepare the EDTA solution for use with this kit, take any
commercially available disodium EDTA or calcium disodium EDTA solution
for injection (for example: Endrate, Edetate disodium injection, USP,
15% solution in water, pH 7, Abbott Laboratories, North Chicago, IL,
60064, USA), and dilute with sterile water for injection to give a
final concentration of 4.4%.

Alternately, an in-house EDTA preparation can be used if desired.
To prepare this, weigh out 4.4 g disodium EDTA or calcium disodium
EDTA (reagent grade) and dissolve in sterile water for injection or
distilled water, under stirring and make up the volume to 100 ml.
Sterilize this solution by autoclaving.

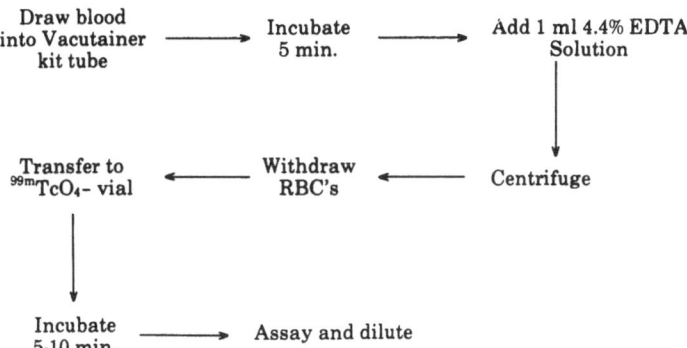

Fig. 5. Schematic representation of steps involved in the
currently used BNL kit method.

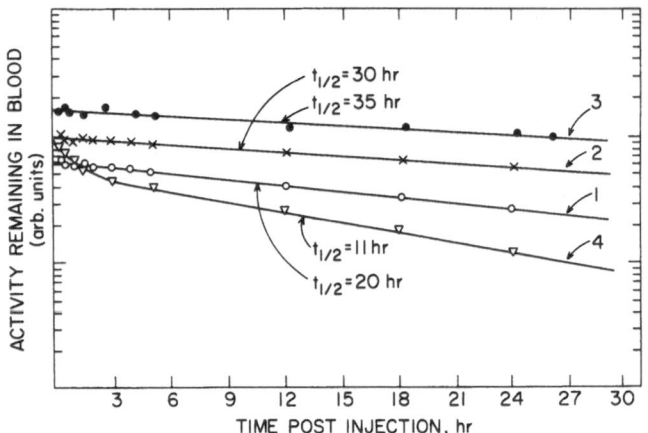

Fig. 6. Blood
clearance curves
of in vitro
labeled Tc-99m-RBC
in asplenic dog.
Curve 1, saline
procedure; curve
2, EDTA procedure;
curve 3, 50 μg
kit, whole blood
procedure; curve
4, heat-damaged
RBC, saline
procedure.

Aliquots of the sterile 4.4% EDTA solution (either in-house or commercial) can be dispensed into a number of individual sterile vials and stored in the refrigerator for subsequent use.

The various steps involved are shown schematically in Figure 5. The determination of labeling yield is carried out as follows:

Draw an aliquot (0.1 - 0.5 ml) of the well-mixed labeled red-blood-cell suspension into a syringe and add to a tube containing 2 ml saline. Mix briefly and centrifuge for 5 minutes at approximately 1300 x G. Withdraw supernatant solution and transfer to another tube. Make volumes in both tubes same with water. Count supernatant solution and RBC and calculate the yield as follows:

$$\text{Percent labeling yield} = \frac{\text{Activity in RBC's x 100}}{\text{Activity in RBC's + Activity in supernatant solution}}$$

The Tc-99m activity should be measured in a dose calibrator unless the sample has been allowed to cool down to less than 1 μCi, in which case a NaI gamma counter can be used.

The procedure now uses 1 ml of 4.4% EDTA instead of 6 ml saline (as previously recommended (52) after the tinning step prior to centrifugation. This improves the labeling yield and the in vitro stability of the label as shown in Table 3. The blood clearance curves are shown in Figure 6. These were obtained in an asplenic dog and indicate the EDTA treatment producing a small but significant improvement in the in vivo stability.

Red cell mass determinations using the BNL kit provide values that are in close agreement with those obtained using Cr-51-RBC. The

Table 3.   Effect of EDTA Addition vs. Saline Addition on the Labeling
Efficiency and Stability of Tc-99m-RBC Using the BNL Kit*

	Saline Addition	EDTA Addition
Labeling yield, percent	97.1 $\pm$ 0.9	98.7 $\pm$ 0.6
Tc-99m washout**		
30 min	2.5 $\pm$ 0.3	1.6 $\pm$ 0.2
24 hr	10.7 $\pm$ 2.4	5.6 $\pm$ 1.4

* After incubating 4 ml blood with the kit for 5 minutes, and prior
to centrifugation, either 6 ml saline or 1 ml 4.4% Na_2EDTa were
added, n=5.

** Washed, labeled cells were incubated with saline at room
temperature, and periodic aliquots removed to determine the loss of
Tc-99m from the cells.

Table 4.   Technetium-99m/Chromium-51 Red Cell Mass Ratios

Patient No.	Sampling Time after Injection (min)			
	15	30	60	120
1	1.05	1.01	1.05	1.03
2	0.96	0.93	1.01	1.04
3	1.00	1.01	1.06	1.08
4	1.01	0.97	1.03	1.05
mean	1.01	0.98	1.04	1.05
$\pm$ s.d.	$\pm$ 0.03	$\pm$ 0.03	$\pm$ 0.02	$\pm$ 0.03

data are shown in Table 4(64). Experiments with double-labeled RBC
(Cr-51 labeling followed by labeling same cells with Tc-99m) showed
(Table 5) that it is the Tc-99m label that elutes from the cells and
leaves the circulation, rather than the intact labeled cells
themselves(64). These results suggest that the labeling operation
does not significantly damage the RBC.

A study was recently completed to develop an in vitro kit method
for selectively labeling RBC with Tc-99m in whole blood(54). This new
method eliminates the main drawbacks of the previously-mentioned
methods, namely, the need for separating plasma, centrifugations,
multiple transfers, etc. The overall effectiveness of such a procedure was thought to be greatly dependent on maximizing the availa-

Table 5.   Technetium-99m/Chromium-51
            Ratios in Blood in Dog Using
            Double-Labeled RBC

Time (hr) after Injection	Ratio
0.25	1.00
0.5	0.99
2	0.96
4	0.89
21.5	0.59
25.5	0.54

Table 6.   Effect of Sodium Hypochlorite (NaOCl), EDTA, and Plasma on
            the Tc-99m Labeling of Red Blood Cells Using BNL Tin Citrate
            (15 µg Sn2+) Kits

Sample and Treatment	Tc-99m Activity, %		
	RBC	Supernatant	2, 2 ml Saline Washes
Whole blood			
No NaOCl	1.8 ± 0.3	87.1 ± 1.6	11.3 ± 0.5
NaOCl (0.6 ml, 0.1%) added	93.3 ± 0.4	5.2 ± 0.4	1.5 ± 0.2
NaOCl (as above) + EDTA (1 ml, 4.4%) added	98.0 ± 1.2	1.5 ± 0.3	0.5 ± 0.1
RBC (plasma removed after tinning)			
NaOCl (as above) added	99.6 ± 0.3	0.4 ± 0.3	--

bility of tin in the stannous form within the cells and at the same
time effectively removing all the extracellular tin(II). Reduced
Technetium in most cases does not pass in or out of the cells, and
thus any premature extracellular reduction of pertechnetate results in
considerably poor labeling yields. It was believed that the addition
of an oxidizing agent which is not transported into the cells will
render extracellular tin(II) ineffective by oxidizing it to tin(IV)
(NaOCl was found most effective). The use of hypochlorite as an
oxidant for tin(II) in plasma was suggested earlier by Narra and

Kuczynski(55). It was found(54) that hypochlorite alone did not
completely oxidize all the tin(II) and that labeling yields greater
than 92 ± 3% could not be obtained. Use of a chelating agent such as
EDTA (in combination with NaOCl) was found to increase the labeling
yields to consistently higher values (98 ± 2%) either by effectively
sequestering the remaining tin(II) and thus making it more accessible

Table 7. Effect of Adding EDTA after NaOCl and Prior to Pertechnetate
Addition on the Tc-99m Labeling of Red Blood Cells

| | | Tc-99m Activity, % |
Conditions	RBC	Supernatant (Including 2, 2 ml Saline Washes)
Tin citrate kits, n=10 (15 µg Sn^{2+})		
No EDTA	93.3 ± 0.4	6.7 ± 0.4
1.0 ml EDTA, pH 4.6	97.7 ± 1.2	2.3 ± 1.2
1.0 ml EDTA, pH 7.2	98.0 ± 1.2	2.0 ± 1.2
Tin glucoheptonate kits, n=15 (15 µg Sn^{2+})		
No EDTA	95.2 ± 1.0	4.8 ± 1.0
1.0 ml EDTA, pH 4.6	98.0 ± 0.2	2.0 ± 0.2
1.0 ml EDTA, pH 7.2	97.6 ± 1.0	2.4 ± 1.0

Table 8. Effect of Carrier Tc-99 on RBC Labeling Yields (3 ml whole
blood, n=4)

| Tc-99 Added, Equivalent to x mCi of Mo-99 decay,* x = | %Tc-99m Activity | | | | | |
| | Tin Glucoheptonate Kits (15 µg Sn^{2+}) | | | Tin Citrate Kits (15 µg Sn^{2+}) | | |
	0	500	600	0	500	600
RBC	97.3 ± 2.4	97.3 ± 0.8	72.7 ± 10.8	97.9 ± 1.8	96.8 ± 2.1	90.8 ± 7.3
Supernatant	2.7 ± 2.4	2.7 ± 0.8	27.3 ± 10.8	2.1 ± 1.8	3.2 ± 2.1	9.2 ± 7.3

* Approximately 1.27×10^{16} atoms or 2.09 µg of Tc-99 are produced
upon the decay of 1 Ci Mo-99.

to the hypochlorite or by other possible mechanisms. Representative
data are summarized in Tables 6 and 7.

Comparable labeling efficiencies at the tracer level are achieved
regardless of whether glucoheptonate (GH), citrate (Cit), or other
suitable ligands are used as complexing agents for tin in the kit.
However, tin uptake into the RBC is higher with GH and Cit, and the
use of these ligands may thus be more advantageous, especially in
situations where excessive amounts of Tc-99 are present in Tc-99m
solutions (e.g., when using instant Technetium, or the first generator
milking following an overly long in-growth period). The effect of
carrier on labeling yields is shown in Table 8.

The data in Tables 6-8 were obtained using kits containing 15 µg
stannous tin. Similar results are obtained using a later version of
tin citrate kits that contains 50 µg tin.

The new kit for labeling RBC in whole blood consists of the
following: One Vacutainer reagent tube 100 x 15/16 mm - 10 ml
capacity, or a 10 ml multi-injection bottle, evacuated to draw up to 6
ml and containing 50.0 µg tin, 3.67 mg sodium citiate, 50.0 mg
dextrose, and 1.40 mg sodium chloride (maximum).

The other reagents required during the labeling procedure (sodium
hypochlorite, EDTA) can be prepared or obtained as follows. These may
also become available as part of the kit in the future.

Sodium hypochlorite. To prepare the sodium hypochlorite (NaOCl)
solution, dilute reagent grade NaOCl (for example, J. T. Baker,
reagent NaOCl, 5.25%) with saline to give a 0.1% final concentration.
This solution should be prepared fresh and used the same day.

Directions for preparing and using the 4.4% EDTA solution are
identical to those provided in an earlier section.

The labeling protocol is as follows (use aseptic techniques):

1. Draw 1 ml of patient blood (0.5 to 6 ml may be used) into a
    heparinized syringe and add to the kit tube.
2. Mix immediately to dissolve the freeze-dried solids in the blood
    and incubate for 5 minutes at room temperature.
3. Add 0.6 ml of 0.1% sodium hypochlorite solution. Mix by gently
    inverting the tube 3-4 times.
4. Add 1 ml of a 4.4% EDTA solution. Mix by gently inverting the
    tube 3-4 times.
5. Store the tube in a lead shield and add the desired quantity of
    Technetium-99m pertechnetate in a volume of 0.5-3 ml.
6. Incubate the Technetium-RBC mixture for 15 minutes at room
    temperature with occasional gentle mixing.

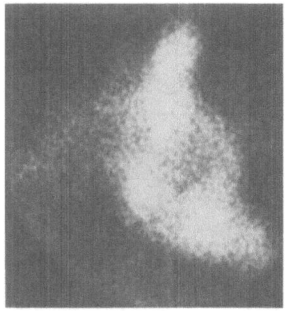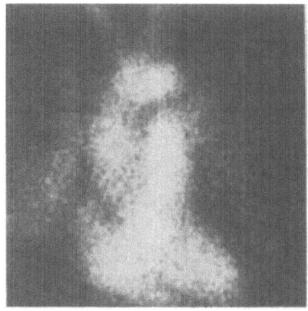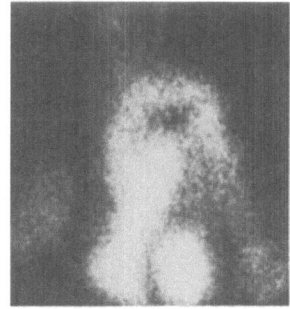

Fig. 7. Typical images (right anterior oblique, anterior, and left
         anterior oblique) of the heart obtained following the
         administration of in vitro labeled Tc-99m-RBC. Note the
         excellent visualization of the interventricular septum,
         space between the liver and heart, and the aorta and
         pulmonary artery. Data from Atkins, et al(51), reproduced
         with permission.

AG   99m Tc RBC

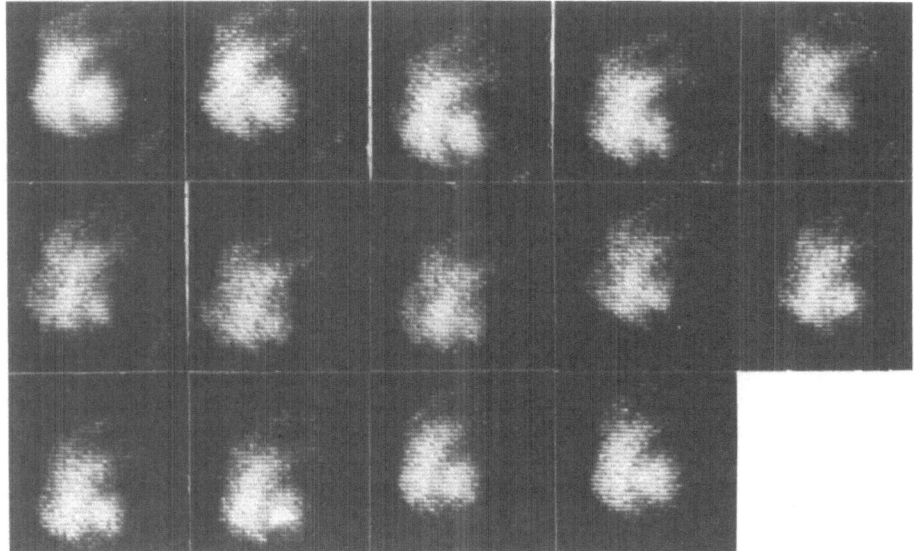

Fig. 8. Dynamic views of the heart obtained using in vitro labeled
         Tc-99m-RBC. Note the emptying and the filling of the heart
         chambers.

7.   The above procedure yields an excellent agent for blood pool
      imaging and red cell mass studies. Substitution of the following

for Step 6 produces an ideal splenic agent: incubate the
Technetium-blood mixture 15 minutes at 49°C with gentle mixing.

In summary, the new kit method for selective labeling of RBC in
whole blood 1) eliminates the need for separating plasma, and thus
cells remain in their native plasma environment; 2) requires no
centrifugation and thus greatly reduces handling of RBC; 3) tolerates
greater quantities of Tc-99 contamination in Tc-99m eluates; 4)
involves one vessel operation and no transfers, and thus is more
convenient to carry out in practice; and 5) provides consistently high
labeling efficiencies (98 $\pm$ 2%). Preliminary studies in dogs have
demonstrated that the in vivo survival of the label in blood is equal
to or somewhat superior than that using other in vitro labeling
techniques.

Typical images of the heart obtained following the injection of
in vitro kit labeled RBC are shown in Figure 7. A representative
dynamic study using gated data acquisition is shown in Figure 8.

In vivo method. Labeling of red cells with Tc-99m can also be
carried out in vivo. The mothed is based on the observation that
prior administration of stannous compounds causes alterations in the
subsebquent in vivo distribution of pertechnetate(56). Pertechnetate
alone does not bind strongly with red cells, but in vivo, because of
the prior introduction of tin(II) in the red cell compartment, after
finding entry into the cells, the administered pertechnetate gets
reduced and firmly bound. Pavel and coworkers (53) introduced this
method for RBC labeling in vivo and found stannous pyrophosphate to be
most effective. Labeling efficiency (fraction of the total
administered pertechnetate incorporated into the red cells), however,
is variable, usually ranging anywhere between 60 and 90%. The exact
role of stannous ion in the labeling process has not been completely
elucidated. It has been proposed that tin(II) selectively activates
the redox mechanisms in the choroid plexus and red cells, thus causing
in situ reduction of the pertechnetate and its retention primarily at
these two sites. Compared to the in vitro method, which routinely
provides quantitative labeling of the cells (a necessary requirement
for many applications), the in vivo method results in generally poor
and frequently irreproducible labeling efficiencies. Its usefulness,
however, cannot be disputed, especially because of the convenience.
The process requires only two injections, and no outside handling of
blood is involved. When higher labeling efficiencies are required,
and for splenic studies, in vitro labeling is the method of choice.

Various parameters, such as the optimum quantity of the tin(II)
preparation to be injected and the appropriate time delay between tin
injection and the pertechnetate administration, have been studied by
various investigators(57-59). Administration of 10 to 20 µg of
stannous ion per kilogram of body weight is thought to be adequate for
optimal labeling. A commercial stannous pyrophosphate kit containing
2 mg stannous ion is reconstituted with saline, and a suitable aliquot

Table 9. In Vitro RBC Labeling with Tc-99m of Blood Samples Obtained
          at Various Intervals Following Sn-PYP Injection in Normal
          Human Volunteers (n=4)

Time After Sn-PYP Injection	Percent Labeling Yield		
	Time (min) of Incubation, RBC's + $Tc\text{-}99mO_4^-$		
	15	60	300
30 min	98.5 + 0.7	96.7 + 1.2	--
24 hr	98.6 + 0.3	98.4 + 0.6	94.6 + 2.3
7 d	55.7 + 3.7	85.4 + 5.8	96.5 + 2.0
21 d	29.5 + 4.3	47.6 + 7.1	89.2 + 5.2
42 d	21.4 + 3.5	27.3 + 6.8	61.7 + 10.8
63 d	20.7 + 13.7	31.7 + 28.2	57.1 + 30.1

Table 10. In Vivo RBC Labeling with Tc-99m in Normal Human Volunteers
           at Various Intervals Following a Single Sn-PYP Injection
           (n=3)

Time of Tc-99mO4- Injection (Post Sn-PYP)	Time (min) of Blood Sampling (post $Tc\text{-}99mO_4^-$ Injection)	Tc-99m in Blood (% of Total Injected, Normalized)	Percent of Tc-99m (total)	
			RBC	Plasma (Plus 1, 2 ml Saline Wash)
30 min	30	100.0	94.5 + 2.5	5.5 + 2.5
	1440	71.6 + 1.5	92.9 + 1.0	7.1 + 1.0
7 d	60	41.2 + 9.5	63.6 + 3.3	36.4 + 3.3
	300	39.4	90.3	9.7
21 d	60	27.5 + 2.8	27.9 + 3.6	72.1 + 3.6
	300	20.2 + 1.2	43.8 + 2.9	56.2 + 2.9
42 d	60	25.4 + 2.7	16.5 + 0.8	83.5 + 0.8
	300	16.6 + 3.3	16.7 + 2.5	83.3 + 2.5

(depending upon the patient's weight) is injected into the patient.
After 30 minutes, the desired quantity of $Tc\text{-}99mO_4^-$ (usually 10 to 30
mCi) is injected intravenously. The red cells get labeled almost
immediately, and the Technetium activity incorporated into the cells

has a clearance half-time of about 30 hours or more(60). The stannous
ion taken up by the cells appears to have a quite slow clearance.
Thus, it appears that following pertechnetate injection up to several
days after the patient has had a Tc-99m-Sn-pyrophosphate bone study
performed, significant in vivo cell labeling with Tc-99m can occur.

The long-term retention of tin following in vivo RBC labeling was
the subject of a recent study(60). The data from this study are
summarized in Tables 9-10. It was found as follows: 1) There is a
significant retention of tin in the RBC, even after a period of 2
months following a single Sn-PYP injection. 2) early blood samples
(following tin administration) give high labeling yields (in vitro)
with Tc-99m; some labeling is achieved even with 2-month samples. 3)
The kinetics of the Tc-99m labeling of RBC (in vitro) slow down
considerably with the later samples. This may be due to the slow loss
of Sn(II) from the cells or its oxidation to Sn(IV). Normal loss of
RBC (and thus of tin) from the circulation may also be an important
contributing factor. 4) Significant in vivo labeling of RBC results
when $Tc-99mO_4^-$ is injected up to 42 days after a single Sn-PYP
administration. 5) Blood samples obtained 60 minutes after the
injection of $Tc-99mO_4^-$ showed that the activity in blood (% injected
dose) was high for early periods after the Sn-PYP injection (30 min,
98.5) and dropped slowly with time (7 d, 41; 21 d, 27.5; 42 d, 25.4).
The ratio of RBC to plasma activity also decreased with time (30 min,
19; 7 d, 1.75; 21 d, 0.39; 42 d, 0.20).

Combined in vivo/in vitro ("in vivitro") methods. Some investigators claim to overcome the variable tagging achieved with the in
vivo procedure with the adaptation of the "in vivitro" method(61-63).
The method basically is an in vivo "tinning" procedure followed by
presentation of Tc-99m-pertechnetate activity to a smaller number of
RBC in vitro and reinjecting the labeled RBC in plasma into the
patient. A typical procedure(61) is presented below in view of its
use at various centers.

Modified in vivo labeling of red blood cells

1.   Pretreatment with 0.5-1 mg stannous ion as stannous pyrophosphate
     (I.V. injection not through indwelling catheter).
2.   Place 19 gauge butterfly needle into antecubital vein.
3.   Attach 4-way stopcock to butterfly; place 10 ml syringe
     containing 4 ml heparin (10 units/ml) and 4 ml 0.9% NaCl on free
     port of stopcock.
4.   Flush butterfly and tubing with heparin saline solution.
5.   Twenty minutes after injection of stannous reagent, attach a
     shielded 5 ml syringe containing 20 mCi of Tc-99m-pertechnetate
     to free port of stopcock.
6.   Withdraw 3 ml of blood into Tc-99m syringe.
7.   Flush butterfly and tubing with heparin saline solution.
8.   Invert the Tc-99m-RBC syringe every 1 minute for 10 minutes; then
     inject labeled red cells via the indwelling butterfly needle.

Table 11. Effect of Blood Volume and Amount of Tin(II) in Kit on RBC Tin Uptake (n=5)

Blood Used (ml)	% Tin Uptake into RBC	
	15 µg Kit	50 µg Kit
1	12.3 ± 1.2	9.5 ± 0.1
2	19.3 ± 1.2	12.5 ± 0.2
3	25.7 ± 0.3	15.9 ± 1.1
4	28.9 ± 3.7	20.9 ± 1.1

Note: Lyophilized Kits labeled with Sn-117m or Sn-113 were used. The kits contained the stated amount of tin (II), 3.67 mg trisodium citrate, and 5.5 mg dextrose. The Kits were incubated for 5 minutes with the indicated volume of blood, and then the cells were separated and washed (2x) to determine radioactivity uptake.

Table 12. Percent Distribution of Tin in Blood Components Following In Vitro Labeling (Blood Volume 3 ml; n=10)

Tin Content of Kit, µg	Red Cell Bound		Plasma (Including Wash)
	Membrane	Nonmembrane	
2	4.0 ± 2.0	17.5 ± 9.7	77.1 ± 12.1
15	5.7 ± 2.9	18.3 ± 5.1	77.1 ± 4.2

Table 13. Percent Distribution of Tc-99m in Blood Components Following In Vitro Labeling (n=10)

Tin Content of Kit, µg	Red Cell Bound		Plasma	Wash
	Membrane	Nonmembrane		
2	1.7 ± 0.3	94.8 ± 2.0	2.7 ± 1.7	0.6 ± 0.3
15	1.3 ± 0.2	93.3 ± 3.2	6.1 ± 3.0	0.7 ± 0.1

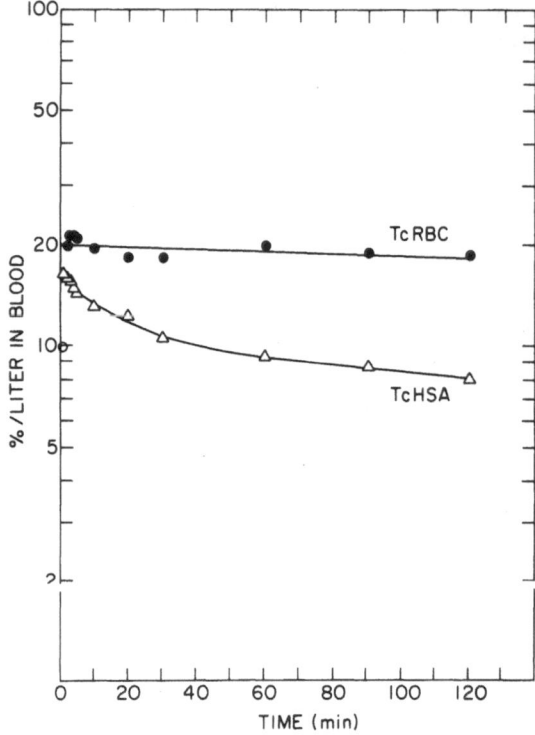

Fig. 9. Change in blood
concentration with time of
Tc-99m-RBC and Tc-99m-HSA
(average of 5 patients).
The average blood clearance
t1/2 were 28.7 hours for
Tc-99m-RBC and 5.3 hours
for Tc-99m-HSA. (Data from
Atkins, et al(73), repro-
duced with permission.)

Mechanistic studies. The mechanism involved in the Tc-99m-RBC
labeling process is not completely understood. However, some evidence
has been accumulated (64-67) to support the following conclusions:
1) stannous ior complexed with citrate or other suitable agents
diffuses into the cell and becomes bound to a cellular component; 2)
pertechnetate ion diffuses freely in and out of the cells; 3)
pertechnetate, once inside the cell and if tin(II) is already present
there, gets reduced and bound mainly to the globin part of hemoglobin;
4) the binding of Technetium with globin is predominantly to the
β-chain; 5) reduced forms of Technetium cannot be transported across
the cell membrane; and 6) any tin(II) remaining outside of the cells
prematurely reduces the pertechnetate, thus forbidding the entry of
Technetium into the cells and thereby causing low labeling yields. It
thus becomes necessary to remove excess tin before adding pertechnetate. Data on the uptake and distribution of Tc-99m and tin
(Sn-117m or Sn-113 was used) in blood and within the red cell components (64) are described in Tables 11-14.

Alterations in the binding of Tc-99m by red cells have been
reported as a result of various patient medications(68, 69) as well as
from diseases or medications associated with RBC antibody formation
(70). The exact mechanisms responsible for these effects are not
known.

Table 14.   Percent Distribution of Tc-99m and Tin in Hemoglobin
(n=10)

Fraction	Technetium (Tc-99m)	Tin (Sn-113 or Sn-117m)
Heme	18.3 ± 9.8	90.1 ± 4.5[1]
Globin	80.5 ± 10.1	12.9 ± 4.2

[1] Results not reliable due to high solubility of tin in acid.
The method for separating heme and globin involved HCl/acetone
treatment.

Comparison of Tc-99m-RBC with Tc-99m-HSA.   The activity concentration of Tc-99m in blood is significantly higher with Tc-99m-RBC as
compared with Tc-99m-HSA (human serum albumin) for several hours after
injection. For a number of reasons(71), Tc-99m-HSA preparations with
favorable blood clearance characteristics are not easily achievable.
For blood pool imaging, especially of the cardiac chambers, labeled

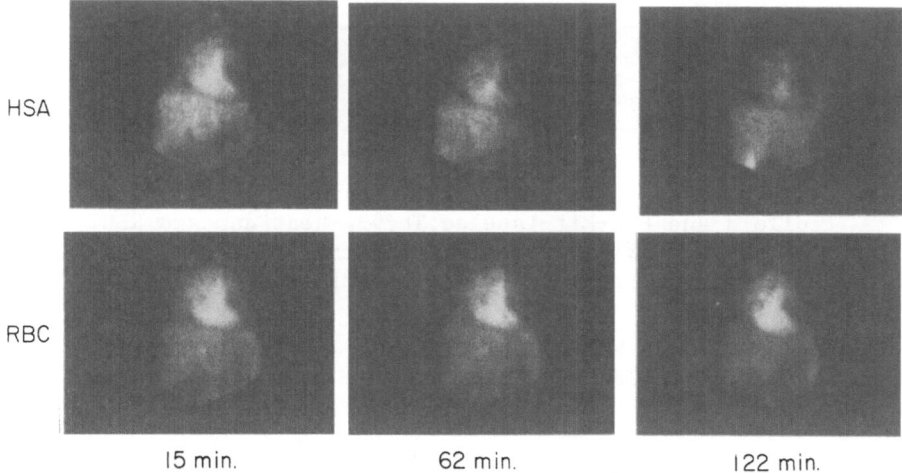

HSA

RBC

15 min.   62 min.   122 min.

Fig. 10.   Comparison scintiphotos of Tc-99m-HSA and Tc-99m-RBC distribution in heart, lungs, and liver. Note the relative
decrease in cardiac blood pool activity of Tc-99m-HSA with
time and the biliary excretion of Tc-99m. In the Tc-99m-RBC
study, the relative cardiac blood pool radioactivity remains
high compared with that of the liver. (Data from Atkins, et
al(73), reproduced with permission.)

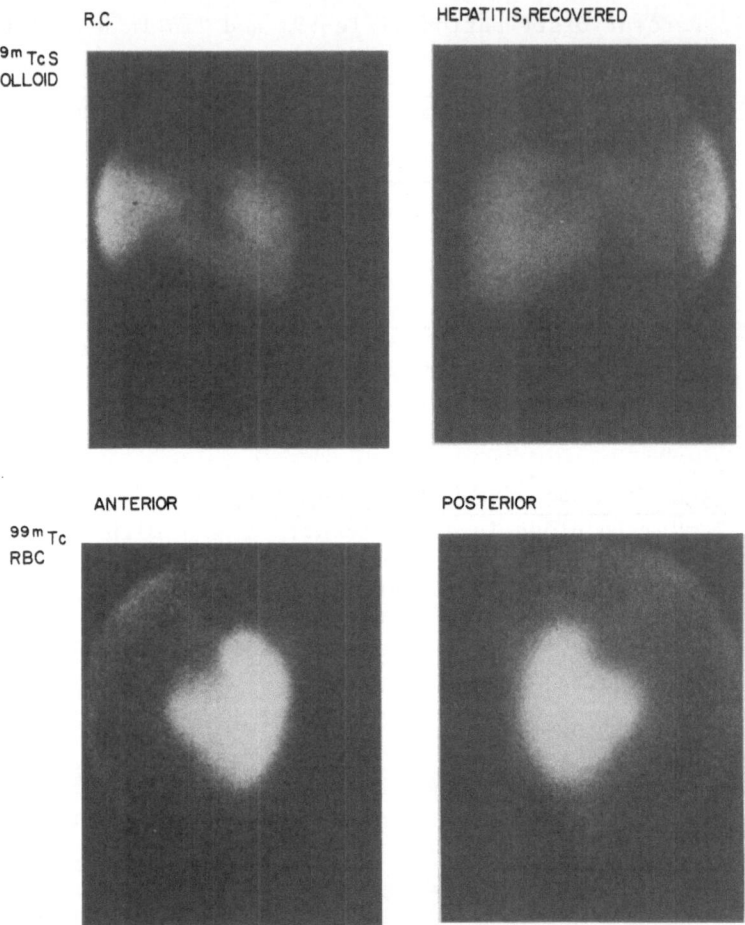

R.C.   HEPATITIS,RECOVERED

^{99m}TcS
COLLOID

ANTERIOR   POSTERIOR

^{99m}Tc
RBC

Fig. 11.   A comparison of spleen images obtained with Tc-99m-sulfur
           colloid and BNL-kit-labeled Tc-99m-heat damaged RBC. Note
           the absence of liver activity in the RBC image. Data from
           Reference 52, reproduced with permission.

RBC have thus proven to be superior to labeled HSA in several recent
studies (72,73). Some of the representative data are shown in Figures
9-10.

    Splenic studies. Heat-damaged Tc-99m RBC (Tc-99m-HDRBC) find
application in the imaging of spleen, which is useful in a number of
clinical situations, including trauma, investigation of left upper
quadrant masses or pain, evaluation of spleen size, splenic infarcts
and space occupying disease, and accessory spleens(74,75).
Specificity higher than Tc-99m-sulfur colloid is obtained by using
Tc-99m-HDRBC. Heating at 49.5°C is a very reliable technique for
inducing splenic sequestration of Tc-99m-RBC. The rapid blood
clearance ($t^{1/2}$ 6 min) and a plateauing of the activity in the spleen

Table 15.   Radiation-Absorbed Dose from Technetium-99m-Labeled
            Red Blood cells

	Dose, rad/mCi	
Organ	Normal RBC	Heat-Damaged-RBC
Whole body	0.019	0.018
Heart	0.078	-
Spleen	0.050	2.87
Liver	0.070	0.011
Blood	-	0.027
Lungs	0.056	-
Kidneys	0.054	-
Red marrow	0.033	-

by 30 minutes make it possible to accomplish imaging soon after the
radionuclide administration(75). High splenic uptake (~70%) allows
the administration of small quantities of the radiotracer (~1 mCi is
adequate for rapid imaging in multiple views), thus reducing the
patient radiation dose considerably. Representative spleen images are
shown in Figure 11, obtained using Tc-99m-S colloid and
Tc-99m-HDRBC(52).

Patient radiation dose. Average radiation doses to the whole
body as well as to various other tissues and organs from the
administration of normal and heat-damaged Tc-99m-RBC are shown in
Table 15(75, 76). The whole body dose is generally lower than that to
many other organs, which is due to the lower average blood content of
whole body.

CONCLUSION

Radiolabeling of red cells and their clinical and research
applications in nuclear medicine imaging and other areas have been a
field of intense interest during the last two decades. Significant
advances have been made so that at the present time sufficiently
stable labels are available for various applications. Technetium-99m-
labeled RBC have revolutionized the field of nuclear cardiology, and
it is now possible to evaluate various heart parameters externally
without significant radiation dose or trauma to the patient.

The ideal radioisotopic label for the determination of red cell life-span in normal and hemolytic states is yet to be developed. The tracers available so far, especially Cr-51-RBC, have provided valuable information but occasionally have given misleading results as well. Caution has to be exercised when interpreting the results.

For various clinical procedures, the choice of label varies. Red cells labeled with long (∿30 d) as well as intermediate (2-3 d) half-life nuclides with sufficient in vivo stability are highly desirable. Future work will perhaps fulfill this need now that we have acquired a strong base of useful knowledge on radiolabeled red cells. Approaches such as using labeled antibodies to red cell antigens or receptors or enzymes in the cell may provide encouraging results in terms of labeling RBC with the least damage.

ACKNOWLEDGMENTS

The secretarial assistance of Mrs. R. Bailey during various stages of the preparation of this manuscript is gratefully acknowledged. This work was supported by the United States Department of Energy under Contract #DE-ACO2-76CH00016.

REFERENCES

1. C. Bishop and D. M. Surgenor, eds., "The Red Blood Cell," Academic Press, New York (1964).
2. R. Whittam, "Transport and Diffusion in Red Blood Cells," Edward Arnold, London (1964).
3. M. Pollycove and M. Tono, Studies of the erythron, Semin Nucl Med 5:11-61 (1975).
4. N. I. Berlin, Determination of red blood cell life-span, JAMA 188:375 (1964).
5. M. S. Wheby and W. H. Crosby, The gastrointestinal tract and iron absorption, Blood 22:416 (1963).
6. L. Saylor and C. A. Finch, Determination of iron absorption using two isotopes of iron, Am J Physiol 172:372 (1953).
7. M. Pollycove, Iron metabolism and kinetics, Semin Hematol 3:235 (1966).
8. M. Kesse-Elias, E. Gyfteki, B. Malamos, Fe-59 and Cr-51 studies in aplastic anemia and myelosclerosis, Acta Haematol 39:139 (1968).
9. J. A. Cohen and M. G. P. Warringa, The fate of P32-labeled diisopropylfluorophosphate in the human body and its use as a labeling agent in the study of the turnover of blood plasma and red cells, J Clin Invest 33:459 (1954).
10. L. E. Bratteby, L. Garby, B. Wadman, Studies on erythrokinetics in infancy. XII. Survival in adult recipients of cord blood red cells labeled in vitro with di-isopropyl fluorophosphate (DF-32-P), Acta Pediatr Scand 57:305 (1968).

11. S. J. Gray and K. Sterling, The tagging of red cells and plasma
     proteins with radioactive chromium, J Clin Invest 29:1604
     (1950).

12. F. G. Ebaugh, Jr., C. P. Emerson, J. F. Ross, The use of radio-
     active Chromium-51 as an erythrocyte-tagging agent for the
     determination of red cell survival in vivo, J Clin Invest
     32:1260 (1953).

13. R. C. Read, G. W. Wilson, F. H. Gardner, The use of radioactive
     sodium chromate to evaluate the life span of the red cell in
     health and in certain hematologic disorders, Am J Med Sci
     228:40 (1954).

14. H. A. Pearson, The binding of Cr-51 to hemoglobin. I. In vitro
     studies, Blood 22:218 (1963).

15. H. N. Wagner, Jr., I. M. Weiner, J. G. McAfee, et al, 1-mercuri-
     2-hydroxypropane (MHP): a new pharmaceutical for visualization
     of the spleen by radioisotope scanning, Arch Intern Med
     113:696 (1964).

16. K. Mayer, A. Dwyer, J. S. Laughlin, Spleen scanning using ACD-
     damaged red cells tagged with Cr-51, J Nucl Med 11:455 (1970).

17. P. M. Johnson, E. H. Wood, S. L. Morring, Splenic scintillation
     scanning, Am J Roentgen 86:575 (1961).

18. F. Spinelli-Ressi, Scintillation scanning of the spleen with red
     cells labeled with Chromium-51, in: "Medical Radioisotope
     Scanning", vol II, IAEA, Vienna (1964).

19. P. M. Johnson, J. C. Herion, S. L. Mooring, Scintillation
     scanning of the normal human spleen utilizing sensitized
     radioactive erythrocytes, Radiology 74:99 (1960).

20. M. L. Thakur, D. Dees, S. S. L. Harwig, et al, Labeling blood
     components with 8-hydroxyquinoline chelates: simplified
     procedure and mechanism of labeling, J Labeled Compds
     Radiopharm 13:177 (1977).

21. M. L. Thakur, Gallium-67 and Indium-111 radiopharmaceuticals, Int
     J Appl Radiat Isot 28:183 (1977).

22. M. L. Thakur, M. J. Welch, J. H. Joist, et al, Indium-111-labeled
     platelets: studies on preparation and evaluations of in vitro
     and in vivo functions, Thromb Res 9:345 (1976).

23. M. L. Thakur, J. P. Lavender, R. N. Arnot, et al, Indium-111-
     labeled autologous leukocytes in man, J Nucl Med 18:1014
     (1977).

24. M. J. Welch, M. L. Thakur, R. E. Coleman, et al, Gallium-68-
     labeled red cells and platelets: new agents for positron
     tomography, J Nucl Med 18:558 (1977).

25. M. Murrell, U. Scheffel, J. M. Whipple, et al, In-111-oxine as a
     red blood cell label, Proceedings of Second International
     Congress World Federation of Nuclear Medicine Biology,
     Washington, D.C., p. 130 (1978).

26. G. G. Winzelberg, F. P. Castronovo, R. J. Callahan, et al,
     In-111-oxine-labeled red cells for detection of simulated
     lower gastrointestinal bleeding in an animal model, Radiology
     135:455 (1980).

27. H. Sinn, P. Georgi, J. Clorius, et al, Die markierung von erythrozyten mit radioactiven indiumisotopen, Nuclear medizin 13:180 (1974).

28. H. Sinn and D. J. Silvester, Simplified cell labeling with Indium-111-acetylacetone, Br J Radiol 52:758 (1979).

29. C. J. Mathias, W. A. Heaton, M. J. Welch, et al, Comparison of In-111-oxine and In-111-acetylacetone for the labeling of cells: in vivo and in vitro biological testing, Int J Appl Radiat Isot 32:651 (1981).

30. A. Ferrant, N. Dehasque, N. Leners, et al, Scintigraphy with In-111-labeled red cells in intermittent gastrointestinal bleeding, J Nucl Med 21:844 (1980).

31. R. L. Beckman, G. L. Pittenger, D. P. Swenson, et al, Blood loss measured with Indium-111-labeled red blood cells in dogs, Radiology 148:243 (1983).

32. D. J. Hnatowich, A method for the preparation and quality control of Ga-68 radiopharmaceuticals, J Nucl Med 16:764 (1975).

33. D. D. Pant, J. J. Coupal, W. J. Shih, et al, A new approach to Gallium-67 labeling of human erythrocytes and platelets, J Nucl Med 24:P123 (1983).

34. M. K. Dewanjee, S. A. Rao, J. A. Rosemark, et al, Indium-111-tropolone, a new tracer for platelet labeling, Radiology 145:149 (1982).

35. H. I. Glass, A. Brant, J. C. Clark, et al, Measurement of blood volume using red cells labeled with radioactive carbon monoxide, J Nucl Med 9:571 (1968).

36. G. Hevesy and G. Nylin, Application of K-42-labeled red corpuscles in blood volume measurements, Acta Physiol Scand 24:285 (1952).

37. S. A. Berson and R. S. Yalow, The use of K-42- or P-32-labeled erythrocytes and I-131-tagged human serum albumin in simultaneous blood volume determinations, J Clin Invest 31:572 (1952).

38. D. M. Ackery, J. Singh, P. Wyeth, Enzyme-inhibitor mediated red cell labeling, Proceedings Fourth International Symposium Radiopharmicological Chemistry, Julich, KFA (1982).

39. H. Jackson, Studies with erythrocytes labeled with radioactive p-iodophenylhydroxylamine, Nature 172:80 (1953).

40. D. Van Dyke, H. O. Anger, M. Pollycove, The effect of erythropoeitic stimulation on marrow distribution in man, rabbit, and rat as shown by Fe-59 and Fe-52, Blood 24:356 (1964).

41. J. A. Penner, Selenomethionine incorporation into hemoglobin, Clin Res 12:228 (1964).

42. S. C. Srivastava and P. Richards, Technetium-labeled compounds, in: "Radiotracers for Medical Applications," vol. I, G. V. S. Rayudu, ed., CRC Press, Boca Raton, FL (1983).

43. J. Fischer, R. Wolf, A. Leon, Technetium-99m as a label for erythrocytes, J Nucl Med 8:229 (1967).

44. U. Haubold, H. W. Pabst, G. Hor, Scintigraphy of the placenta
    with Tc-99m-labeled erythrocytes, in: "Symposium on Medical
    Radioisotope Scintigraphy," vol. 2, Vienna, IAEA (1969).
45. M. B. Weinstein and W. M. Smoak, Technical difficulties in
    Tc-99m-labeling of erythrocytes, J Nucl Med 11:41 (1970).
46. R. Berger, B. Johanssen, Markierung von Erythrozyten mit Tc-99m,
    Math-Naturwiss R/18:634 (1960).
47. J. P. Nouel and P. Brunelle, Le marquage des hematies par le
    Technetium-99m, Presse Med 78:73 (1970).
48. W. Eckelman, P. Richards, W. Hauser, et al, Technetium-labeled
    red blood cells, J Nucl Med 12:22 (1971).
49. K. D. Schwartz and M. Kruger, Improvement in labeling
    erythrocytes with Tc-99m-pertechnetate, J Nucl Med 12:323
    (1971).
50. W. C. Eckelman, R. C. Reba, S. N. Albert, A rapid simple improved
    method for the preparation of Tc-99m red blood cells for the
    determination of red cell volume, Am J Roentgenol Radium Ther
    Nucl Med 118:861 (1973).
51. H. L. Atkins, W. C. Eckelman, J. F. Klopper, et al, Vascular
    imaging with Tc-99m red blood cells, Radiology 106:357 (1973).
52. T. D. Smith and P. Richards, A simple kit for the preparation of
    Tc-99m-labeled red blood cells, J Nucl Med 17:126 (1976).
53. D. G. Pavel, A. M. Zimmer, V. N. Patterson, In vivo labeling of
    red blood cells with Tc-99m: a new approach to blood pool
    visualization, J Nucl Med 18:305 (1977).
54. S. C. Srivastava, J. B. Babich, P. Richards, A new kit method for
    the selective labeling of erythrocytes in whole blood with
    Tc-99m, J Nucl Med 24:P128 (1983).
55. R. K. Narra and B. L. Kuczynski, Kit for in vitro labeling of red
    blood cells with Tc-99m, in: "Applications of Nuclear and
    Radiochemistry," R. M. Lambrecht and N. Morcos, eds., Pergamon
    New York (1982).
56. J. McRae, R. M. Sugar, B. A. Shipley, et al, Alterations in
    tissue distribution of Tc-99m-pertechnetate in rats given
    stannous tin, J Nucl Med 15:151 (1974).
57. R. G. Hamilton and P. O. Alderson, A comparative evaluation of
    techniques for rapid and efficient in vivo labeling of red
    cells with Tc-99m, J Nucl Med 18:1010 (1977).
58. A. M. Zimmer, D. G. Pavel, S. M. Karesh, Technical parameters of
    in vivo red blood cell labeling with Tc-99m, Nucl Med 18:241
    (1979).
59. M. W. Billinghurst, D. Jette, D. Greenberg, Determination of the
    optimal concentrations of stannous pyrophosphate for in vivo
    RBC labeling with Tc-99m, Int J Appl Radiat Isot 31:499
    (1980).
60. S. C. Srivastava, P. Richards, Y. Yonekura, et al, Long-term
    retention of tin following in vivo RBC labeling, J Nucl Med
    23:P91 (1982).
61. G. G. Winzelberg, K. A. McKusick, J. W. Froelich, et al,
    Detection of gastrointestinal bleeding with Tc-99m-labeled red
    blood cells, Semin Nucl Med 12:139 (1979).

62. R. Armas, M. L. Thakur, A. Gottschalk, A simple method of spleen
     imaging with Tc-99m-labeled erythrocytes, <u>Radiology</u> 132:215
     (1979).
63. R. J. Callahan, J. W. Froelich, K. A. McKusick, et al, A modified
     method for the in vivo labeling of red blood cells with
     Tc-99m: concise communication, <u>J Nucl Med</u> 23:315 (1982).
64. S. C. Srivastava, R. Straub, P. Richards, Unpublished data, 1983.
65. M. K. Dewanjee, Binding of Tc-99m to hemoglobin, <u>J Nucl Med</u>
     15:702 (1974).
66. M. K. Dewanjee and S. A. Rao, Mechanism of in vivo red cell
     labeling with Tc-99m-pertechnetate and role of Sn2+ pump at
     RBC membrane on the distribution of Sn2+ ion and Tc-99m, <u>in</u>:
     <u>Nucl Med Biol</u>, Proceedings of Third World congress of Nuclear
     Medicine and Biology, Paris, 1982, C. Raynaud, ed., Pergamon,
     Paris (1982).
67. M. M. Rehani and S. K. Sharma, Site of Tc-99m binding to the red
     blood cell, <u>J Nucl Med</u> 21:676 (1980).
68. L. R. Chervu, J. J. Castronuovo, S. S. Huq, et al, Alterations in
     red cell tagging with sulfonamides, <u>J Nucl Med</u> 22:P70 (1981).
69. H. B. Lee. J. P. Wexler, S. C. Scharf, et al, Pharmacologic
     alterations in Tc-99m binding by red blood cells. coneise
     communication, <u>J Nucl Med</u> 24:397 (1983).
70. G. P. Leitl, H. M. Drew, M. E. Kelly, et al, Interference with
     Tc-99m labeling of red blood cells (RBCs) by RBC antibodies,
     <u>J Nucl Med</u> 21:P44 (1980).
71. G. Meinken, S. C. Srivastava, T. D. Smith, et al, Is there a
     "good" Tc-99m-albumin?, <u>J Nucl Med</u> 17:537 (1976).
72. J. A. Dahlstrom, S. Carlsson, B. Lilja, et al, Cardiac blood pool
     imaging - a clinical comparison between RBC labeled with
     Tc-99m in vivo and in vitro and Tc-99m-HSA, <u>J Nucl Med</u> 18:271
     (1979).
73. H. L. Atkins, J. F. Klopper, A. N. Ansari, et al, A comparison of
     Tc-99m-labeled HSA and in vitro-labeled RBC for blood pool
     studies, <u>Clin Nucl Med</u> 5:166 (1980).
74. H. L. Atkins, W. C. Eckelman, W. Hauser, et al, Splenic
     sequestration of Tc-99m-labeled red blood cells, <u>J Nucl Med</u>
     13:811 (1972).
75. H. L. Atkins, A. G. Goldman, R. F. Fairchild, et al, Splenic
     sequestration of Tc-99m-labeled heat-treated RBC, <u>Radiology</u>
     136:501 (1980).
76. H. Malamud, Dosimetry of Tc-99m-labeled blood pool scanning
     agents, <u>Clin Nucl Med</u> 3:420 (1978).

QUALITY CONTROL OF NEUTROPHIL LABELING METHODS

D. English and K. P. Gunter

Kuzell Institute for Arthritis Research
Medical Research Institute at Pacific
Medical Center
San Francisco, CA 94115

Department of Pediatrics
Vanderbilt University School of Medicine
Nashville, TN 37232

INTRODUCTION

Almost without exception, evaluations of new cell-labeling
techniques document the labeling yield obtained and offer data
comparing some function of labeled cells to that of cells not exposed
to the radioisotope. However, the possibility that the majority of
the label associates with cells of diminished function cannot be
dismissed. At least two conditions can lead to such a result. First,
it is possible that the isotope is preferentially concentrated in
nonviable cells, a small number of which are present in any cell
preparation. Secondly, the radiopharmaceutical may be toxic, and
uneven distribution of the radioisotope may preferentially damage
those cells which become most heavily labeled. It is also possible
that deleterious radiation effects may preferentially disrupt the
function of cells most heavily labeled by the radioisotope.

Recent interest in the diagnostic utility of labeled autologous
neutrophils has led to the description of a variety of methods for
labeling these cells (1-10). We undertook this investigation to
comparatively assess the functional integrity of labeled neutrophils
and to determine the amount of radioactivity confined to viable cells.
To accomplish this, we exploited the ability of chemotactically viable
neutrophils in labeled cell suspensions to migrate toward a chemoattractant, isolated the responsive cells, and compared the extent to
which the chemotactically viable cells held radioactivity.

393

MATERIALS AND METHODS

Neutrophils were isolated from heparinized blood obtained from healthy adults by centrifugation over cushions of ficoll-hypaque washed three times with Hanks' balanced salt solution and resuspended in phosphate buffered saline at a concentration of 3.0×10^7 cells/ml prior to incubation with the radioisotopes (11). Washed neutrophils were exposed to 10 to 100 μCi of the various radioisotopes according to previously described protocols for 30 minutes at room temperature. These procedures involved incubation with Ga-67-citrate (4), Tc-99m, SC (1), Cr-51-Na$_2$CrO$_4$ (3), In-111-oxine (10), and P-32-DFP (12) with continuous gentle agitation. After incubation, the non cell-associated radioactivity was removed by washing the cells three times with BSS. Cells were resuspended in BSS at 2.0×10^7/ml for use in chemotaxis assays.

Chemotaxis

The granulocyte chemotaxis radioassay was performed as described by Gallin et al (13) using endotoxin activated serum as the chemoattractant in the lower compartment of the chemotaxis chamber and labeled PMN's in the upper compartment. Two, 5 μm Millipore filters separated the lower from the upper compartments. After 3 hours incubation at 37°C, the two filters were removed. The upper filter was stained with hematoxylin, and the average number of cells responding to the chemoattractant (viewed on the underside of the filter) per high-power field (HPF, 420x) was determined, giving a comparative estimate of the labeled neutrophils' functional competence. Lower filters were rinsed in saline, placed in polypropylene tubes, and their radioactivity determined (CPM), reflecting the amount of radioactivity carried by migrating neutrophils. The chemotactic response of the cells exposed to radioisotopes determined by microscopic examination of upper filters reflects "viability" and is expressed as the percent of the response obtained with cells not exposed to isotope. The relative radioactivity of labeled cells responding to the chemoattractant (determined by counting lower filters) is represented by a chemotactic index (CI), where

$$CI = \frac{CPM \text{ of lower filter}}{CPM \text{ of cells originally placed in chamber}} \times 10^6$$

By using the corrective factor of 10^6, the CI reflects the CPM that would have been recovered on lower filters if one million CPM were contained on the cells placed in each chamber's upper compartment (13). Chemotactic indices of filters from chambers containing chemoattractants (CI+) were compared with those from chambers without attractant (which may be influenced by spontaneous isotope elution and other nonspecific factors) to determine their relation to migration of labeled cells.

RESULTS AND DISCUSSION

Table 1. Chemotactic Characteristics of Netrophils Labeled with
          Tc-99m-SC, Cr-51, In-111-oxine, Ga-67 and P-32-DFP.

Isotope	Labeling Yield	Chemotactic Viability	CI+	CI-
Tc-99m-SC	15-34%	96%	62,000	1,300
In-111-oxine	31-65%	94%	55,000	3,700
Cr-51	3.2-8.9%	93%	37,050	9,500
Ga-67	0.3-2.5%	103%	1,290	1,450
P-32-DFP	30-55%	98%	78,000	2,100

Chemotactic viability was determined by microscopic examination
of the undersurface of upper filters. Chemotactic indices (CI+, CI-)
were determined by radiometric analysis of lower filters.

Table 1 describes the chemotactic responses of cells in
suspensions of neutrophils incubated with Tc-99m-sulfur colloid,
In-111-oxine, Cr-51, Ga-67, and P-32-diisopropylfluorophosphate to
effect cell labeling. Optimal results were achieved using DFP,
In-111-oxine and Tc-99m-SC. None of these isotopes substantially
decreased chemotactic viability as determined by microscopic
enumeration of cells migrating through the upper filters toward the
chemoattractant. Moreover, Cr-51 and Ga-67 did not decrease
chemotactic viability determined microscopically. The data clearly
indicate, however, that cells migrating from neutrophil suspensions
labeled with Ga-67 and Cr-51 contained a lower portion of the total
radioactivity of the suspension than did cells migrating from
suspensions labeled with In-111, P-32 or Tc-99m-SC. For Cr-51, the
CI- was higher than values obtained with the three other isotopes,
indicating isotope elution (reflected by CI-) may have been higher in
unstimulated chambers. With Ga-67, little difference was observed
between the CI+ and CI- even though chemotaxis (determined
microscopically) was not inhibited.

For Ga-67, the results indicate that little or no radioactivity
became associated with chemotactically viable cells. The explanation
for this finding is unclear but may reflect exclusion of the isotope
by viable cells. Whatever the reason, the results with Ga-67
illustrate the utility of the chemotaxis radioassay for quality
control. In the absence of this type of analysis, it would have been

difficult to determine that the radioactivity confined to neutrophils
in cell suspension exposed to Ga-67 was confined to nonviable cells
with respect to migratory activity. This approach should permit
comparative analysis for quality control, for improvement of existing
technology, and for development of new methods of neutrophil labeling.
This study was supported by NIH grant #AI-17950.

REFERENCES

1. D. English and B. R. Andersen, Labeling of phagocytes from human
   blood with Technetium-99m sulfur colloid, J Nucl Med 16:5
   (1975).

2. D. English and B. R. Andersen, The organ distribution of canine
   leukocytes labeled with Technetium-99m sulfur colloids, J Nucl
   Med 18:289 (1977).

3. W. C. Harvey and J. Silva, Cr-51 labeling of concentrated
   phagocytes, J Nucl Med 14:890 (1973).

4. R. L. Burleson, N. C. Johnson, H. Head, In vivo and in vitro
   labeling of rabbit blood leukocytes with Ga-67-citrate, J Nucl
   Med 15:98 (1974).

5. T. Usbida and R. C. Vincent. In vitro studies of leukocyte
   labeling with Technetium-99m, J Nucl Med 17:730 (1976).

6. H. J. Glenn, N. Ruksawin, T. P. Harvine, T. Konikowski, Leukocyte
   labeling with technetium-99m, Int J Nucl Med Biol 3:9 (1976).

7. J. G. McAfee and M. L. Thakur, Survey of radioactive agents for
   in vitro labeling of phagocytic leukocytes. I. Soluble
   agents, J Nucl Med 17:480 (1976).

8. J. G. McAfee and M. L. Thakur, Survey of radioactive agents for
   in vitro labeling of phagocytic leukocytes. II. Particles, J
   Nucl Med 17:488 (1976).

9. M. L. Thakur, R. E. Coleman, and M. J. Welch, Indium-111-labeled
   leukocytes for the localization of abscesses: preparation,
   analysis, tissue distribution and comparison with Gallium-67
   citrate in dogs, J Lab Clin Med 89:217 (1977).

10. B. Zakhireh, M. L. Thakur, H. L. Malech, M. S. Cohen, A.
    Gottaschalk, R. K. Root, Indium-111-labeled human
    polymorphonuclear leukocytes: viability, random migration,
    chemotaxis, bactericidal capacity, and ultrastructure, J Nucl
    Med 20:741 (1979).

11. R. S. Weening, D. Roos, and J. A. Loos, Oxygen consumption of
    phagocytizing cells in human leukocyte and granulocyte
    preparations: a comparative study, J Lab Clin Med 83:570
    (1974).

12. J. W. Athens, A. M. Maurer, H. Ashenbruker, G. E. Cartwright, M.
    M. Wintrobe, Leukokinetic studies. I. A method for labeling
    leukocytes with diisoopropyl-flourophosphate (DFP^{32}), Blood
    14:303 (1959).

13. J. I. Gallin, R. A. Clark, H. R. Kimball, Granulcoyte chemotaxis:
    an improved in vitro assay employing Cr-51-labeled
    granulocytes, J Immunol 110:233 (1973).

INDIUM-111-LABELED LEUKOCYTE IMAGING:

A 5-YEAR EXPERIENCE

L. A. Forstrom, R. L. Morin, R. Carpenter,
J. McCullough, and M. K. Loken

The Nuclear Medicine Service
Department of Radiology
University of Minnesota/
VA Medical Center
Minneapolis, MN

INTRODUCTION

Indium-111-labeled leukocytes (In-111-WBC) have been used in the diagnosis of inflammatory disease (1-4), renal graft rejection (5, 6), and studies of granulocyte kinetics (7, 8). In this article we report results from a review of 1,178 In-111-WBC studies performed during a 5-year period (1977-82). Of these, 819 studies were carried out at the University of Minnesota Hospital, and 359 studies were performed in affiliated community hospitals. Results were analyzed to determine possible differences in diagnostic accuracy of the test between groups of patients separated by presence or absence of leukocytosis (WBC $\geq$ 10,000/mm³), sex (male/female), referring service (medicine/surgery), cell source (autologous/homologous), or hospital (U. of Minn./Community). Results were also analyzed in relation to certain primary clinical diagnoses and major sites of inflammatory disease.

METHODS

Patient Population

All patients selected for study had clinically suspected sepsis. The study population included pediatric as well as adult patients. Of the 1,178 cases reviewed, 819 studies were done at the University of Minnesota Hospital and 359 studies were performed in affiliated community hospitals. In 41 leukopenic patients, cells for labeling were obtained from ABO-compatible normal donors. Informed consent was obtained in all cases. The studies were performed under an investigational protocol (IND-13572) approved by the appropriate review committees at the University of Minnesota.

Preparation and Labeling of Cells

Leukocytes were harvested from whole blood and were prepared and labeled according to previously reported methods (9).

In leukopenic patients, cells for labeling were obtained from ABO-compatible normal donors. In most cases, relatively pure granulocyte suspensions were prepared by the double-density gradient centrifugation method which we previously reported for use in granulocyte kinetic studies (7). In a few instances, leukocytes from donor blood were harvested by the method used in nonleukopenic patients.

In studies performed in community hospitals affiliated with the University of Minnesota, an appropriate sample of whole blood from the patient was transported to our laboratory for leukocyte preparation and labeling. The dose of In-111-WBC was then returned for administration to the patient. The usual interval between drawing of the blood sample and reinfusion of the labeled cells was approximately 4-5 hours.

Patient Procedure

Indium-111-WBC doses were administered by intravenous injection in amounts of 0.3-0.6 mCi for adult patients, with proportionately smaller doses in pediatric patients. Imaging was routinely performed at approximately 24 hours after injection, typically including both whole body scans and camera views of selected regions of interest. Approximately 100,000 counts were obtained per image, using both photon peaks of In-111. Occasionally, additional images were obtained at either earlier or later times after injection.

Data Analysis

All studies were retrospectively reviewed by Nuclear Medicine physicians, and comparison was made between scan findings and the presence or absence of clinical sites of focal inflammatory disease. Scan and clinical outcomes were classified as positive, negative, or equivocal for each site of interest. Other pertinent clinical information was also recorded, including age, sex, primary clinical diagnoses, leukocyte count, postoperative interval (where applicable), and source of the leukocytes used for labeling. A computer-assisted analysis of these data was then carried out to determine sensitivity and specificity values for the test and the possible effect on these values of the variables mentioned above.

RESULTS

Labeling efficiency of the leukocyte suspensions was uniformly high, with average values of approximately 90%. Similarly, the images obtained were of consistently good quality and were considered satisfactory in nearly all cases. No adverse reactions to the test were observed in any of the patients studied.

Of the 1,178 In-111-WBC studies reviewed, there were 187 cases in which clinical and/or scan diagnoses were considered indeterminate. In most of these cases, the clinical findings were interpreted as equivocal for a possible site of inflammatory disease. These studies were excluded from further analysis.

In the remaining 991 studies, a total of 363 sites of inflammatory disease was diagnosed by tissue examination (153 cases) or by other radiographic and clinical findings. Indium-111-WBC images showed abnormal uptake in 294 of these sites, for a test sensitivity of 80%. There were 19 false-positive sites of In-111-WBC uptake, representing a specificity of 97%. These values yield an overall diagnostic accuracy of 91% and are comparable to values we previously reported from a smaller number of patient studies (9).

Although primary clinical diagnoses included a great variety of diseases, several diagnoses predominated among the patients studied. For example, there were 206 patients with some type of neoplastic disease, 188 patients with disease of the gastrointestinal system, and 114 patients who had received an organ transplant (usually kidney). No major differences were noted in diagnostic accuracy of the test among these patient groups.

Data were also analyzed for various sites of inflammatory disease. These most commonly involved the abdomen (144 cases) or the lung (65 cases). Although uptake of In-111-WBC in the lung showed less specificity for inflammatory disease (46% versus 88% in the abdomen), diagnostic accuracy of the test in the lung remained quite high at 83%.

These data are summarized in Table 1. This table includes data from cases in which leukocyte count at the time of the test was recorded by the reviewing physician. Although not analyzed by formal statistical methods, it is evident from these data that the presence or absence of leukocytosis (white blood cell count $\geq$ 10,000/mm^3) did not significantly affect test accuracy.

Statistical analysis was performed to determine the possible effect of several other variables on test sensitivity and specificity. These data are summarized in Table 2. No significant differences were observed between male and female patients. Patients referred from surgical services showed a significantly higher test sensitivity than

Table 1. Results of In–111–WBC Imaging (991 Cases)

Variable	Category	No. of Cases	Sensitivity (%)	Specificity (%)	Accuracy (%)
Primary clinical diagnosis	Neoplasia	206	90	98	95
	GI disease	188	81	96	90
	Transplant Recipient	114	93	94	93
	Ill-defined condition	92	50	96	90
	Other	391	69	98	88
Site of inflammatory disease	Abdomen	144	88	88	88
	Lung	65	89	46	83
	Other	154	72	76	72
WBC/mm3	< 10,000	187	74	97	89
	≥ 10,000	228	75	96	87
All non-equivocal cases		991	80	97	91

Table 2. Dichotomous Comparisons for In-111-WBC Imaging (991 Cases)

Variable	Possible Categories	Sensitivity (%)	Specificity (%)	Accuracy (%)
Sex	Male/ Female	79/82	97/97	90/92
Referring Service	Medicine/ Surgery	71/85*	98/95*	91/90
Source	Autologous/ Homologous	80/94*	97/92	91/93
Clinic	Univ. of MN/ Community	84/69*	96/98	91/91

*Results are significantly different (p < 0.05)

medical patients (85% versus 71%), but a slightly lower specificity
(95% versus 98%). Test sensitivity was also significantly higher in
patients studied with homologous cells than in those with autologous
cells (94% versus 80%) and in patients studied at the University of
Minnesota as compared with those in community hospitals (84% versus
69%). These differences were all fairly small and were not reflected
in significant differences in overall diagnostic accuracy for any of
these variables.

DISCUSSION

      We have previously reported on the usefulness of In-111-WBC in
the diagnosis of occult inflammatory disease, in smaller numbers of
patients (2, 9). Similar results have been reported by other
investigators (3, 4, 10, 11) and are confirmed in the present review
of 1,178 patient studies. In this study, the effects of several
variables on test accuracy were examined and were found to be
generally small or absent. The sensitivity of the test was somewhat
higher in patients studied by means of homologous than with autologous
cells (94% versus 80%). This may reflect a difference in the
chemotactic responsiveness of these cell types,
because many of the patients studied had illnesses or were on
medications which could have affected granulocyte function. However,
this difference did not significantly affect overall diagnostic
accuracy of the test.

      The test sensitivity was also somewhat less in patients studied
in community hospitals as compared with University of Minnesota
Hospital patients (69% versus 84%). This difference is probably
accounted for by the longer interval between drawing of the patient's
blood and reinfusion of the labeled cells and by difficulties in
controlling temperature of these products during transportation. Both
of these conditions could adversely affect the viability and function
of labeled cells. In spite of this difference, overall diagnostic
accuracy in these groups did not significantly differ.

      There are certain disadvantages in In-111-WBC imaging, including
the inconvenience of leukocyte preparation and labeling. There is
also significant patient radiation exposure, especially to the liver
and spleen (1, 12). False-negative tests may result from use of cells
with diminished function and in patients having inflammatory foci with
decreased blood flow or poor chemotactic stimulus. The test has been
shown to be less sensitive in older lesions than in acute lesions
(2, 13). Occasionally, cell clumping may result in small foci of
uptake in the lungs (14), although this occurs infrequently in our
experience. A more serious problem for interpretation of images is
the occasional occurrence of bowel activity due to swallowing of
In-111-WBC in patients with uptake in the respiratory system (15, 16).

Finally, In-111-WBC does not distinguish between abscesses and other types of inflammatory foci. We do not regard this as a "pitfall" in the use of this test, as localization of In-111-WBC in sites other than abscess (e.g., sinusitis, pneumonitis, osteomyelitis) often provides clinically useful information. In the presence of abnormal In-111-WBC uptake, moreover, the differentiation between abscess and other inflammatory processes can usually be made clinically or by other radiographic techniques. Our results demonstrate the usefulness of In-111-WBC imaging in the detection of localized inflammatory disease in a wide variety of clinical settings.

REFERENCES

1. M. L. Thakur, J. P. Lavender, R. N. Arnot, et al, Indium-111-labeled autologous leukocytes in man, J Nucl Med 18:1014 (1977).

2. N. L. Ascher, D. H. Ahrenholz, R. L. Simmons, et al, Indium-111 autologous tagged leukocytes in the diagnosis of intraperitoneal sepsis, Arch Surg 114:386 (1979).

3. I. R. McDougall, J. E. Baumert, R. L. Lantieri, Evaluation of In-111 leukocyte whole body scanning, Am J Radiol 133:849 (1979).

4. R. E. Coleman, R. E. Black, D. M. Welch, et al., Indium-111-labeled leukocytes in the evaluation of suspected abdominal abscesses, Amer J Surg 139:99 (1980).

5. M. P. Frick, C. E. Henke, L. A. Forstrom, et al, Use of In-111-labeled leukocytes in evaluation of renal transplant rejection: a preliminary report, Clin Nucl Med 4:24 (1979).

6. L. A. Forstrom, M. K. Loken, A. Cook, et al, In-111-labeled leukocytes in the diagnosis of rejection and cytomegalovirus infection in renal transplant patients, Clin Nucl Med 6:146 (1981).

7. B. J. Weiblen, L. A. Forstrom, J. McCullough, Studies of the kinetics of Indium-111-labeled granulocytes J Lab Clin Med 94:246 (1979).

8. J. McCullough, B. J. Weiblen, M. E. Clay, et al, Effect of leukocyte antibodies on the fate in vivo of Indium-111-labeled granulocytes, Blood 58:164 (1981).

9. L. A. Forstrom, B. J. Weiblen, L. Gomez, et al, Indium-111-labeled leukocytes in the diagnosis of occult inflammatory disease, in: "Indium-111-Labeled Neutrophils, Platelets, and Lymphocytes," M. L. Thakur and A. Gottschalk, ed., p. 123, Trivirum, New York, (1980).

10. D. A. Goodwin, P. W. Doherty, I. R. McDougall, Clinical use of Indium-111-labeled white cells: an analysis of 312 cases, in: "Indium-111-Labeled Neutrophils, Platelets, and Lymphocytes," M. L. Thakur and A. Gottschalk, eds., p. 131, Trivirum, New York, (1980).

11. J. Q. Knochel, P. R. Koehler, T. G. Lee, et al, Diagnosis of
     abdominal abscesses with computed tomography, ultrasound, and
     In-111 leukocyte scans, <u>Radiology</u> 137:425 (1980).
12. L. E. Williams, L. A. Forstrom, B. J. Weiblen, et al, Human
     dosimetry of In-111 granulocytes, <u>J Nucl Med</u> 21:86 (1980).
13. G. N. Sfakianakis, W. Al-Sheikh, A. Heal, et al, Comparisons of
     scintigraphy with In-111 leukocytes and Ga-67 in the diagnosis
     of occult sepsis, <u>J Nucl Med</u> 23:618 (1982).
14. R. E. Coleman and D. Welch, Possible pitfalls with clinical
     imaging of Indium-111 leukocytes: concise communication, <u>J
     Nucl Med</u> 21:122 (1980).
15. J. R. Crass, P. L'Heureux, M. Loken, False-positive In-111-
     labeled leukocyte scan in cystic fibrosis, <u>Clin Nucl Med</u> 4:291
     (1979).
16. S. Mikhail, M. K. Loken, L. A. Forstrom, Abdominal localization
     of Indium-111-labeled leukocytes in the study of consecutive
     patients with apparent abdominal sepsis, <u>Minn Med</u> (in press).

USE OF INDIUM-111-OXINE LABELED AUTOLOGOUS LYMPHOCYTES IN

THE DIAGNOSIS OF KIDNEY GRAFT REJECTION

J. Martin-Comin, M. Roca, J. M. Griño*
C. L. Paradell, and A. Caralps*

Departments of Nuclear Medicine and Nephrology*
Hospital de Bellvitge, "Principes de España"
Hospitalet de Llobregat
Barcelona, Spain

INTRODUCTION

Kidney graft rejection is mediated through immunological consequences. Lymphocytes, granulocytes, and platelets are involved in the rejection process (1-3). Their deposition in the transplanted kidney indicates specific pathophysiological mechanisms and represents a rejection episode.

Distribution of radiolabeled cellular blood elements in the body may be studied by external imaging. Labeled leukocytes have been used for abscess localization (4). Labeled platelets are used for the diagnosis of graft rejection and for detection of thrombosis (5-8). This article presents our experience in the use of labeled lymphocytes in the detection of a rejection episode of a transplanted kidney.

MATERIAL AND METHODS

We studied 26 transplant recipients, (20 males and 6 females). Autologous lymphocytes from the patient were labeled with In-111-oxine, (Radiochemical Center, Amersham), as follows:

1. Eighty ml of venous blood are mixed with 20 ml of Hank's solution and 100 I.U. of sodium heparine.

2. Four, 25 ml alliquots are transferred to 60 ml Falcon tubes which contain 12 ml of Ficol-sodium-diatrizoate gradient.

3. Each tube is centrifuged at 400 g for 30 minutes at room temperature.

4. Five ml of supernatant are kept for final lymphocyte resuspension (lymphocyte poor plasma, LPP).

5. Lymphocytes are obtained from the interphase by careful aspiration and centrifugated at 100 g for 10 minutes.

6. Lymphocytes pellet is resuspended in 10 ml of Hank's solution and centrifuged at 100 g for 10 minutes.

7. Lymphocytes pellet is again resuspended in 10 ml of Hank's solution.

8. 100 µCi of In-111-oxine are added to the lymphocyte suspension.

9. The mixture is incubated at 37° C for 10 minutes and centrifuged at 100 g for 10 minutes.

10. Labeled lymphocytes are resuspended in LPP and reinjected to the patient.

The entire procedure is carried out under sterile conditions in a laminar flow hood.

Two groups of patients were studied and are described below.

Control Group: Three patients who have undergone kidney transplantation 3 months before this procedure. Renal function at examination time was normal in all cases. No rejection signs were observed.

Study Group: Twenty-three transplant recipients with clinically suspected acute graft rejection.

Rejection diagnosis was based on clinical, biochemical, isotopic (I-131-hippuran), and, at times histological data.

Once labeled lymphocytes had been reinjected, daily scintigrams of graft area, contralateral iliac fossa, and spleen were obtained with a gamma camera. A 410 KeV, parallel hole, medium resolution collimator was used in all studies. Ten minute images were obtained. Data were stored in a DEC-PDP 11/34 computer.

The following indices of activity were calculated:

i : graft/contralateral iliac fossa

a : graft/iliac crest

  b : graft/spine

  c : spleen/graft

RESULTS

Control Group

   No lymphocyte deposition in the graft area was seen in any of
the patients in the control group (Fig. 1). Most lymphocytes were
spleen trapped; In-111 activity was also detected over spine,
iliac crest, and liver. Mean index values in nine studies
performed in the three patients are shown in Table 1. Index i
range was 0.94-1.27. Index a was in all patients but one lower
than 0.9. Index b was in all cases under 0.82. Finally, index c
range was 9.1-17.27.

Study Group

   Lymphocyte graft trapping was observed in all patients.
Twenty-one cases were diagnosed to have suffered from an acute
graft rejection, (2 interstitial type, 1 vascular type, and 1
mixed type). In the remaining two patients activity was due to
surgically demonstrated abdominal wall abscess over graft area.
Mean index values calculated in the first exploration performed
(2-4 hours post-reinjection) are shown in Table 2.

Normal

i=1.04, a=0.7, b=0.6, c=17.3

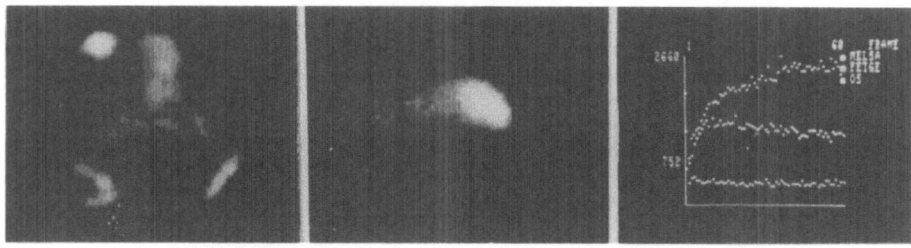

Fig. 1.   Normal study at 3 hours. Left: graft area image.
          Center: spleen. Right: time. Activity curves 0-60
          minutes post-injection. Upper: spleen. Middle:
          liver. Down: spine.

   Index i: Was always over 1.35. The highest value was 2.42.
In no rejection patients its value was 1.23 and 1.8.

   Index a: Could not be calculated in 8 patients due to no
visualization of homolateral iliac crest. The lowest value found
was 0.92. In all patients but 2 its value was greater than 1.23.

Table 1.

Control Group

	N	$\overline{X}$ (range)	S
i	9	1.05 (0.94 − 1.27)	0.1
a	9	0.87 (0.80 − 1.16)	0.15
b	9	0.64 (0.54 − 0.82)	0.11
c	9	13.7 (9.10 − 17.2)	3.59

N: Number of explorations $\overline{X}$: Mean   S: Standard deviation

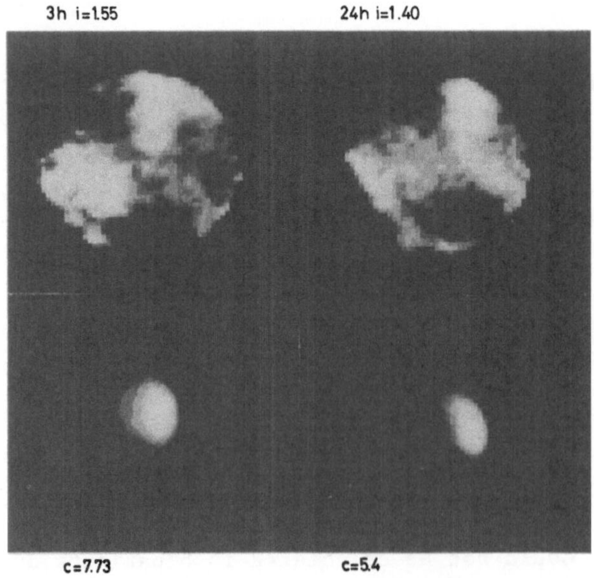

Fig. 2.   Acute graft rejection. Two and 24 hours images.

Table 2.

Study Group

	N	X̄ (range)	S
i	21	1.86 (1.35 - 2.42)	0.3
a	13	1.37 (0.88 - 2.02)	0.42
b	14	1.18 (0.79 - 1.80)	0.28
c	20	4.35 (3.42 - 14.5)	2.70

N: Number of explorations   X̄: Mean   S: Standard

In the patients with abdominal wall abscess its value was 1. and 1.8.

Index b: Could not be calculated in 7 patients due to no visualization of spine activity. In 2 patients its value was under 0.81. Its value in no rejection patients was 0.7 and 1.2.

Index c: In all patients who showed graft rejection but 2 c value was under 8. In no rejection patients its value was 12.5 and 4.4.

DISCUSSION

We have previously shown the use of In-111-oxine labeled platelets in kidney transplantation (6, 9). Sinzinger et al (7), have demonstrated that platelet deposition in graft area occurs before plasma creatinine increase is evident. Buckel et al (8) have also used In-111 labeled platelets in the management of patients with renal transplants. The possible usefulness of labeled Cr-51-lymphocytes in graft rejection was pointed out by Williams et al in 1975 (10). Oluwole et al (11), have demonstrated in experimental cardiac transplants in rats that lymphocyte infiltrate preceded platelet deposition.

Our results in human kidney transplants agree with those reported by Lennard et al (12), and show that In-111-oxine labeled

lymphocytes deposit in rejected kidney graft. The activity found
in spleen, liver, and bone has been already discussed by Wagstaff
et al (13).

Indium-111-lymphocyte deposition in grafts was observed in
all patients from the study group. In all but 2 the diagnosis was
acute graft rejection. The 2 false-positives resulted from an
abdominal wall abscess lying over the graft area.

Indices i and c were the most useful ones. In patients with
acute graft rejection, index i increased while index c decreased.

Indices a and b were the less useful. Iliac crest and spine
localization was difficult in some cases. Due to no visualization
of activity, both indices could not be calculated in 8 and 7
patients, respectively. Index values were different in the 2
patients with abdominal wall abcess. In the first patient index i
was borderline and c did not decrease (i=1.23; c=12.5). In the
second patient index i increased and c decreased (i=1.8; c=4.4).
The difference may be due to a slight graft rejection mixed with
the wall abcess.

In summary, In-111-oxine labeled lymphocytes may be used for
kidney graft rejection diagnosis by imaging their deposition in
the grafted kidney. Infectious foci over graft area must be
excluded.

The increase of index i (over 1.3) is indicative of acute
graft rejection, especially if index c falls below 9. However,
due to the high lymphocyte irradiation and the high number of
chromosomic aberrations demonstrated by ten Berge et al (14) in
labeled lymphocytes, in our opinion the test must be reserved to
clinical situations in which rejection diagnosis is difficult,
especially if false-positive results may be expected with labeled
platelets.

REFERENCES

1. S. Reitamo, Y. T. Konttinen, A. Ranki, and Hâjry, The relation of
    different inflamatory cells types to the various parenquimal
    components of rejecting kidney allografts, Histopathology 4:
    517 (1980).
2. B. H. Spargo, A. E. Seymour, and N. G. Ordonez, Transplantation,
    in: Renal Biospy with Diagnostic and Therapeutic Implications,
    Spargo, Seymour, and Ordonez, eds., John Wiley & Sons, New York
    (1980).
3. A. Capitanio, P. M. Mannuci, F. Ponticeli, and F. Pareti, Detection
    of circulating released platelets after renal transplantaion,
    Transplantation 33:298 (1982).

4. R. E. Coleman, Radiolabeled leukocytes, in: "Medicine Annual 1982,"
    L. M. Freeman and H.S. Weismann, eds., Raven Press, New York
    (1982).

5. D. A. Cunningham and B. A. Siegel, Radiolabeled platelets, in:
    "Nuclear Medicine Annual 1982". E. L. M. Freeman and H. S.
    Weisman, eds. Raven Press, New York (1982).

6. J. Martin-Comin, M. Roca, J. M. Grino, C. Paradell, and A. Caralps,
    Indium-111-oxine autologous labeled platelets in the diagnosis
    of kidney graft rejection, Clin Nucl Med 98:7, (1983).

7. H. Sinzinger, Ch. Liehner, M. Schwarz, and R. Hofer, Labelling of
    autologous human platelets with In-111-oxine sulfate for
    monitoring of human kidney transplants, in: "Proceedings of the
    Third World Congress of Nuclear Medicine and Biology,"Paris
    (1982).

8. S. A. C. Buckel, S. Chandler, R. Hawker, C. N. McCollum, and A. D.
    Barnes, The early diagnosis of acute renal transplant rejection
    using In-111-labeled platelets, Transplant Proc 15:1192 (1983).

9. J. M. Grino, J. Alsina, J. Martin, Indium-111-labeled autologous
    platelets as a diagnostic method in kidney allograft rejection,
    Transplant Proc 14:198 (1982).

10. L. E. Williams, G.E. Merio, G. Goren, et al. Detection of canine
    kidney allograft rejection with Cr-51 labeled lymphocytes,
    Radiology 115:205 (1975).

11. S. Oluwole, T. Wang, R. Fawz, Use of In-111-labeled cells in
    measurement of cellular dynamics of experimental cardiac
    allograft rejection, Transplatation 31:51 (1981).

12. L. Lennard, J. M. Blair, A. M. Holroyd, M. Fox, Evaluation of
    In-111-labeled lymphocytes in diagnosis of kidney transplant
    rejection, "VIII International Congress on Nephrology," Athens
    (1981).

13. J. Wagstaff, C. Gibson, N. Thatcher, A method for following human
    lymphocyte traffic using In-111-oxine labeling, Clin Exp Immunol
    43:435 (1981).

14. R. J. M. ten Berge, A. T. Natarajan, M. R. Hardeman, Labeling with
    In-111-has detrimental effects on human lymphocytes: concise
    communication, J Nucl Med 24:615 (1983).

ABDOMINAL PATTERNS OF IN-111-LABELED LEUKOCYTES

SCANS IN PATHOLOGIES OTHER THAN ABDOMINAL ABSCESSES

Daniel Picard, Raymonde Chartrand, Lucie Carrier,
and Daniel Dionne

Nuclear Medicine Department
Hospital St-Luc
Université de Montréal
1058 St-Denis
Québec, Canada

Indium-111-labeled leukocytes scintigraphy is now a well-accepted technique for the detection of abdominal infectious processes. High specificity and sensitivity are achieved. However, In-111-labeled leukocyte accumulation does not always suggest the diagnosis of abscess. Higher specificity could be obtained by the knowledge of In-111 leukocyte distribution pattern in the abdomen in other pathologies like 1) inflammation of abdominal wounds, 2) surgical complication without abscess formation, 3) inflammatory or ischemic bowel disease, or 4) swallowing of leukocytes in the intestinal lumen.

One hundred fifty-three labeled leukocyte scans performed over 18 months were reviewed. Ninety-six cases showed no abdominal uptake, and 57 cases had accumulation of leukocytes in the abdomen. Twenty-nine of these patients had proven abdominal abscesses, and the remaining 28 were positive due to other causes. This report will briefly discuss the distribution pattern of In-111-labeled leukocyte accumulation in these last 28 scans.

TECHNIQUE AND METHODS

The labeling technique of Thakur et al was used(1). Labeling efficacy was 75% to 85%. The scintigraphy was performed on a LFOV gamma camera with a medium energy collimator, at 2, 24, and/or 48 hours after injection.

413

RESULTS

Four patterns of abdominal distribution were observed:
1) circumscribed to the abdominal wall, 2) in the abdominal cavity, 3)
in the intestinal walls, 4) in the intestinal lumen.

The circumscribed abnormalities to the abdominal walls were found
in six patients. In all these cases, the uptake was lower than the
liver uptake. The activity was localized at the site of the wound,
did not move, diminished or disappeared when the dressings were
changed. These cases included 1 coecostomy, 1 colostomy, 1 open
wound, 2 Parks pouch, and 1 ileostomy. The uptake was secondary to
inflammatory reaction in the healing process of the wound.

Free activity in the abdominal cavity was found in nine patients
with surgical complication. The uptake had an irregular contour, it
neither moved nor had a bowel shape but was localized in the free
spaces and disappeared or faded with time. Also, activity was seen in
the collecting bags. Four of these patients had a pancreatic
intervention, two had colonic, two had a hepatic, and one had a
gastric surgery.

The third category includes four cases in which the uptake had
the shape of the bowels. The accumulation occured early (4 hours
after the injection) and persisted at the same site 24 hours and 48
hours after the injection. On delayed views, some activity was seen
in the intestinal lumen and progressed with time. This category
included two cases of Crohn disease, one of ulcerative colitis, and
one of ischemic colitis following an aorto-bifemoral bypass.

In the last category, activity was not seen in the abdomen
initially but was seen at 24 hours. The uptake had the shape of the
intestinal lumen, and activity moved completely with time. Clinical
facts suggested leukocyte swallowing. This included patients with
pneumonia(2), shock lung(1), levine and endotracheal intubation(3)
esophagitis(1).

In two cases, a faint uptake was visualized in the abdomen.
Delayed views were not available. No certain etiology could be given
to explain the uptake. One patient had active sarcoidosis, but no
uptake was seen in the lungs. The other patient was treated for
osteomyelitis of the foot.

DISCUSSION

Many authors have already demonstrated the high specificity and
sensitivity of In-111-labeled leukocytes in the search of abdominal
abscess. However, some false-positive cases could occur. Coleman et
al have described different etiologies for these false-positive

results. We have learned not to report In-111-labeled leukocyte scan as being positive for an abscess when the uptake is irregular in distribution, has a bowel shape appearance, or corresponds to the site of a wound. Delayed views are useful to see if the abnormal uptake moves or changes in intensity with time. Localization of wounds, drains, and collecting bags need to be known because of the transient accumulation of leukocytes often seen at these sites. Early views (4 hours) are also helpful in order to identify the involved ischemic or inflammatory segments of the bowel and also to confirm the extra-abdominal origin of the labeled leukocytes.

We conclude that early and delayed views of the abdomen are mandatory to increase the already high specificity of In-111-labeled leukocyte scan.

REFERENCES

1. M. L. Thakur, J. P. Lavender, R. N. Arnot, D. J. Silvester, A. W. Segal, Indium-111 labelled autologous leukocytes in man, J Nucl Med 18:1012 (1977).
2. R. E. Coleman and D. Welch, Possible pitfalls with clinical imaging of Indium-111 leukocytes: concise communication, J Nucl Med 21: 122 (1980).

EVALUATION OF IN-111-LEUKOCYTE IMAGING

IN PATIENTS WITH INTRACRANIAL LESIONS

K. Uno, A. Yamaura, G. Uchiyama and N. Arimizu

Chiba University School of Medicine
Chiba, Japan

This study was undertaken to evaluate usefulness of this leukocyte imaging in patients with suspected intracranial infections or with difficult diagnostic problems, including differentiation of cerebral abscess from other intracranial lesions.

Twenty-three patients, with ages ranging from 7 to 70 years, with signs of infection or with encapsulated low-density areas on X-ray CT were studied. Autologous polymorphonuclear leukocytes were separated, labeled with 0.5-1mCi of In-111-oxine and readministered to the patients. Imaging was performed at 24 hours and occasionally after several days.

Eight cases were positive. Of them five were proved to have intracranial infections and three were having metastases from lung and gastric cancers. A patient with multilocular cerebellar abscess was successfully treated by conservative therapy under monitoring of In-111-leukocytes. A child with brain abscess with tetralogy of Fallot was initially treated with antibiotics, and imaging showed steady decrease of leukocytes. Then an operation was performed because the mass effect remained unchanged. Three times imaging performed in this case was very useful for quantitative monitoring of therapeutic and surgical effects. In case of glioblastoma multiform having a low-density area with marked ring enhancement on CTs, Indium leukocyte scan was negative. Indium imaging was a reliable diagnostic procedure in excluding an abscess from other intracranial lesions. Brain metastasis might be the cause of false-positive results. These lesions could not be proved to have an abscess or any inflammatory process.

It is concluded that Indium leukocytes imaging becomes not only a reliable diagnostic procedure but an effective means for selecting treatment procedure and for monitoring therapeutic effect in an acute intracranial abscess.

REFERENCES

1.   M. L. Thakur, J. P. Lavender, R. N. Arnot, D. J. Silvester,
      A. W. Segal, Indium-111-labeled autologous leukocytes in
      man, J Nucl Med 18: 1012 (1977).
2.   A. M. Peters, J. P. Lavender, J. Macdermot, Diagnosing
      cerebral abscess with Indium-111-labeled leukocytes,
      Lancet 309 (1980).
3.   K. Uno, G. Uchiyama, A. Yamaura, A case of subdural empyema
      with In-111-labeled leukocyte imaging, Radioisotopes Hosp
      Radiol 14:23 (1981).
4.   E. B. Rotheram and L. A. Kessler, Use of computerized
      tomography in nonsurgical management of brain abscess,
      Arch Neurol 36:25 (1979).
5.   E. F. Crocker, A. F. Mclaughlin, J. G. Morris, R. Benn, J. G.
      McLeod, J. L. Allsop, Technetium brain scanning in the
      diagnosis and management of cerebral abscess, Am J Med
      56:192 (1974).

TRANSPULMONARY KINETICS OF IN-111-POLYMORPHONUCLEAR

LEUKOCYTES (PMN) IN ADULT RESPIRATORY DISTRESS SYNDROME (ARDS)

A. A. Driedger, G. Morrissey, F. Warshawski
and W. J. Sibbald
Division of Nuclear Medicine
Department of Medicine
University of Western Ontario

Department of Nuclear Medicine
Critical Care/Trauma Unit
Victoria Hospital, London, Ontario

PMNs may play a role in the pathogenesis of ARDS(1). In 68
patients with ARDS we prospectively performed In-111-PMN scans using
autologous cells labeled by the ethanolic oxine method. The labeled
cells were reinjected with the patient positioned under a computerized
gamma camera, and the continuously changing PMN distributions in the
lungs were monitored for an hour. On the next day analog whole body
and digital lung images were obtained. The analog images were
subsequently scored on a 0-4+ scale as per the concentration of
In-111-PMNs remaining in the lungs in 33 patients. The computerized
transpulmonary kinetic data were analyzed as per the rate of PMN
washout from the lung, and the late digitized lung images were
quantified to measure the percentage of initial PMNs remaining at
18-20 hours. Patients with cardiac edema were used for comparison
purposes.

Of 14 patients with cardiac edema studied in this way, none
demonstrated any visually detectable PMN accumulations in the lungs.
Seventeen of 22 patients with ARDS demonstrated diffuse pulmonary PMN
retention. Another 11 patients with ARDS who were receiving high-dose
corticosteroids demonstrated visible pulmonary PMN retention in only
one instance(2).

Subsequent to these observations, another group of patients was classified by clinical criteria as having active ARDS (Aa), steroid treated active ARDS (As), systemic sepsis without pulmonary manifestations (S), recovering ARDS (Ar) and cardiogenic pulmonary edema (C). The results of the PMN scintigraphy were evaluated independently (by AAD). The washout of In-111-PMN from the lungs after I.V. injection was a monoexponential function, and the t1/2 for the pulmonary transit differed significantly between patient categories (p<0.001). The retained pulmonary activity at 17-20 hours more clearly separated Aa patients from all other categories (p<0.001). Data from 46 patients in whom complete studies were obtained are shown in Table 1.

There are several possible interpretations of these results:

1. That altered transpulmonary PMN kinetics have been demonstrated in patients with ARDS and this is due to alteration of PMN affinity for pulmonary microvascular endothelium. This interpretation could be reconciled with the hypothesis that complement fraction C5a activation leads to PMN clumping and embolization(3).

2. That in active ARDS the PMNs are more susceptible to injury by the cell-labeling procedure. If, as seems likely, the PMNs are metabolically stimulated in response to the septic process, then it might be that removal from plasma as is required in the oxine method may be more deleterious when the cells are from patients with active ARDS. Studies will need to be conducted to compare the behavior of PMNs labeled by the tropolone method(4) in patients with ARDS.

Table 1.

Clinical Classification	Aa	As	S	Ar	C
Transpulmonary t1/2	85 ± 33*	58 ± 25	50 ± 17	40 ± 13	32 ± 9
% Pulmonary Retention (17 - 20 hours)	28 ± 17†	10.5 ± 25	11.5 ± 5.0	12.2 ± 6.4	6.3 ± 1.9
n	13	7	7	9	10

* minutes ± S. D.
† % of maximum within 4 minutes of injection ± S. D.

REFERENCES

1. R. M. Tate and J. E. Repine, Neutrophils and the adult
    respiratory distress syndrome, <u>Am Rev Resp in Dis</u> 128:522
    (1983).
2. F. Warshawski, A. A. Driedger, W. J. Sibbald, Pulmonary
    microvascular localization of labeled polymorphs in human
    ARDS and the influence of corticosteroids, <u>Annals R Coll
    Phys Surg Canada</u> 15:282 (1982).
3. H. S. Jacob, P. R. Craddock, D. E. Hammerschmidt, C. F.
    Moldow, Complement-induced granulocyte aggregation, <u>New Eng
    J Med</u> 302:789 (1980).
4. S. H. Saverymuttu, A. M. Peters, H. J. Dampare, H. J. Reavy,
    S. Osman, J. P. Lavender, Lung transit of In-111-Labelled
    granulocytes, <u>Scand J Haematol</u> 30:151 (1983).

EFFECT OF INDIUM-111 OXINE AND TROPOLONE LABELING ON

HUMAN PLATELET AGGREGATION AND BETA-THROMBOGLOBULIN

H. M. A. Towler, A. P. Bautista, and B. Bennett

Department of Medicine
University of Aberdeen
Aberdeen, Scotland

INTRODUCTION

Much concern has been expressed about the possibly deleterious effect of in vitro platelet labeling techniques on platelet function and viability, thus questioning the validity of studies of their in vivo properties.

Indium-111-oxine has become a popular platelet label over the last six years, but criticism has been made of the necessity for separation of platelets from plasma, the use of nonphysiological additives such as ethanol, and the use of buffers such as HEPES. Indium-111-tropolone has been advocated as an agent which allows platelets to be labeled efficiently in plasma. In this preliminary study we have examined the aggregation response of platelets labeled with In-111-oxine and tropolone to ADP (adenosine diphosphate), adrenaline, collagen, and ristocetin. We have also studied the effects of labeling upon the β-thromboglobulin (BTG) content of platelets and BTG release after aggregation.

METHODS

Platelets were obtained from healthy human donors who were receiving no drugs which might interfere with platelet function. Twenty-six ml blood was obtained via a 19G needle, 17 ml being mixed with 3 ml acid citrate for preparation of platelet-rich plasma (PRP) for labeling, and 9 ml mixed with 1 ml 3.8% trisodium citrate for preparation of platelet-poor plasma (PPP) and control aggregation studies. PRP was then labeled with 250 μCi In-111 by one of three methods:

1. with In-111-oxine by the method of Hawker et al (1)

2. with In-111-tropolone in either

    (a) acid citrate dextrose in saline (ACD-saline)
    (b) acid citrate dextrose in plasma (ACD-PLASMA) by the
         method of Dewanjee et al (2,3). The concentration of
         tropolone was 20 μg/ml.

After labeling and resuspension, the PRP was incubated at 30°C to
allow recovery from the inhibitory effects on platelet aggregation of
prostaglandin E_1 (method 1) or ACD (method 2).

Aggregation studies were then performed with ADP, adrenaline,
collagen, and ristocetin. BTG was measured at each stage of labeling,
and before and after aggregation by radioimmunoassay kit (Amersham
International) as described by Ludlam et al (4).

RESULTS

Platelet Aggregation Studies

Aggregation responses to all four agents were normal compared
with controls with each of the three labeling techniques. The
responses of tropolone-labeled platelets were uniformly less than
oxine, with the exception of ristocetin-induced aggregation.

BTG studies

Tropolone-labeled platelets showed a greater loss of BTG than
oxine-labeled platelets (Table 1), the greatest BTG loss occurring
when the platelet pellet is washed to remove residual plasma.

Table 1.

Preparation	Platelet-Associated BTG (ng/10⁶ Cells)		Labeling
	Pre-labeling	Post-labeling	Efficiency
In-111-oxine	6.14	4.98	80 $\pm$ 13%
In-111-tropolone in ACD-saline	6.14	0.80	88 $\pm$ 10%
In-111-tropolone in ACD-plasma	6.14	0.75	70 $\pm$ 20%

BTG release after aggregation with ADP, adrenaline, and collagen was reduced with each of the three methods, but the sample size was too small to assess the significance of this.

DISCUSSION

We have confirmed that human platelets can be labeled with In-111-oxine and In-111-tropolone in a plasma-free medium without impairment of aggregability when compared with controls. The presence of plasma reduces the efficiency of labeling but does not alter the aggregation responses, nor does it reduce the number of manipulations required during the labeling process. The apparent submaximal aggregation to ADP, adrenaline, and collagen of tropolone-labeled platelets may reflect the longer duration of incubation at $37°C$ found necessary to abolish the inhibitory effects of ACD. Fifteen minutes incubation was sufficient to negate the effects of PGE_1 upon platelets labeled with oxine, whereas tropolone-labeled platelets required incubation for 2 to 6 hours. This prolonged period of incubation may also explain the suboptimal aggregation curve of the control platelets. Thakur noted reduced aggregation to ADP in platelets incubated in saline and citrated saline (6), but did not comment whether this was reversible following incubation in plasma.

The use of PGE_1 in labeling by Hawker's method facilitates platelet resuspension (1), and we did note that when labeled in ACD-saline or plasma the platelet pellet required more vigorous agitation to achieve resuspension. This may be a factor resulting in an increase of BTG loss during labeling as mechanical stress is known to result in BTG loss without actually resulting in platelet lysis or aggregation (7).

During labeling, the greatest loss of BTG occurs when the platelet pellet is washed, and most of this loss is due to the elution of whole platelets. With tropolone-labeled platelets, a greater amount of BTG was found in the supernatant washing fluid after centrifugation to remove intact platelets, and this suggests that platelet activation has occurred with subsequent release. It is thought that ADP loss from platelets is reversible and does not affect platelet life-span (8), and the same may be true for BTG loss.

There are considerable methodological differences between the methods employed, and a direct comparison is therefore difficult.

We feel that at present In-111-tropolone offers no significant advantages over In-111-oxine for routine platelet labeling and that further study of the optimal labeling technique with In-111 is required.

REFERENCES

1. R.J. Hawker, L.M. Hawker, A.R. Wilkinson, Indium (In-111)-
    labeled human platelets: optimal method, <u>Clin</u> <u>Science</u>
    58:243 (1980).
2. M.K. Dewanjee, S.A. Rao, P. Didisheim, Indium-111 tropolone,
    a new high-affinity platelet label: preparation and
    evaluation of labeling parameters, <u>J</u> <u>Nucl</u> <u>Med</u> 22:981
    (1981).
3. M.K. Dewanjee, S.A. Rao, J.A. Rosemark, S. Chowbury, P.
    Didisheim, Indium-111 tropolone, a new tracer for
    platelet labeling, <u>Radiology</u> 145:149 (1982).
4. C.A. Ludlam, A.E. Bolton, S. Moore, J.D. Cash, Evidence for
    the platelet specificity of β-thromboglobulin and
    studies on its plasma concentration in healthy
    individuals, <u>Br</u> <u>J</u> <u>Haematol</u> 41:271 (1979).
5. H.J. Danpure, S. Osman, F. Brady, The labeling of blood cells
    in plasma with In-111-tropolonate, <u>Br</u> <u>J</u> <u>Radiol</u> 55:247
    (1982).
6. M.L. Thakur, L. Walsh, H.L. Malech, A. Gottshalk, Indium-111-
    labeled human platelets. improved methed, officaey and
    evaluation, <u>J</u> <u>Nucl</u> <u>Med</u> 22:381 (1981).
7. M.T. Santos, J. Valles, J. Aznar, P. Villa, Platelet BTG
    release in vitro induced by mechanical and chemical
    stimulus; correlation with the aggregation curve para-
    meters, <u>Scand</u> <u>J</u> <u>Haematol</u> 29:368 (1982).
8. H.J. Reimers, R.A. Kinlough-Rathbone, J.P. Cazenave, A.F.
    Senyi, J. Hirsh, M.A. Packham, J.F. Mustard, In vitro
    and in vivo functions of thrombin-treated platelets,
    <u>Thromb</u> <u>Haemost</u> 35:151 (1976).

CHELATES OF INDIUM-111 FOR GRANULOCYTE LABELING

F. Brady, H. J. Danpure, S. Osman
and D. J. Silvester

MRC Cyclotron Unit
Hammersmith Hospital
Ducane Road
London W12 OHS, U. K.

INTRODUCTION

Chelates of the cyclotron-produced radionuclide In-111 (t1/2 2.8d) with 8-hydroxyquinoline (1) (oxine) and acetylacetone (2, 3) (acac) have been used for labeling human blood cells in diagnostic nuclear medicine (4, 5). However, to obtain efficient labeling with oxine or acac complexes the cells must first be washed free from plasma which may result in damage to the cells. In contrast, In-111-tropolone was found to label cells efficiently in plasma (6, 7). Thus the harmful effects of depriving cells of plasma was eliminated and a more rapid detection of inflammatory lesions was achieved (8).

As part of a study to find more efficient agents for labeling cells in plasma with In-111 we have examined the ligands mentioned above together with substituted tropolone derivatives; N,N'-dialkyldithiocarbamates and 2-mercaptopyridine-N-oxide (9).

EXPERIMENTAL MATERIALS AND METHODS

The following chemicals were used as supplied: tropolone, N,N'-dimethyldithiocarbamate, N,N'-diethyldithiocarbamate (Fluka). Acetylacetone (Sigma) was distilled before use. β-thujaplicin (4-iso-propyltropolone) (Roth) was converted to the more water-soluble sodium salt (10) and recrystallized from water as the dihydrate. Eight-hydroxyquinoline sulphate (Sigma) was recrystallized from water/ethanol. The following compounds were synthesized according to reported methods 7-carboxy-8-carboxy-methyl-tropolone (11); 6-methyltropolone (11); ammonium 5-sulphonyl-tropolone (12).

PREPARATION AND LABELING OF GRANULOCYTES

Separation of granulocytes from fresh samples of venous blood and labeling with In-111 complexes were carried out in a similar manner to that which we have previously described for labeling with In-111-tropolone (6).

RESULTS AND DISCUSSION

The optimum ligand concentration for labeling and the labeling efficiencies obtained with eight neutral In-111 complexes for granuloctyes labeled in plasma or plasma-free medium (Hepes-saline buffer) are shown in Table 1. All the compounds label granulocytes very efficiently in buffer (64 - 86%). However, in plasma the

Table 1. Labeling Efficiencies (LE) and Optimum Ligand concentration for granulocytes (5×10^6/ml) labeled with Various In-111-L_3 Complexes in Plasma and Buffer Solutions.

Ligand L	90% Plasma		Hepes-saline buffer	
	LE	Optimum ligand conc. (M)	LE	Optimum ligand conc. (M)
Acetylacetone	a	a	86 ± 2	1.8×10^{-2}
8-hydroxyquinoline sulphate	a	a	71 ± 1	4×10^{-6}
2-mercaptopyridine-N-oxide	16 ± 2	1×10^{-4}	84 ± 3	4×10^{-5}
Tropolone	17 ± 2	4×10^{-4}	70 ± 5	4×10^{-5}
6-methyltropolone	12 ± 3	2×10^{-4}	64 ± 7	4×10^{-5}
4-iso-propyltropolone	14 ± 1	4×10^{-4}	62 ± 4	2×10^{-5}
N,N'-dimethyldithio-carbamate	22 ± 1	2×10^{-4}	77 ± 4	1×10^{-4}
N,N'-diethyldithio-carbamate	9 ± 1	1×10^{-4}	66 ± 3	1×10^{-5}

a   Labeling could not be determined at this cell concentration. At a granulocyte concentration of 2×10^7/ml[3] the LE was 28% at a ligand concentration of 5×10^{-2}M for acac and 9% at a ligand concentration of 4×10^{-5}M for oxine sulphate.

labeling efficiencies are greatly reduced (9 - 22%); the largest
reductions are for In-111-acac and In-111-oxine for which the labeling
in plasma was negligible at a granulocyte concentration of 5 x 10^6/ml.
Tropolone derivatives with ionizable substituents; 7-carboxy-6-
carboxymethyltropolone, and 5-sulphonyltropolone (not shown in table)
failed to label cells in either plasma or buffer.

These results support and extend our previous observations of the
existence of an optimum ligand concentration for cell-labeling (6),
which is greater in plasma than in buffer.

The marked variation in the labeling efficiencies of the eight
neutral complexes (0 - 22%) is probably a consequence of differences
in the interaction of the ligands with plasma proteins. In two cases
where the lipophillicity of the ligand was increased, methyltropolone
and iso-propyltropolone compared with tropolone and
N,N'-dimethyldithiocarbamate compared with
N,N'-diethyldithiocarbamate, the labeling efficiency was not improved.

While In-111 complexes of all the ligands presented in Table 1
label cells efficiently in plasma-free medium, only N,N'-di-
methyldithiocarbamate labels cells better in plasma than tropolone,
but the increase in efficiency is small.

REFERENCES

1.   M. L. Thakur, R. E. Coleman, M. Welch, J Lab Clin Med 89:217
     (1977).
2.   H. S. Sinn and D. J. Silvester, Br J Radiol 52:758 (1979).
3.   H. J. Danpure and S. Osman, Br J Radiol 54:587 (1981).
4.   M. L. Thakur, J. P. Lavender, R. N. Arnot, D. J. Silvester, A.
     W. Segal, J Nucl Med 18:1014 (1977).
5.   A. M. Peters, S. Karimjee, S. H. Saverymuttu, J. P. Lavender, Br
     J Radiol 55:827 (1982).
6.   H. J. Danpure, S. Osman, F. Brady, Br J Radiol 55:247 (1982).
7.   M. K. Dewanjee, S. A. Rao, P. Didisheim, J Nucl Med 22:981
     (1981).
8.   A. M. Peters, S. H. Saverymuttu, H. J. Reavy, H. J. Danpure, S.
     Osman, J. P. Lavender, J Nucl Med 24:39 (1983).
9.   M. L. Thakur and M. J. Barry, J Labeled Comp Radiopharm 19:1410
     (1982).
10. H. Iinuma, J Chem Soc Japan 64:742 (1943).
11. R. D. Haworth and J. D. Hobson, J Chem Soc 561 (1951).
12. T. Nozoe, S. Seto, T. Ikemi, T. Arai, Proc Japan Acad Sci 27:24
     (1951).